DESIGN AND ANALYSIS OF HEAT SINKS

WILEY SERIES IN THERMAL MANAGEMENT OF MICROELECTRONIC AND ELECTRONIC SYSTEMS
Allan D. Kraus and Avram Bar-Cohen, Series Editors

An Introduction to Heat Pipes: Modeling, Testing, and Applications
G.P. Peterson

Design and Analysis of Heat Sinks
Allan D. Kraus
Avram Bar-Cohen

DESIGN AND ANALYSIS OF HEAT SINKS

Allan D. Kraus
Allan D. Kraus Associates
Pacific Grove, CA

Avram Bar-Cohen
Department of Mechanical Engineering
University of Minnesota
Minneapolis, MN

A Wiley-Interscience Publication
JOHN WILEY & SONS, INC.
New York / Chichester / Brisbane / Toronto / Singapore

This text is printed on acid-free paper.

Library of Congress Cataloging in Publication Data:

Kraus, Allan D.
Design and analysis of heat sinks / Allan D. Kraus, Avram Bar-Cohen.
p. cm. — (Wiley series in thermal management of microelectronic and electronic systems)
Includes index.
ISBN 0-471-01755-8 (cloth: alk. paper)
1. Heat sinks (Electronics) — Design and construction.
I. Bar-Cohen, Avram. 1946 – II. Title. III. Series.
TK7872.H4K69 1995
821.381'046 – dc20 I 95-32934

Printed in the United States of America

10 9 8 7 6 5 4 3 2

To our teachers and mentors

A. E. Bergles
K. A. Gardner
A. L. London
and
W. M. Rohsenow

CONTENTS

PREFACE

Heat sinks are the most common thermal management hardware in use in the electronics industry. They are used to improve the thermal control of electronic components, assemblies, and modules by enhancing their exterior surface area through the use of fins or spines. The presence of the heat sink has a positive impact on both the reliability and the functional performance of electronic, telecommunication, and power conversion systems.

In the present day competitive environment, superior thermal performance must be attained simultaneously with lower manufacturing costs, minimum added mass, and shortened product development cycles. Thus, the successful design and/or selection of cost-effective heat sinks for electronic equipment requires a detailed understanding of the governing heat transfer phenomena, mastery of thermal modeling tools for geometrically complex fin structures, and an appreciation for thermal optimization opportunities.

In this book, the authors have attempted to provide a comprehensive treatment of the "Analysis and Design of Heat Sinks," including development of the mathematical tools, discussion of the prevailing heat transport processes, articulation of the tools for a variety of fin geometries, and presentation of simple "transfer function" relationships for single fin, cascaded fin, and fin array heat sinks. Indeed, many of the fin geometries examined in this book cannot be thermally characterized by conventional analytical techniques. Thus, this self contained book can serve as a text for the engineer interested in learning the subject of heat sink design and analysis in its entirety. And yet, it can be used as a source book for the accomplished thermal specialist seeking to hone his/her skills.

The book opens with an extensive introduction, presenting the fundamentals of heat transfer, thermal modeling of electronic packages, and the first-order thermal analysis of finned arrays. The next two chapters review the mathematics of linear transformations and prepare the reader for the application of this methodology to fin structures. Chapters 4 through 9, the core of this heat sink book, systematically explore the use of linear transformations in the thermal parameterization of a wide variety of single fin, cascaded fin, and fin array designs encountered in commercial heat sinks, as well as the least material optimization of specific fin shapes. The last chapter of the book is devoted to the thermal characterization and optimization of the ubiquitous plate-fin heat sink, including both convective and radiative cooling and presentation of empirical results, as well as theoretical formulations. The reader will find a useful review of matrices and determinants, along with the list of nomenclature and references in the appendices.

Mastery of the contents of this book will equip the reader with a most powerful thermal analysis technique which is uniquely suited to the analysis, design and optimization of convective heat sinks. This will provide a rich understanding of the thermal performance and optimization potential of the most commonly used plate-fin heat sinks and a deep appreciation of the thermal context of heat sink applications in the electronics industry.

The authors are indebted to Abdul Aziz, Jim Welty, Bud Peterson, Dave Snider, and Mike Yovanovich for their continued encouragement, to Frank Cerra whose vision brought about the series in thermal management, and to Howard and Jo Aksen who handled the production of the book so ably.

ALLAN D. KRAUS
Pacific Grove, CA

AVRAM BAR-COHEN
Minneapolis, MN

Chapter 1

INTRODUCTION

1.1 INTRODUCTION

This book is devoted to the heat sink which is used to enhance heat transfer from the surface of an electronic component or package to the surrounding environment. Although the presence of the heat sink (and its bonding layer) produces a thermal resistance, when properly designed, the thermal resistance attributed to the heat sink will be substantially lower than the resistance associated with convection and radiation from the bare surface of the component. Thus, thermal design of electronic equipment often requires detailed consideration of the thermal resistance of the heat sink as part of the thermal path between the component junction and the "ground" temperature (usually the environment). The heat sink represents just one element along this thermal path, but, because it may be the controlling resistance, thermal analysis and design of heat sinks are of considerable importance in the development of electronic products. Some typical electronic equipment heat sinks are shown in Fig. 1.1.

This chapter prepares the reader for the study of the methods used for expeditious design and analysis of heat sinks for electronic equipment and electronic components. Section 1.2 deals with the heat-transfer phenomena in terms of the thermal resistances attributed to individual mechanisms. Section 1.3 introduces the concept of the chip package resistance between the component junction and the surroundings and provides the context for the design and use of heat sinks. Section 1.4 provides a summary of the considerations and procedures involved in the design and analysis of a typical array of fins. This introductory chapter concludes with a general discussion of extended surfaces that will serve to introduce the detailed considerations that follow in Chapters 2 through 10.

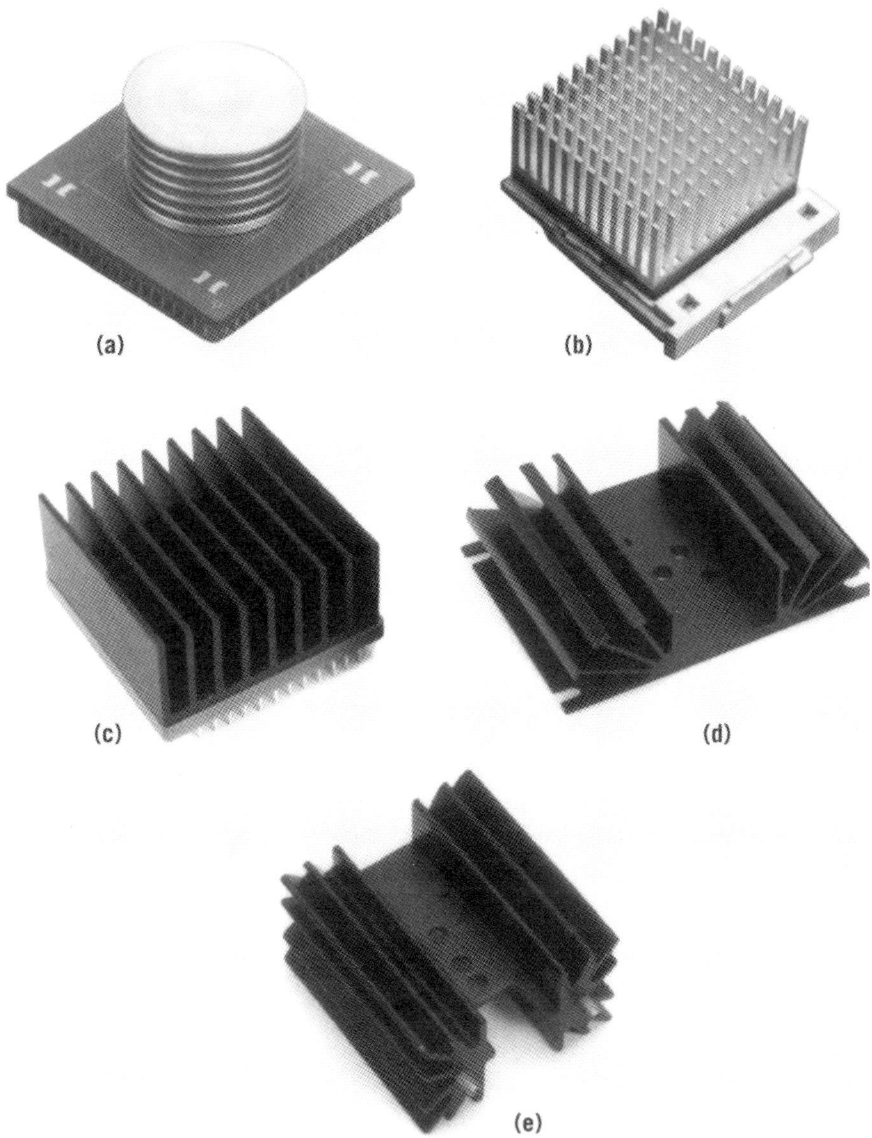

Figure 1.1 (*a*) Radial fin cooler, (*b*) clamp-on pin fin cooler/heat sink, (*c*) plate fin heat sink, (*d*) and (*e*) finned array heat sinks (photographs courtesy of Wakefield Engineering Co., Wakefield, MA).

1.2 HEAT-TRANSFER FUNDAMENTALS

1.2.1 Introduction

To determine the temperature differences encountered in the flow of heat within electronic systems, it is necessary to recognize the relevant heat-transfer mechanisms and their governing relations. In a typical system, heat removal from the active regions of the microcircuit(s) or chip(s) may require the use of several mechanisms, some operating in series and others in parallel, to transport the generated heat to the coolant or ultimate heat sink. Practitioners of the thermal arts and sciences generally deal with four basic thermal transport modes: conduction, convection, phase change, and radiation.

The process by which heat diffuses through a solid or a stationary fluid is termed *heat conduction*. Situations in which heat transfer from a wetted surface is assisted by the motion of the fluid give rise to *heat convection*, and when the fluid undergoes a liquid/solid or liquid/vapor state transformation, at, or very near, the wetted surface, attention is focussed on this *phase-change heat transfer*. The exchange of heat between surfaces, or between a surface and a surrounding fluid, by long-wavelength electromagnetic radiation is termed *heat radiation*.

1.2.2 Conduction Heat Transfer

One-Dimensional Conduction.
Steady thermal transport through solids is governed by the Fourier equation, which in one-dimensional form is expressible as

$$q = -kA\frac{dT}{dx} \qquad [\mathrm{W}] \tag{1.1}$$

where q is the heat flow, k is the thermal conductivity of the medium, A is the cross-sectional area for the heat flow, and dT/dx is the temperature gradient in the direction of heat flow. In this book, heat flow produced by a negative temperature gradient will be considered positive. This convention requires the insertion of the minus sign in Eq. (1.1) to ensure a positive heat flow, q. The temperature difference resulting from the steady-state diffusion of heat is thus related to the thermal conductivity of the material, the cross-sectional area, and the path length, L (Fig. 1.2), according to

$$(T_1 - T_2)_{cd} = q \cdot \frac{L}{kA} \qquad [\mathrm{K}] \tag{1.2}$$

The form of Eq. (1.2) suggests that, by analogy to Ohm's Law governing electrical current flow through a resistance, it is possible to define a thermal resistance for conduction, R_{cd}, as

$$R_{cd} \equiv \frac{(T_1 - T_2)}{q} = \frac{L}{kA} \qquad [\mathrm{K/W}] \tag{1.3}$$

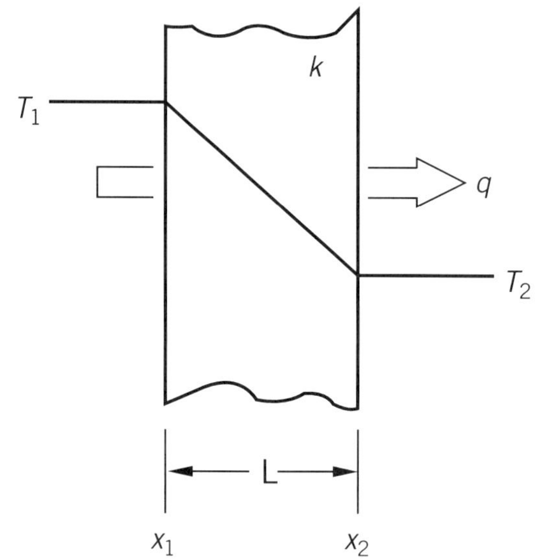

Figure 1.2 Heat transfer by conduction through a slab.

One-Dimensional Conduction with Internal Heat Generation. Situations in which a solid experiences internal heat generation, such as that produced by the flow of an electric current, give rise to more complex governing equations and require greater care in obtaining the appropriate temperature differences. The axial temperature variation in a slim, internally heated, conductor whose edges (ends) are held at a temperature, T_o (Fig. 1.3) is found to equal

$$T = T_o + q_g \frac{L^2}{2k} \left[\left(\frac{x}{L}\right) - \left(\frac{x}{L}\right)^2 \right]$$

when the volumetic heat generation rate, q_g, in W/m^3, is uniform throughout. The peak temperature is developed at the center of the solid and is given by

$$T_{max} = T_o + q_g \cdot \frac{L^2}{8k} \qquad [K] \tag{1.4}$$

Alternatively, because q_g is the volumetric heat generation $q_g = q/LW\delta$, the center-to-edge temperature difference can be expressed as

$$T_{max} - T_o = q\frac{L^2}{8kLW\delta} = q\frac{L}{8kA} \tag{1.5}$$

where the cross-sectional area, A, is the product of the width, W, and the thickness, δ. Comparison of Eq. (1.5) with Eq. (1.3), for a path length of $L/2$, reveals that the thermal resistance of a conductor with a distributed heat input

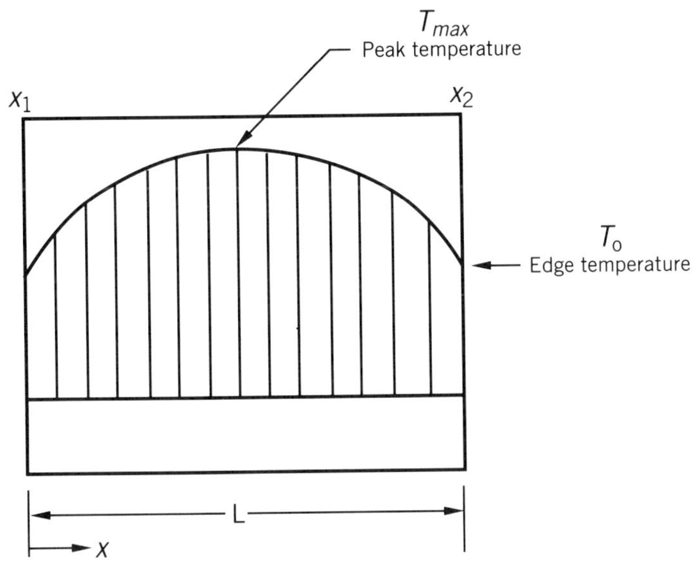

Figure 1.3 Temperature variation in an internally heated conductor.

is only one-quarter that of a structure in which all of the heat is generated at the center.

Spreading Resistance

In chip packages that provide for lateral spreading of the heat generated in the chip, the increasing cross-sectional area for heat flow at successive "layers" below the chip reduces the internal thermal resistance. Unfortunately, however, there is an additional resistance associated with this lateral flow of heat. This, of course, must be taken into account in the determination of the overall chip package temperature difference.

For the circular and square geometries common in microelectronic applications, Negus et al. (1989) provide an engineering approximation for the spreading resistance for a small heat source on a thick substrate or heat spreader. This can be expressed as

$$R_{sp} = \frac{0.475 - 0.62\epsilon + 0.13\epsilon^3}{k\sqrt{A_c}} \quad [\mathrm{K/W}] \tag{1.6}$$

where ϵ is the square root of the heat source area divided by the *substrate* area, k is the thermal conductivity of the substrate, and A_c is the area of the heat source. It is to be noted that the use of Eq. (1.6) requires that the substrate be 3 to 5 times thicker than the square root of the heat source area.

For relatively thin layers on thicker substrates, such as encountered in the use of thin lead frames, or heat spreaders interposed between the chip and substrate, Eq. (1.6) cannot provide an acceptable prediction of R_{sp}. Instead, use can be made of the numerical results plotted in Fig. 1.4 to obtain the requisite value of the spreading resistance.

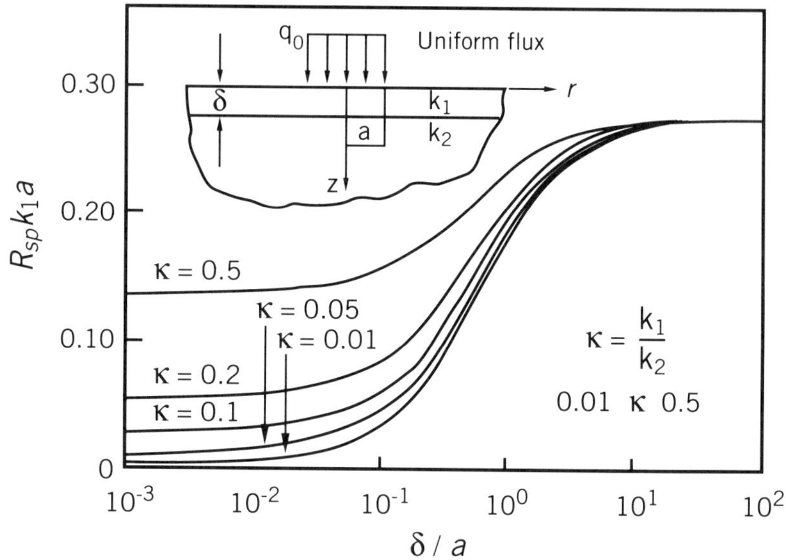

Figure 1.4 The thermal resistance for a circular heat source on a two layer substrate (Yovanovich and Antonetti, 1988).

Interface/Contact Resistance.
Heat transfer across the interface between two solids is generally accompanied by a measurable temperature difference, which can be ascribed to a contact or interface thermal resistance. For perfectly adhering solids, geometrical differences in the crystal structure (lattice mismatch) can impede the flow of phonons and electrons across the interface, but this resistance is generally negligible in engineering design. However, when dealing with real interfaces, the asperities present on each of the surfaces, as shown in an artist's conception in Fig. 1.5, limit actual contact between the two solids to a very small fraction of the apparent interface area. The flow of heat across the gap between two solids in nominal contact is, thus, seen to involve solid conduction in the areas of actual contact and fluid conduction across the "open" spaces. Radiation across the gap can be important in a vacuum environment or when the surface temperatures are high.

The heat transferred across an interface can be found by adding the effects of the solid-to-solid conduction and the conduction through the fluid and recognizing that the solid-to-solid conduction, in the contact zones, involves heat flowing sequentially through the two solids. With the total contact conductance, h_{co}, taken as the sum of the solid-to-solid conductance, h_c, and the gap conductance, h_g,

$$h_{co} = h_c + h_g \qquad [\mathrm{W/m^2\text{-}K}] \qquad (1.7a)$$

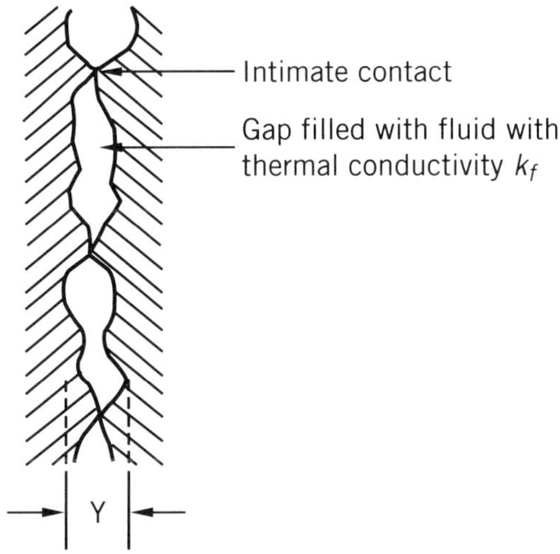

Figure 1.5 Physical contact between two nonideal surfaces.

the contact resistance based on the apparent contact area, A_a, may be defined as

$$R_{co} \equiv \frac{1}{h_{co} A_a} \qquad \text{[K/W]} \tag{1.7b}$$

In Eq. (1.7*a*), h_c is given by Yovanovich and Antonetti (1988) as

$$h_c = 1.25 k_s \left(\frac{m}{\sigma}\right)\left(\frac{P}{H}\right)^{0.95} \tag{1.8a}$$

where k_s is the harmonic mean thermal conductivity for the two solids with thermal conductivities, k_1 and k_2,

$$k_s = \frac{2k_1 k_2}{k_1 + k_2} \qquad \text{[W/m-K]}$$

σ is the effective rms surface roughness developed from the surface roughnesses of the two materials, σ_1 and σ_2,

$$\sigma = \sqrt{\sigma_1^2 + \sigma_2^2} \qquad [\mu\text{-m}]$$

and m is the effective absolute surface slope composed of the individual slopes of the two materials, m_1 and m_2,

$$m = \sqrt{m_1^2 + m_2^2}$$

and where P is the contact pressure, and H is the microhardness of the softer material, both in N/m^2. In the absence of detailed information, the m/σ ratio can be taken equal to 0.111 to 0.200 microns^{-1} for relatively smooth surfaces (Yovanovich, 1990).

In Eq. (1.7*a*), h_g is given by

$$h_g = \frac{k_g}{Y + M} \tag{1.8b}$$

where k_g is the thermal conductivity of the gap fluid, Y is the distance between the mean planes (Fig. 1.5) given by

$$\frac{Y}{\sigma} = 1.185 \left[-\ln \left(3.132 \frac{P}{H} \right) \right]^{0.547}$$

and M is a gas parameter used to account for rarified gas effects:

$$M = \alpha \beta \mathbf{\Lambda}$$

In this relationship, α is an accomodation parameter (approximately equal to 2.4 for air and clean metals), $\mathbf{\Lambda}$ is the mean free path of the molecules (equal to approximately 0.06 μm for air at atmospheric pressure and 15°C), and β is a fluid property parameter (equal to approximately 1.7 for air and other diatomic gases).

Returning to Eq. (1.7a) and inserting the expressions from Eqs. (1.8*a*) and (1.8*b*), the contact resistance becomes

$$R_{co} \equiv \left\{ \left[1.25 k_s \left(\frac{m}{\sigma} \right) \left(\frac{P}{H} \right)^{0.95} + \frac{k_g}{Y + M} \right] A_a \right\}^{-1} \tag{1.9}$$

Lumped-Capacity Heating and Cooling.
An internally heated solid of relatively high thermal conductivity, which is experiencing no external cooling, will undergo a constant rise in temperature according to

$$\frac{dT}{dt} = \frac{q}{mc} \qquad \text{[K/s]} \tag{1.10}$$

where q is the rate of internal heat generation, m is the mass of the solid, and c is the specific heat of the solid. Equation (1.10) assumes that all of the mass can be represented by a single temperature. This relation is frequently termed the *lumped-capacity* solution for transient heating.

Expanding upon the analogy between thermal and electrical resistances suggested previously, the product of mass and specific heat can be viewed as analogous to electrical capacitance and to thus constitute the *thermal capacitance*.

When this same solid is externally cooled, the temperature rises asymptotically toward the steady-state temperature, which is itself determined by the external resistance to heat flow, R_{ex}. Consequently, the time variation of the temperature of the solid, with the initial temperature equal to the fluid temperature, is expressible as

$$T(t) = T(t=0) + qR_{ex}[1 - e^{-t/R_{ex}mc}] \qquad \text{[K]} \qquad (1.11)$$

where the product of the external resistance, R_{ex}, and the thermal capacitance, mc, is seen to constitute the *thermal time constant* of the system.

1.2.3 Convective Heat Transfer

The Heat Transfer Coefficient.
Convective thermal transport from a surface to a fluid in motion can be related to the heat-transfer coefficient, h, the surface-to-fluid temperature difference, and the "wetted" surface area, S, in the form

$$q = hS(T_s - T_{fl}) \qquad \text{[W]} \qquad (1.12)$$

The differences between convection to a rapidly moving fluid, a slowly flowing fluid, or a stagnant fluid, as well as variations in the convective heat-transfer rate among various fluids, are reflected in the values of h. For a particular geometry and flow regime, h may be found from available empirical correlations and/or theoretical relations. Moreover, use of Eq. (1.12) makes it possible to define the convective thermal resistance, as

$$R_{cv} \equiv \frac{1}{hS} \qquad \text{[K/W]} \qquad (1.13)$$

Dimensionless Parameters.
Common dimensionless parameters that are used in the correlation of heat-transfer data are the *Nusselt number*, Nu, which relates the convective heat-transfer coefficient to conduction in the fluid [1]

$$\text{Nu} \equiv \frac{h}{k_{fl}/L} = \frac{hL}{k_{fl}}$$

the *Prandtl number*, Pr, which is a fluid property parameter relating the diffusion of momentum to the conduction of heat

$$\text{Pr} \equiv \frac{c_p \mu}{k_{fl}}$$

the *Grashof number*, Gr, which accounts for the bouyancy effect produced by the volumetric expansion of the fluid

$$\text{Gr} \equiv \frac{\rho^2 \beta g L^3 \Delta T}{\mu^2}$$

[1]At this point, it becomes necessary to use the subscript fl whenever there is a chance for confusion between solid and fluid properties.

and the *Reynolds number*, Re, which relates the momentum in the flow to the viscous dissipation

$$\text{Re} \equiv \frac{\rho_{fl} V L}{\mu}$$

Natural Convection.
In natural convection, fluid motion is induced by density differences resulting from temperature gradients in the fluid. The heat-transfer coefficient for this regime can be related to the buoyancy and the thermal properties of the fluid through the *Rayleigh number*, which is the product of the Grashof and Prandtl numbers,

$$\text{Ra} = \frac{\rho^2 \beta g c_p}{\mu k_{fl}} L^3 \Delta T$$

where the fluid properties, ρ, β, c_p, μ, and k, are evaluated at the fluid bulk temperature, and ΔT is the temperature difference between the surface and the fluid.

Empirical correlations for the natural convection heat-transfer coefficient generally take the form

$$h = C\left(\frac{k_{fl}}{L}\right)(\text{Ra})^n \qquad [\text{W/m}^2\text{-K}] \tag{1.14}$$

where n is found to be approximately 0.25 for $10^3 < \text{Ra} < 10^9$, representing laminar flow, 0.33 for $10^9 < \text{Ra} < 10^{12}$, the region associated with the transition to turbulent flow, and 0.4 for $\text{Ra} > 10^{12}$, when strong turbulent flow prevails. The precise value of the correlating coefficient, C, depends on the fluid, the geometry of the surface, and the Rayleigh number range. Nevertheless, for common plate, cylinder, and sphere configurations in air, it is found to vary in the relatively narrow range of 0.45 to 0.65 for laminar flow and 0.11 to 0.15 for turbulent flow past the heated surface (Kraus and Bar-Cohen, 1983).

Natural convection in vertical channels, such as those formed by arrays of longitudinal fins, is of major significance in the analysis and design of heat sinks. Elenbaas (1942) was the first to document a detailed study of this configuration. His experimental results for isothermal plates were later confirmed numerically by Bodoia and Osterle (1964) and have subsequently been extended to many other configurations.

These many studies have revealed that the value of the Nusselt number lies between two extremes associated with the separation between the plates or the channel width. For wide spacing, the plates appear to have little influence on one another, and the Nusselt number in this case achieves its *isolated plate limit*. On the other hand, for closely spaced plates or for relatively long channels, the fluid attains its *fully developed* velocity and temperature profiles, and the Nusselt number reaches its *fully developed limit*. Intermediate values of the Nusselt number can be obtained from a correlating expression following the form suggested by Churchill and Usagi (1972) for smoothly varying processes, as has been verified by detailed experimental and numerical studies (Bar-Cohen and Rohsenow, 1984).

Thus, the correlation for the average value of h along isothermal vertical plates separated by a spacing, z, is given by

$$h = \frac{k_{fl}}{z}\left[\frac{576}{(\mathrm{El})^2} + \frac{2.873}{(\mathrm{El})^{1/2}}\right]^{1/2} \tag{1.15}$$

where El is the *Elenbaas number,*

$$\mathrm{El} \equiv \frac{\rho^2 \beta g c_p z^4 \Delta T}{\mu k_{fl} L}$$

and $\Delta T = T_s - T_{fl}$. The use of this correlation will be explored in some detail in an example in Sec. 1.4, and further details concerning its use will be provided in Chap. 10.

Forced Convection.
For forced flow in long, or very narrow, parallel-plate channels, the heat-transfer coefficient attains an asymptotic value (a fully developed limit), which for symmetrically heated channel surfaces is equal approximately to

$$h = \frac{4k_{fl}}{d_e} \qquad [\mathrm{W/m^2\text{-}K}] \tag{1.16}$$

where d_e is the *hydraulic diameter* defined in terms of the flow area, A, and the wetted perimeter of the channel, P_w,

$$d_e \equiv \frac{4A}{P_w}$$

In the inlet zones of such parallel-plate channels and along isolated plates, the heat-transfer coefficient varies with the distance from the leading edge. The low velocity, or laminar flow, average convective heat-transfer coefficient along a surface of length L, for $\mathrm{Re} \leq 3 \times 10^5$ is given for example, in Kraus and Bar-Cohen(1983) by:

$$h = 0.664\left(\frac{k_{fl}}{L}\right)\mathrm{Re}^{0.5}\mathrm{Pr}^{0.333} \qquad [\mathrm{W/m^2\text{-}K}] \tag{1.17}$$

where k_{fl} is the fluid thermal conductivity, L is the characteristic dimension of the surface, and Re is the Reynolds number.

A similar relation applies to a flow in tubes, pipes, annuli, or channels with the hydraulic diameter, d_e, serving as the characteristic dimension in both the Nusselt and Reynolds numbers. For laminar flow, $\mathrm{Re} \leq 2100$,

$$\frac{hd_e}{k_{fl}} = 1.86\left[(\mathrm{Re})(\mathrm{Pr})\left(\frac{d_e}{L}\right)\right]^{1/3}\left(\frac{\mu}{\mu_w}\right)^{0.14} \tag{1.18}$$

which is attributed to Sieder and Tate (1936), and where μ_w is the viscosity of the convective medium at the channel wall temperature. Observe that this relationship shows the heat-transfer coefficient to attain its maximum value at the inlet to the channel and decreases towards the asymptotic fully developed value given by Eq. (1.16) as d_e/L decreases.

In higher-velocity turbulent flow along plates, the dependence of the convective heat-transfer coefficient on the Reynolds number increases and, in the range $\mathrm{Re} \geq 3 \times 10^5$, is typically given by (Kraus and Bar-Cohen 1983):

$$h = 0.036\left(\frac{k_{fl}}{L}\right)(\mathrm{Re})^{0.8}(\mathrm{Pr})^{0.333} \qquad [\mathrm{W/m^2\text{-}K}] \tag{1.19}$$

In pipes, tubes, annuli, and channels, fully turbulent flow occurs at an equivalent diameter-based Reynolds number of 10,000, with the regime bracketed by $2100 \leq \mathrm{Re} \leq 10,000$ usually referred to as the *transition region*. For the transition region, Hausen (1943) has provided the correlating equation

$$\frac{hd_e}{k_{fl}} = 0.116\,[\mathrm{Re} - 125]\,(\mathrm{Pr})^{1/3}\left[1 + \left(\frac{d_e}{L}\right)^{2/3}\right]\left(\frac{\mu}{\mu_w}\right)^{0.14} \tag{1.20}$$

The Sieder and Tate (1936) relation

$$\frac{hd_e}{k_{fl}} = 0.023(\mathrm{Re})^{0.8}(\mathrm{Pr})^{1/3}\left(\frac{\mu}{\mu_w}\right)^{0.14} \tag{1.21}$$

is commonly used for turbulent flow.

Phase Change Heat Transfer.

Boiling heat transfer displays a complex dependence on the temperature difference between the heated surface and the saturation temperature (boiling point) of the liquid. Following Rohsenow (1952), the heat-transfer rate in nucleate boiling can be approximated by a relation of the form

$$q_\phi = C_{sf}A(T_s - T_{sat})^3 \qquad [\mathrm{W}] \tag{1.22}$$

where C_{sf} is a function of the surface/fluid combination and various fluid properties. For comparison purposes, it is possible to define a boiling heat-transfer coefficient, h_ϕ,

$$h_\phi = C_{sf}(T_s - T_{sat})^2 \qquad [\mathrm{W/m^2\text{-}K}]$$

which, however, will vary strongly with surface temperature.

Finned Surfaces.

In the thermal design of electronic equipment, frequent use is made of finned or "extended" surfaces in the form of *heat sinks* or *coolers*. While such finning can substantially increase the surface area in contact with the coolant, resistance to heat flow in the fin reduces the average temperature of the exposed surface relative to the fin base. In the analysis of such finned surfaces, it is, thus, common to define a fin efficiency, η, equal to the ratio of the actual heat dissipated by the fin to the heat that would be dissipated if the fin possessed an infinite thermal conductivity (as discussed in Section 1.3.4). Using this approach, heat transferred from a fin or a fin structure can be expressed in the form

$$q_f = hS_f\eta(T_b - T_a) \qquad [\mathrm{W}] \tag{1.23}$$

where T_b is the temperature at the base of the fin and where, here, T_a is the surrounding temperature [2] and q_f is the heat entering the base of the fin which, in the steady state, is equal to the heat dissipated by the fin.

The thermal resistance of a finned surface is given by

$$R_f \equiv \frac{1}{hS_f\eta} \tag{1.24}$$

where η, the fin efficiency, can be expected to lie between 0.5 and 0.7 for a well designed fin cross-section fin (Kraus and Bar-Cohen, 1983).

Flow Resistance.
The transfer of heat to a flowing gas or liquid that is not undergoing a phase change results in an increase in the coolant temperature from an inlet temperature of T_{in} to an outlet temperature of T_{out}, according to

$$q = mc_p(T_{out} - T_{in}) \qquad [\mathrm{W}] \tag{1.25}$$

Based on this relation, it is possible to define an effective flow resistance, R_{fl}, as

$$R_{fl} \equiv \frac{1}{\dot{m}c_p} \qquad [\mathrm{K/W}] \tag{1.26}$$

where $\dot{m}$ is in kg/sec.

1.2.4 Radiative Heat Transfer

Unlike conduction and convection, radiative heat transfer between two surfaces or between a surface and its surroundings is not linearly dependent on the temperature difference and is expressed instead as

$$q = \sigma S\mathcal{F}(T_1^4 - T_2^4) \qquad [\mathrm{W}] \tag{1.27}$$

where $\mathcal{F}$ includes the effects of surface properties and geometry and σ is the Stefan-Boltzman constant, $\sigma = 5.67 \times 10^{-8}$ W/m^2-K^4. For modest temperature differences, this equation can be linearized to the form

$$q = h_rS(T_1 - T_2) \qquad [\mathrm{W}] \tag{1.28}$$

[2]The reader is reminded that T_s is used in this chapter to represent *surface temperature*. Here, the subscript a is used to represent the *ambient temperature*.

where h_r is the effective "radiation" heat-transfer coefficient

$$h_r = \sigma \mathcal{F}(T_1^2 + T_2^2)(T_1 + T_2) \qquad [\text{W/m}^2\text{-K}] \tag{1.29a}$$

and, for small $\Delta T = T_1 - T_2$, h_r is approximately equal to

$$h_r = 4\sigma \mathcal{F}(T_1 T_2)^{3/2} \qquad [\text{W/m}^2\text{-K}] \tag{1.29b}$$

It is of interest to note that for temperature differences of the order of 10 K, the radiative heat-transfer coefficient, h_r, for an ideal (or "black") surface in an absorbing environment is approximately equal to the heat-transfer coefficient in natural convection of air.

Noting the form of Eq. (1.28), the radiation thermal resistance, analogous to the convective resistance, is seen to equal

$$R_r \equiv \frac{1}{h_r S} \qquad [\text{K/W}] \tag{1.30}$$

1.2.5 Thermal Resistance Network

The expression of the governing heat-transfer relations in the form of thermal resistances greatly simplifies the first-order thermal analysis of electronic systems. Following the established rules for resistance networks, thermal resistances that occur sequentially along a thermal path can be simply summed to establish the overall thermal resistance for that path. In similar fashion, the reciprocal of the effective overall resistance of several parallel heat-transfer paths can be found by summing the reciprocals of the individual resistances. In refining the thermal design of an electronic system, prime attention should be devoted to reducing the largest resistances along a specified thermal path and/or providing parallel paths for heat removal from a critical area.

While the thermal resistances associated with various paths and thermal transport mechanisms constitute the "building blocks" in performing a detailed thermal analysis, they have also found widespread application as "figures of merit" in evaluating and comparing the thermal efficacy of various packaging techniques and thermal management strategies.

1.3 EXTENDED SURFACE AND HEAT SINKS

1.3.1 Introduction

A growing number of engineering disciplines are concerned with energy transitions requiring the rapid movement of heat. They produce an expanding demand for high-performance heat-transfer components with progressively smaller weights, volumes, costs, or more accommodating shapes. The electonic component heat sink is a primary example of these trends. *Extended surface heat transfer* is the study of high-performance heat-transfer components with respect to these parameters and of their behavior in a variety of thermal environments. Typical components are found in air-land-space vehicles and their power sources; chemical, refrigeration, and cryogenic processes; electric and electronic assemblies; conventional furnaces and gas turbines; process heat dissipaters and waste-heat boilers; nuclear-fuel modules, direct energy conversion systems and many more.

In the design and construction of various types of heat-transfer equipment, simple shapes such as cylinders, bars, and plates are used to implement the flow of heat between a source and a sink. They provide heat-absorbing or heat-rejecting surfaces, and each is known as a *prime surface*. When a prime surface is extended by appendages intimately connected with it, such as an electronic component to which additional heat-transfer surface is appended, the additional surface is known as *extended surface*. In a heat-rejecting system for electronic equipment, the component to be temperature controlled is referred to as the *prime surface*, and the extended surface is often called either a *cooler* or a *heat sink*. Moreover, in some cases, the entire configuration of component and cooler/heat sink are considered collectively as extended surfaces to distinguish them from the use of a prime surface component alone. The elements used to extend prime surfaces are referred to as *fins*, *finned arrays* or *arrays of fins*, and when the fin elements are conical or cylindrical, they may be referred to as *spines*, *pins*, or *pegs*. Some typical semiconductor heat sinks and coolers are displayed in Fig. 1.1 at the beginning of this chapter.

1.3.2 Surface Augmentation with Extended Surface

Inspection of the *heat transfer-rate* equation,

$$q = hS\Delta T$$

shows that for a fixed rate of heat dissipation, q, the magnitude of the temperature difference (or *temperature excess*), ΔT, is governed by the product, hS. To obtain a high dissipation at a relatively low temperature excess, one requires either a component with substantial surface area for dissipation or a coolant possessing superior heat-transfer properties.

The trend in component design has been, and most probably will continue to be, toward microminiaturization. Ordinarily, miniaturized electronic equipment is quite small, and there is not much component surface for heat dissipation. Moreover, air, which is inexpensive and often readily available, does not have outstanding attributes as a heat-transfer medium. Thus, the designer of the cooling system is often faced with the problem of cooling small heat-dissipating components to precise temperature limits while using a fluid with definite heat-transfer limitations. This dilemma can be summarized as being one of low hS product.

Increase of the hS product is not easily achieved by increasing the heat-transfer coefficient. A considerably more effective method for obtaining this increase is by augmenting the surface through the use of extended surface or fins. Such augmentation can result in as much as an order-of-magnitude increase in the hS product without an appreciable degradation in the heat-transfer coefficient. While the addition of fins is a straight forward step to increasing hS, it must be recognized that, because they are remotely located from the source of heat, the fins will not perform as well as the prime surface. This consideration can be handled by a parameter known as the *fin efficiency*. However, before discussing the fin efficiency, (Sec. 1.3.4), fin types and the applicable terminology should be considered.

1.3.3 Types of Fins, Terminology, and Nomenclature

Figure 1.6 shows three typical fin structures that can be (and are) used to form electronic-equipment heat sinks. The *longitudinal fin*, often called the *straight fin*, is shown in Fig. 1.6*a*, where the cross-sectional and profile areas may be noted. The longitudinal fins are characterized by the dimensions that form the cross-sectional area; one dimension is very much greater than the other.

The *spine* (or peg or pin fin) is displayed in Fig. 1.6*b*. Its position on the mounting surface is similar to that of the longitudinal fin. However, for the spine, the two dimensions that form the cross-sectional area possess comparable dimensions.

Figure 1.6*c* shows the *radial fin*, which is placed *around* a curved surface such as a tube or a cylindrical component. This fin is often referred to as an *annular* or *circumferential fin*, and, in its radial coordinate system, its cross-sectional area is a function of the radial distance from the origin of the coordinate system.

The extended surface literature, in general, and the electronic heat sink literature, in particular, is burgeoning. There are many investigators publishing the results of their work and, because of this, a standard terminology and associated nomenclature does not exist. The user of this book must have an instinctive grasp of what is meant by *fin length, fin thickness*, and *fin height* when considering individual fins and combinations of fins. For this reason, a discussion of these quantities is now presented.

- The term *height* applies to longitudinal and radial fins and to spines. It is the direction that is associated with the flow of heat and can be easily determined by the so-called height coordinate, b. The height, b, of a longitudinal fin or spine is the difference between the dimensions at the fin or spine tip $x(a)$ and base $x(b)$. It is always a positive quantity:

  $$b = |x(a) - x(b)|$$

  The use of the term *height* is not affected by fin, spine, or heat sink orientation. It applies to horizontal base plates in which the fins or spines are vertical, as well as to vertical base plates in which the fins or spines are horizontal. It is the dimension of the protrusion from the base plate surface.

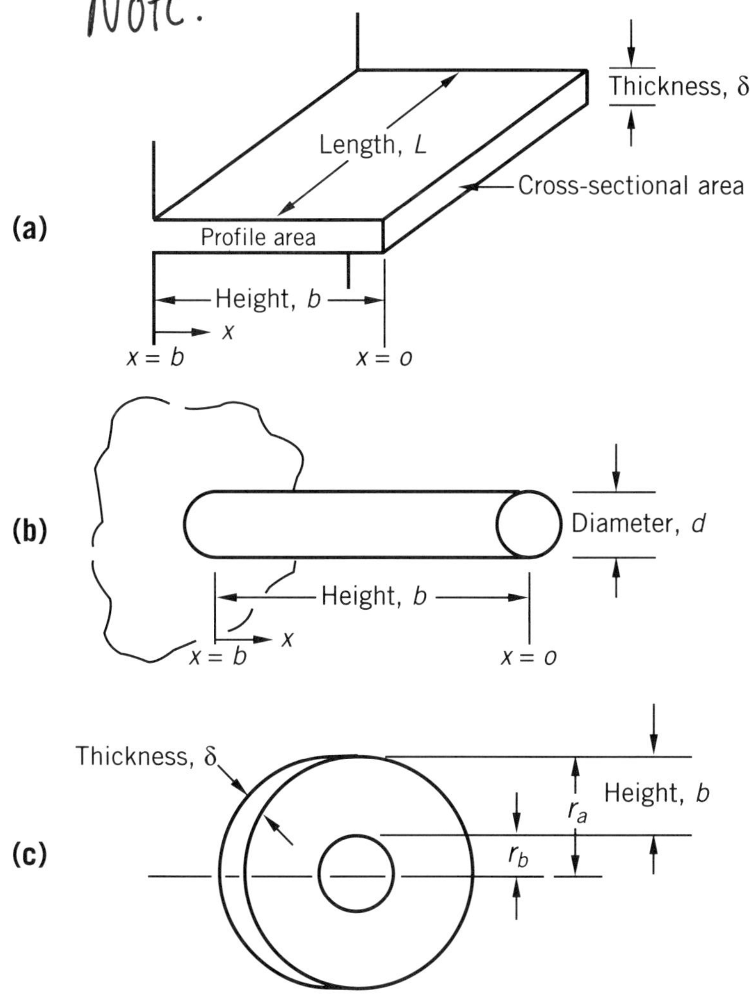

Figure 1.6 Types of fins, terminology, and nomenclature: (*a*) longitudinal fin, (*b*) spine, and (*c*) radial fin.

For the radial fin of rectangular profile, the height, b, is equal to the difference between the tip and base radii:

$$b = r_a - r_b$$

- The fin length, L, applies only to longitudinal fins and is the dimension that is in the direction of flow of the convective medium. It was selected in this manner to permit the use of the terminology *flow length*. Observe that it is, most often, the largest of the three dimensions that pertain to the individual fin.

- The fin thickness applies to both the longitudinal and radial fins of rectangular profile. It is designated by δ and, for the longitudinal fin of rectangular profile, it is the dimension that is perpendicular to the plane formed by the fin height, b, and the fin length, L. For radial fins, where the term *fin length*, has no meaning, it is perpendicular to the radial coordinate. For spines, the thickness is usually designated by a diameter[3] and the term *thickness* does not often apply.

- Observe that the term *width*, is never used as a synonym for *thickness*. The term *width* is reserved for base or mounting plates that support fins or spines. In this case, the base, mounting, or support plate will have a length, L, a width, W, and its own thickness, δ_p.

- In the event that the total height between the interior of the mounting plate and the tip of the fin is of interest, it is designated as

$$H = \delta_p + b$$

where δ_p is the mounting plate thickness.

With the terminology of *fin height, width,* and *thickness*, as well as the cross-sectional and profile areas in hand, attention can now turn to the fin efficiency.

1.3.4 The Fin Efficiency

It can be shown quite readily that when a fin and its prime surface are exposed to a uniform thermal environment, a unit of fin surface will be less effective than a unit of prime surface. Consider the plate equipped with a longitudinal fin of rectangular profile in Fig. 1.7. Let the inner plate surface remove heat from a source with a uniform heat-transfer coefficient and temperature T_1, and let the outer plate and fin surfaces reject it to colder surroundings at temperature T_s, with a uniform heat-transfer coefficient. The colder surface of the plate is at some intermediate temperature T_p, and heat from the source leaves the plate because of the temperature potential, $T_p - T_s$. Similarly, the fin surface is at some temperature, T, and heat leaves it because of the temperature potential, $T - T_s$. The heat enters the fin through its base where it joins the plate and moves continuously through it by conduction. In most cases the temperature at the base of the fin will be very nearly the same as T_p. Heat absorbed by the fin through its base can flow toward its extreme tip only if there is a temperature gradient within the fin such that T_p is greater than T. For this condition, the temperature potential, $T - T_s$, will be smaller than $T_p - T_s$, and a unit of fin surface will be less effective than a unit of plate or prime surface. The temperature, T, of the fin surface varies from the base to the extreme tip of the fin.

[3]Elliptical spines will have a major and a minor diameter, and rectangular spines will have two thickness dimensions, δ_1 and δ_2.

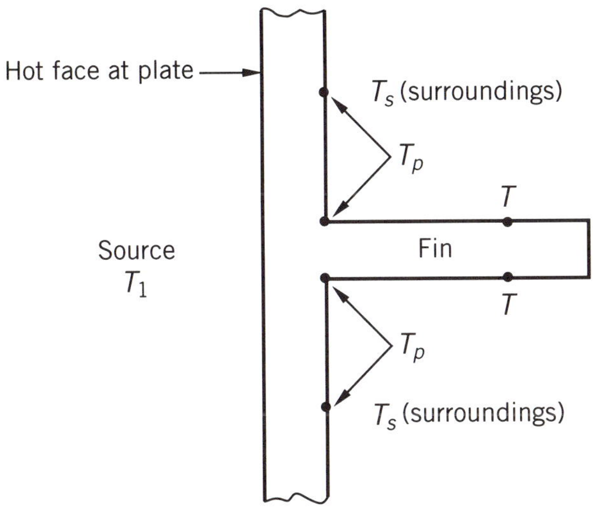

Figure 1.7 Plate equipped with longitudinal fin of rectangular profile.

The inescapable loss in performance of a square meter of fin surface compared with a square meter of prime surface is the inefficiency of the fin. The *fin efficiency* is defined consistently throughout this book as the ratio of *the actual heat dissipation of a fin to its ideal heat dissipation if the entire fin surface were at the same temperature as its base.*[4] Other indexes of performance are also employed, such as *fin effectiveness, weighted fin efficiency*, and *fin resistance*. Fins of given size, shape, and material possess different fin efficiencies, and the efficiency of any fin will also vary with the mode of heat transfer with respect to the environment and the thermal conductivity of the fin material.

1.3.5 Modes of Heat Transfer Involving Fins and Surroundings

The study of extended surface heat transfer in most cases involves two factors that may be conveniently separated. One factor considers only the movement of heat within the fin by conduction. The other considers how the fin exchanges heat with its surroundings, which usually involves convection and radiation acting alone or simultaneously.

1.3.6 Limiting Assumptions

Reference has been made to extended surfaces comprising several types of prime surfaces and several types of fins. Much insight on the heat flows, temperature profiles, efficiencies, and optimization of fin parameters can be obtained from the analysis of three fundamental fin geometries. These are (1) longitudinal fins, (2) radial fins, and (3) spines.

Fins of various geometries and thermal conductivities respond differently

[4]It is to be noted that the fin efficiency is, thus, equal to the ratio of the average fin excess temperature to the base excess temperature.

to identical and uniform heat sources and sinks. Similarly, there are numerous ways in which the temperatures and heat-transfer coefficients of sources and sinks may vary. Important to the analysis of fin geometries are the constraints or assumptions that are employed to define and limit the problem and often to simplify its solution. The analyses of the three fundamental fin geometries given here employ the assumptions of Murray (1938) and Gardner (1945). These limiting assumptions, known as the Murray-Gardner assumptions are:

1. The heat flow in the fin and its temperatures remain constant with time.
2. The fin material is homogeneous, its thermal conductivity is the same in all directions, and it remains constant.
3. The heat-transfer coefficient to the fin is constant and uniform over the entire surface of the fin.
4. The temperature of the medium surrounding the fin is uniform.
5. The fin thickness is so small compared with its height that temperature gradients *across* the fin thickness may be neglected.
6. The temperature at the base of the fin is uniform.
7. There is no contact resistance where the base of the fin joins the prime surface.
8. There are no heat sources within the fin itself.
9. The heat transferred through the outermost edge of the fin is negligible compared with that leaving the fin through its lateral surface.
10. Heat transfer to or from the fin is proportional to the temperature excess between the fin and the surrounding medium.

1.4 CHIP MODULE THERMAL RESISTANCES

1.4.1 Definition

The thermal performance of alternative chip and packaging techniques is commonly compared on the basis of the overall (junction-to-coolant) thermal resistance, R_T. This packaging figure of merit is generally defined in a purely empirical fashion:

$$R_T \equiv \frac{T_j - T_{fl}}{q_c} \qquad [\mathrm{K/W}] \tag{1.31}$$

where T_j and T_{fl} are the junction and coolant (fluid) temperatures, respectively, and q_c is the chip heat dissipation.

Unfortunately, however, most measurement techniques are incapable of detecting the actual junction temperature, that is, the temperature of the small

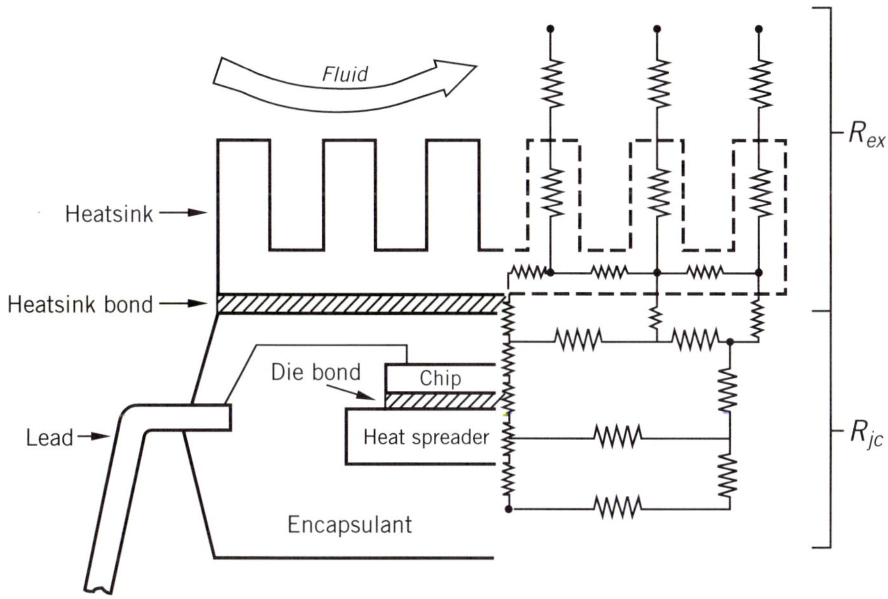

Figure 1.8 Primary thermal resistances in a single chip package.

volume at the interface of p-type and n-type semiconductors. Hence, this term generally refers to the average temperature or a representative temperature on the chip. To lower chip temperature at a specified power dissipation, it is clearly necessary to select and/or design a chip package with the lowest thermal resistance.

Examination of various packaging techniques reveals that the junction-to-coolant thermal resistance is, in fact, composed of an internal, largely conductive, resistance and an external, primarily convective, resistance. As shown in Fig. 1.8, the internal resistance, R_{jc}, is encountered in the flow of dissipated heat from the active chip surface through the materials used to support and bond the chip and on to the case of the integrated circuit package. The flow of heat from the case directly to the coolant, or indirectly through a fin structure and then to the coolant, must overcome the external resistance, R_{ex}.

The thermal design of single-chip packages, including the selection of die-bond, heat spreader, substrate, and encapsulant materials, as well as the quality of the bonding and encapsulating processes, can be characterized by the internal, or so-called *junction-to-case resistance*. The convective heat-removal techniques applied to the external surfaces of the package, including the effect of finned heat sinks and other thermal enhancements, can be compared on the basis of the external thermal resistance. The complexity of heat flow and coolant flow paths in a multichip module generally requires that the thermal capability of these packaging configurations be examined on the basis of overall, or chip-to-coolant, thermal resistance.

1.4.2 Internal Thermal Resistance

As discussed in Sec. 1.2, conductive thermal transport is governed by the Fourier equation, which can be used to define a conduction thermal resistance, as in Eq. (1.3). In flowing from the chip to the package surface or case, the heat encounters a series of resistances associated with individual layers of materials such as silicon, solder, copper, alumina, and epoxy, as well as the contact resistances which occur at the interfaces between pairs of materials. Although the actual heat flow paths within a chip package are rather complex and may shift to accommodate varying external cooling situations, it is possible to obtain a first-order estimate of the internal resistance by assuming that power is dissipated uniformly across the chip surface and that heat flow is largely one-dimensional. To the accuracy of these assumptions,

$$R_{jc} = \frac{T_j - T_c}{q_c} = \sum \frac{\Delta x}{kA} \qquad [\mathrm{K/W}] \tag{1.32}$$

can be used to determine the internal chip module resistance where the summed terms represent the conduction thermal resistances posed by the individual layers, each with thickness, Δx.

As the thickness of each layer decreases and/or the thermal conductivity and cross-sectional area increase, the resistance of the individual layers decreases. R_{cd} values for packaging materials with typical dimensions can be found via Eq. (1.3) or Fig. 1.9, to range from 2 K/W for a 1000 mm^2 by 1 mm thick layer of a typical epoxy encapsulant to 0.0006 K/W for a 100 mm^2 by 25 micron (1 mil) thick layer of copper. Similarly, the values of conduction resistance for typical "soft" bonding materials are found to lie in the range of approximately 0.1 K/W for solders and 1 – 3 K/W for epoxies and thermal pastes for typical $\Delta x/A$ ratios of 0.25 to 1.0 m^{-1}.

Commercial fabrication practice in the mid-1990s yields internal chip package thermal resistances varying from approximately 80 K/W for a plastic package with no heat spreader to 15 – 20 K/W for a plastic package with heat spreader, and to 5 – 10 K/W for a ceramic package or an especially designed plastic chip package. Large and/or carefully designed chip packages can attain even lower values of R_{jc}, down perhaps to 2 K/W.

Comparison of theoretical and experimental values of R_{jc} reveals that the resistances associated with compliant, low thermal conductivity bonding materials and the spreading resistances, as well as the contact resistances at the lightly loaded interfaces within the package, often dominate the internal thermal resistance of the chip package. It is, thus, not only necessary to determine correctly the bond resistance but also to add the values of R_{sp}, obtained from Eq. (1.6) and/or Fig. 1.4, and R_{co} from Eqs. (1.7*b*) or (1.9) to the junction-to-case resistance calculated from Eq. (1.32). Unfortunately, the absence of detailed information on the voidage in the die-bonding and heat-sink attach layers and the present inability to determine, with precision, the

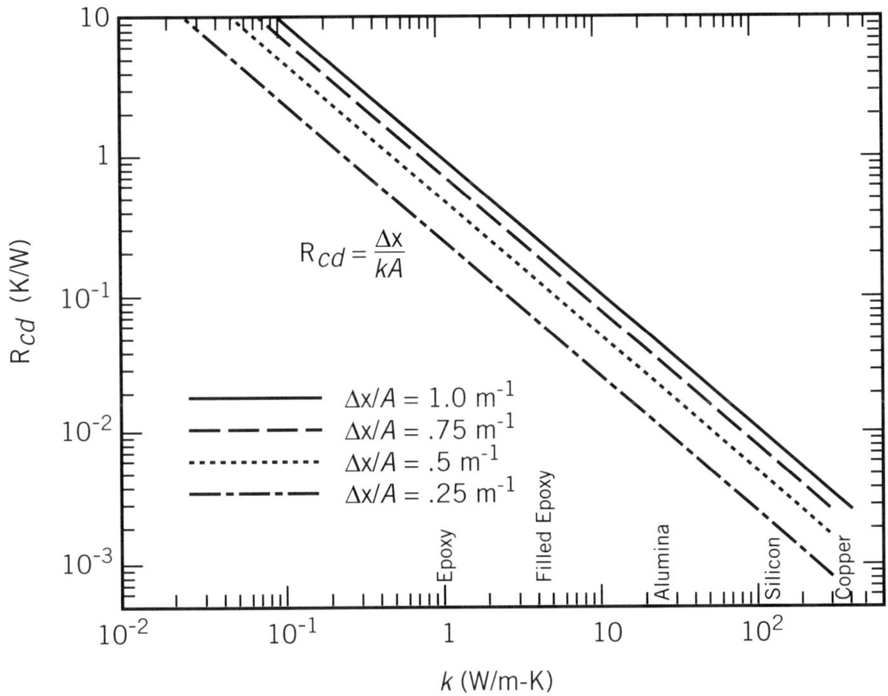

Figure 1.9 Conductive thermal resistance for packaging materials.

contact pressure at the relevant interfaces, conspire to limit the accuracy of this calculation.

1.4.3 Substrate or PCB Conduction

In the design of airborne electronic systems and equipment to be operated in a corrosive or damaging environment, it is often necessary to conduct the heat dissipated by the components down into the substrate or printed circuit board and, as shown in Fig. 1.10, across the substrate/PCB to a cold plate or sealed heat exchanger. For a symmetrically cooled substrate/PCB with approximately uniform heat dissipation on the surface, a first estimate of the peak temperature, at the center of the board, can be obtained by use of Eq. (1.5).

Setting the heat generation rate equal to the heat dissipated by all the components and using the volume of the board in the denominator, the temperature difference between the center at T_{ctr} and the edge of the substrate/PCB at T_o is given by

$$T_{ctr} - T_o = \left(\frac{Q}{LW\delta}\right)\left(\frac{L^2}{8k_e}\right) = \frac{QL}{8W\delta k_e} \tag{1.33}$$

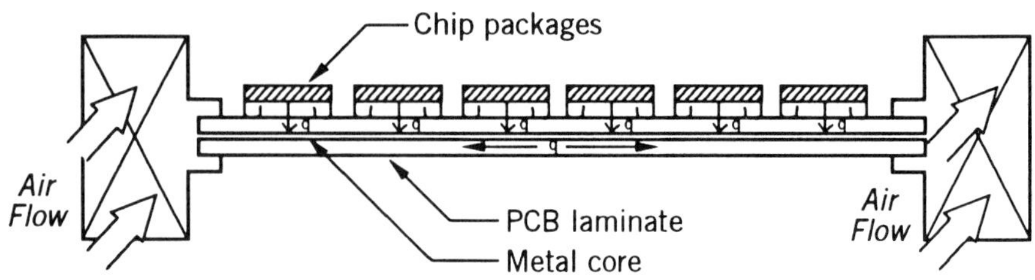

Figure 1.10 Edge cooled printed circuit board populated with components.

where Q is the total heat dissipation, W, L, and δ are the width, length, and thickness respectively, and k_e is the effective thermal conductivity of the board.

This relation can be used effectively in the determination of the temperatures experienced by conductively cooled substrates and conventional printed circuit boards, as well as PCB's with copper lattices on the surface and metal cores or heat sink plates in the center. In each case it is necessary to evaluate or obtain the effective thermal conductivity of the conducting layer. As an example, consider an alumina substrate, 0.20 m long, 0.15 m wide, and 0.005 m thick, with a thermal conductivity of 20 W/m-K, whose edges are cooled to 35°C by a cold-plate. Assuming that the substrate is populated by 30 components, each dissipating 1 W, use of Eq. (1.33) reveals that the substrate center temperature will equal 85°C.

1.4.4 External Resistance

To determine the resistance to thermal transport from the surface of a component to a fluid in motion, i.e., the convective resistance as in Eq. (1.13), it is necessary to quantify the heat-transfer coefficient, h. In the natural convection air cooling of printed circuit board arrays, isolated boards, and individual components, it has been found possible to use smooth-plate correlations, such as Eqs. (1.14) and (1.15), to obtain a first estimate of the peak temperature likely to be encountered on the populated board (Bar-Cohen, 1985). Examination of these relations suggests that an increase in the component/board temperature and a reduction in its length will serve to increase modestly the convective heat-transfer coefficient and thus to decrease modestly the resistance associated with natural convection. To achieve a more dramatic reduction in this resistance, it is necessary to select a high-density coolant with a large thermal expansion coefficient — typically a pressurized gas or a liquid.

When components are cooled by forced convection, the laminar heat-transfer coefficient, given by Eq. (1.17), is found to be directly proportional to the square root of fluid velocity and inversely proportional to the square root of the characteristic dimension. Increases in the thermal conductivity of the fluid and in Pr, as are encountered in replacing air with a liquid coolant, will also result in higher heat-transfer coefficients. In studies of low-velocity convective air cooling of simulated integrated circuit packages, the heat-transfer coefficient, h, has been found to depend somewhat more strongly on Re (using channel height as the characteristic length) than suggested in Eq. (1.17), and to display a Re exponent of 0.54 to 0.72 (Sparrow et al., 1982; Wirtz and Dykshoorn, 1984; and Godsell et al., 1990). When the fluid velocity and the Reynolds number increase, turbulent flow results in higher heat-transfer coefficients which, following Eq. (1.19), vary directly with the velocity to the 0.8 power and inversely with the characteristic dimension to the 0.2 power. The dependence on fluid conductivity and Pr remains unchanged.

An application of Eqs. (1.17) or (1.19) to the transfer of heat from the case of a chip module to the coolant shows that the external resistance, $R_{ex} = 1/hS$, is inversely proportional to the wetted surface area and to the coolant velocity to the 0.5 to 0.8 power, and directly proportional to the length scale in the flow direction to the 0.5 to 0.2 power. It may, thus, be observed that the external resistance can be strongly influenced by the fluid velocity and package dimensions and that these factors must be addressed in any meaningful evaluation of the external thermal resistances offered by various packaging technologies.

Values of the external resistance, for a variety of coolants and heat-transfer mechanisms, are shown in Fig. 1.11 for a typical component wetted area of $10\,\text{cm}^2$ and a velocity range of 2 to 8 m/s. They are seen to vary from a nominal 100 K/W for natural convection in air, to 33 K/W for forced convection in air, to 1 K/W in fluorocarbon liquid forced convection, and to less than 0.5 K/W for boiling in fluorocarbon liquids. Clearly, larger chip packages will experience proportionately lower external resistances than the displayed values. Moreover, conduction of heat through the leads and package base into the printed circuit board or substrate will serve to further reduce the effective thermal resistance.

As previously noted, in the event that the direct cooling of the package surface is inadequate to maintain the desired chip temperature, it is common to attach finned heat sinks, or compact heat exchangers, to the chip package. These heat sinks can considerably increase the wetted surface area, but may act to reduce the convective heat-transfer coefficient by obstructing the flow channel. Similarly, the attachment of a heat sink to the package can be expected to introduce additional conductive resistances in the adhesive used to bond the heat sink and in the body of the heat sink. Typical air-cooled heat sinks can reduce the external resistance to approximately 15 K/W in natural convection and to as low as 5 K/W for moderate forced convection velocities.

When a heat sink or compact heat exchanger is attached to the package, the external resistance accounting for the bond-layer conduction and the total resistance of the heat sink, R_{sk}, can be expressed as

$$R_{ex} = \frac{T_c - T_{fl}}{q_c} = \sum \left(\frac{x}{kA}\right)_b + R_{sk} \qquad [\text{K/W}] \qquad (1.34)$$

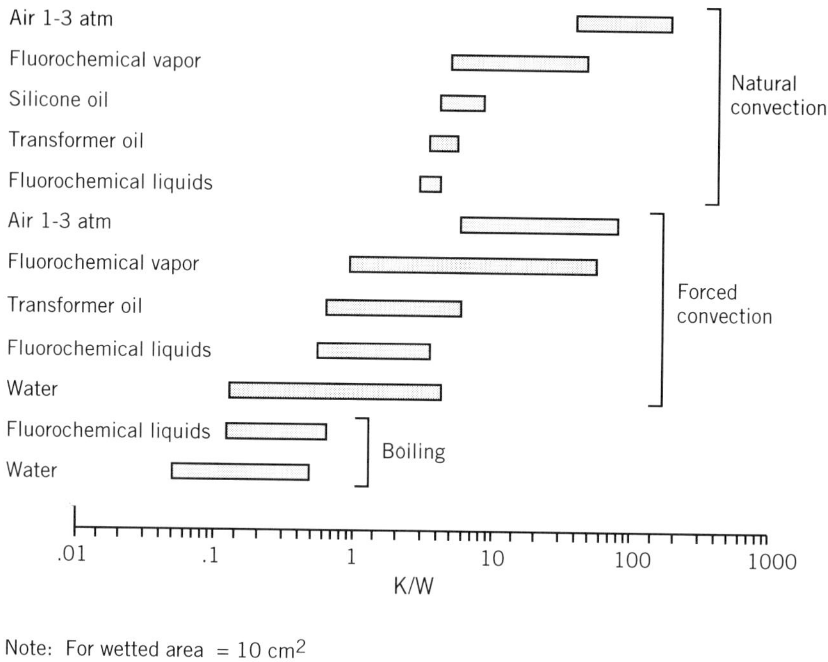

Figure 1.11 Typical external (convective) thermal resistances for various coolants and cooling nodes.

where

$$R_{sk} = \left[\frac{1}{nhS_f\eta} + \frac{1}{h_b S_b} \right]^{-1}$$

is the the parallel combination of the resistance of the n fins

$$R_f = \frac{1}{nhS_f\eta}$$

and the *exposed* or base surface not occupied by the fins

$$R_b = \frac{1}{h_b S_b}$$

Here, the base surface is $S_b = S - nS_f$, and the heat-transfer coefficient, h_b, is used because the heat-transfer coefficient that is applied to the base surfaces is not necessarily equal to that applied to the fins.

An alternative expression for R_{sk} involves an *overall surface efficiency*, η_o, defined by

$$\eta_o = 1 - \frac{nS_f}{S}(1 - \eta)$$

where S is the total surface composed of the base surface and the finned surfaces of n fins

$$S = S_b + nS_f$$

In this case, it is presumed that $h_b = h$ so that

$$R_{sk} = \frac{1}{h\eta_o S}$$

In an optimally designed fin structure, η can be expected to fall in the range of 0.50 to 0.70 (Kraus and Bar-Cohen, 1983). Relatively thick fins in a low-velocity flow of gas are likely to yield fin efficiencies approaching unity. This same unity value would be appropriate, as well, for an unfinned surface and, thus, serve to generalize the use of Eq. (1.34) to all package configurations.

1.4.5 Flow Resistance

In convectively cooled systems, determination of the component temperature requires knowledge of the fluid temperature adjacent to the component. The rise in fluid temperature relative to the inlet value can be expressed in a flow thermal resistance, as done in Eq. (1.26). When the coolant flow path traverses many individual components, care must be taken to use R_{fl} with the total heat absorbed by the coolant along its path, rather than the heat dissipated by an individual component. For system-level calculations, aimed at determining the average component temperature, it is common to base the flow resistance on the average rise in fluid temperature, that is, one-half the value indicated by Eq. (1.26).

1.4.6 Total Resistance–Single-Chip Packages

To the accuracy of the assumptions employed in the preceding development, the overall single-chip package resistance, relating the chip temperature to the inlet temperature of the coolant, can be found by summing the internal, external, and flow resistances to yield

$$R_T = R_{jc} + R_{ex} + R_{fl} = \sum \frac{\Delta x}{kA} + R_{int} + R_{sp}$$

$$+ \frac{1}{\eta h S} + \left(\frac{Q}{q}\right)\left(\frac{1}{2\dot{m}}\right) \qquad [\text{K/W}] \qquad (1.35)$$

In evaluating the thermal resistance by this relationship, care must be taken to determine the effective cross-sectional area for heat flow at each layer in the module and to consider possible voidage in any solder and adhesive layers.

As previously noted in the development of the relationships for the external and internal resistances, Eq. (1.35) shows R_T to be a strong function of the convective heat-transfer coefficient, the flowing heat capacity of the coolant, and geometric parameters (thickness and cross-sectional area of each layer). Thus, the introduction of a superior coolant, use of thermal enhancement techniques that increase the local heat-transfer coefficient, or selection of a heat transfer mode with inherently high heat transfer coefficients (e.g., boiling) will all be reflected in appropriately lower external and total thermal resistances. Similarly, improvements in the thermal conductivity and reduction in the thickness of the relatively low-conductivity bonding materials (e.g., soft solder, epoxy, silicone) would act to reduce the internal and total thermal resistances.

Frequently, however, even more dramatic reductions in the total resistance can be achieved simply by increasing the cross-sectional area for heat flow, within the chip module (e.g., chip, substrate, heat spreader), as well as along the wetted, exterior surface. The implementation of this approach to reducing the internal resistance generally results in a larger package footprint or volume but is rewarded with a lower thermal resistance. The use of heat sinks is, of course, the embodiment of this approach to the reduction of the external resistance.

1.5 THE HEAT SINK DESIGN PROCEDURE

The foregoing sections have attempted to show that the heat sink possesses a thermal resistance and that the heat sink resistance is but one of many resistances in the overall heat flow path between the component and the environment. However, particularly in packages constrained by the external resistance, the selection and/or design of an appropriate heat sink is crucial to the successful thermal management of an integrated circuit package. It is the intent of this section to amplify on the development of subsection 1.2.3 and to set the stage for the detailed treatment of the heat sink art that begins in Chapter 2.

The design or synthesis of a heat sink begins with the selection of a configuration of base surface and fins. The configuration is then analyzed to determine whether its performance meets the required specifications, and adjustments are made until the comparison between results and specifications is favorable. This procedure, called *synthesis by analysis*, is common in the synthesis (design) of heat-transfer equipment.

The actual procedure involves:

- The determination the base surface area, S_b, and the total finned surface area, S_f
- The evaluation of the heat-transfer coefficient, h
- The calculation of the fin efficiency, η
- The computation of the total surface, $S = S_b + \eta S_f$
- The evaluation of the heat sink resistance

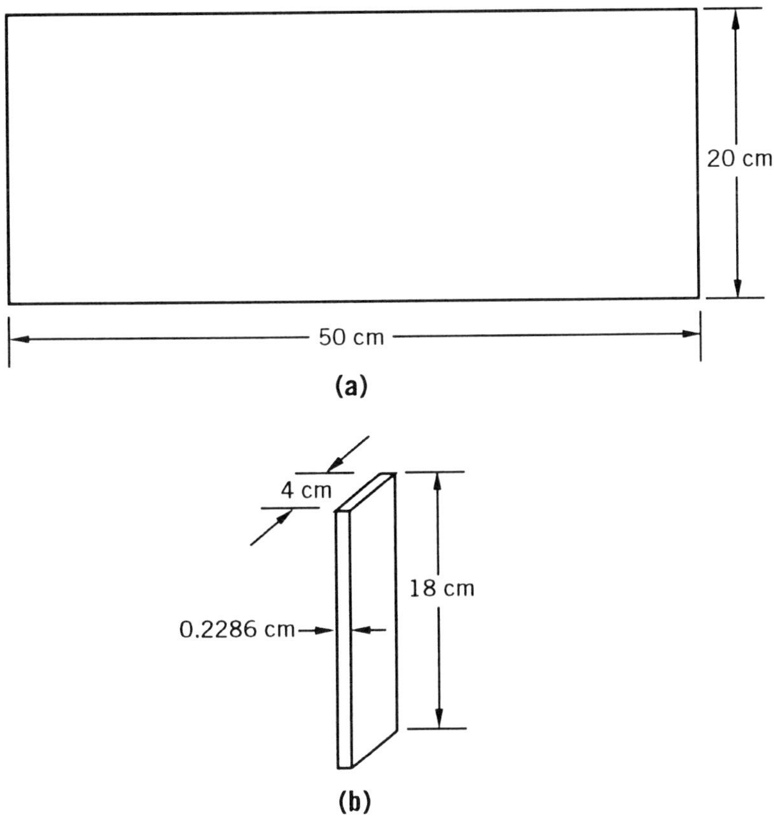

Figure 1.12 (*a*) Plane wall at rear of electronic enclosure; (*b*) one of n aluminum fins that can be attached to the wall.

Consider the rear wall of an enclosure or box containing electronic components, and suppose that, in order to provide an appropriate *microclimate* for the components, the rear wall must passively dissipate 175 W to an environment at 25°C at atmospheric pressure without exceeding a surface temperature of 75°C. The wall measures 50 cm by 20 cm, as indicated in Fig. 1.12*a*. The box under consideration is one of several such enclosures in an equipment bay, and the rear panel is located directly across an air gap from another electrical panel.

In the event that the plane wall fails to provide a low enough thermal resistance, aluminum fins (Fig. 1.12*b*) with a thermal conductivity of $k = 196\ \text{W/m-K}$, a length of 18 cm, a height of 4 cm, and a thickness of 0.2286 cm can be used. The designer must decide whether to use the fins and, if they are to be used, how many fins should be used.

Preliminaries: First, determine if the fins should be employed. The required thermal resistance between the panel and the ambient air is

$$R_{req} = \frac{\Delta T}{q} = \frac{75°\text{C} - 25°\text{C}}{175\,\text{W}} = 0.286\,\text{K/W}$$

The actual thermal resistance without the fins is a convection resistance in the form of Eq. (1.13), $R_{cv} = 1/hS$. Due to the proximity of the neighboring box with identical power dissipation, radiation from the rear surface will be neglected. For a panel surface area of

$$S = (50\,\text{cm})(20\,\text{cm}) = 1000\,\text{cm}^2 \text{ or } 0.10\,\text{m}^2$$

the heat-transfer coefficient needed to yield the required thermal resistance of 0.286 K/W would be

$$h = \frac{1}{R_{cv}S} = \frac{1}{(0.286\,\text{K/W})(0.1\,\text{m}^2)} = 35\,\text{W/m}^2\text{-K}$$

For the specified temperature difference of 50 K, use of Eq. (1.15) will reveal that natural convection coefficients at atmospheric pressure for heat transfer from a vertical surface above 5 or 6 W/m^2-K are not possible. Thus, fins must be used.

Base and Fin Surface Areas: It has been determined that fins must be used. The question now becomes: How many fins? The starting point is the determination of the optimum fin spacing. The Bar-Cohen and Rohsenow (1984) relationships, based on the earlier Elenbaas (1942) investigation of relatively higher fins where in-flow from the open ends may be neglected, can serve as a starting point for this example. Bar-Cohen and Rohsenow give for the optimum fin spacing,

$$z_{opt} = 2.714P \tag{1.36}$$

where P is called the *plate-air parameter*[5], with the thermal properties of the air evaluated at the average temperature between the plate and the surroundings:

$$P = \left[\frac{\mu k_{fl} L}{c_p \rho^2 g \beta \Delta T}\right]^{\frac{1}{4}}$$

Assuming that the fin efficiency is close to unity

$$\Delta T_{av} = \frac{75°\,\text{C} - 25°\,\text{C}}{2} = 50°\,\text{C}\,(323\text{K})$$

$$L = 0.18\,\text{m}$$

[5]This definition differs from the one used in the original paper by Bar-Cohen and Rohsenow.

Then setting

$$\Delta T = 75^\circ\,\mathrm{C} - 25^\circ\,\mathrm{C} = 50^\circ\,\mathrm{C}$$

and taking the thermal properties of air at $T_{av} = 50^\circ$ C as

$$\begin{aligned} c_p &= 1007\,\mathrm{J/kg\text{-}K} \\ \rho &= 1.095\,\mathrm{kg/m^3} \\ k_{fl} &= 0.0274\,\mathrm{W/m\text{-}K} \\ \mu &= 1.95 \times 10^{-5}\,\mathrm{kg/s\text{-}m} \end{aligned}$$

and

$$\beta = \frac{1}{T_{av}} = \frac{1}{323} = 0.003096\,\mathrm{K}^{-1}$$

The plate air parameter is, hence, found to equal

$$\begin{aligned} P &= \frac{\mu k L}{c_p \rho^2 g \beta \Delta T} \\ &= \left[\frac{(1.95 \times 10^{-5})(0.0274)(0.18)}{1007(1.095)^2(9.81)(0.003096)(50)}\right]^{\frac{1}{4}} \\ &= 0.00269\,\mathrm{m} \end{aligned}$$

so that

$$z = 2.714(0.00269\,\mathrm{m}) = 0.007304\,\mathrm{m}$$

or

$$z = 0.7304\,\mathrm{cm}\ \ (7.304\,\mathrm{mm})$$

The centerline spacing of the fins would then be $z + \delta$, and allowing n spaces for n fins gives

$$n = \frac{W}{z + \delta} = \frac{50\,\mathrm{cm}}{0.7304\,\mathrm{cm} + 0.2286\,\mathrm{cm}} = \frac{50\,\mathrm{cm}}{0.9590\,\mathrm{cm}} \approx 52\ \text{fins}$$

The base and fin surface areas may now be determined. In consideration of the fin temperature gradient, which can substantially reduce the fin tip temperature relative to the base, it is prudent to neglect heat transfer from the fin tip in the initial evaluation of heat sink performance. Thus, the total base surface area is expressed as WL minus the *footprints* of the 52 fins

$$\begin{aligned} S_b &= WH - nL\delta \\ &= (0.50\,\mathrm{m})(0.20\,\mathrm{m}) - 52(0.18\,\mathrm{m})(0.002286\,\mathrm{m}) \\ &= 0.1000\,\mathrm{m}^2 - 0.0214\,\mathrm{m}^2 \\ &= 0.0786\,\mathrm{m}^2 \end{aligned}$$

and

$$\begin{aligned} S_f &= 2bL \\ &= 2(0.040\,\text{m})(0.18\,\text{m}) \\ &= 0.0.0144\,\text{m}^2 \end{aligned}$$

The Heat-Transfer Coefficient: The Bar-Cohen-Rohsenow (1984) correlation for natural convection between vertical isothermal plates may be used for this consideration. First, the Elenbaas number is formed:

$$\begin{aligned} \text{El} &= \frac{\rho^2 \beta g c_p z^4 \Delta T}{\mu k_f L} \\ &= \frac{(1.095)^2(0.003096)(9.81)(1007)(0.007304)^4(50)}{(1.95 \times 10^{-5})(0.0274)(0.18)} \\ &= 54.26 \end{aligned}$$

and then Eq. (1.15) may be employed:

$$\begin{aligned} h &= \frac{k_{fl}}{z}\left[\frac{576}{(\text{El})^2} + \frac{2.873}{(\text{El})^{1/2}}\right]^{-1/2} \\ &= \frac{0.0274}{0.007304}\left[\frac{576}{(54.26)^2} + \frac{2.873}{(54.26)^{1/2}}\right]^{-1/2} \\ &= (3.751)(1.3608) \\ &= 4.90\,\text{W/m}^2\text{-K} \end{aligned}$$

Obtain the Fin Efficiency: The fin efficiency, η, is treated in detail elsewhere in this book because it is one of the fundamental parameters employed in the design and analysis of extended surface systems in general and heat sinks in particular. For the fins considered here, it is given by

$$\eta = \frac{\tanh mb}{mb}$$

where

$$m = \left(\frac{2h}{k\delta}\right)^{1/2}$$

Here

$$m = \left(\frac{2h}{k\delta}\right)^{1/2} = \left[\frac{2(4.90)}{(196)(0.002286)}\right]^{1/2} = 4.68\,\text{m}^{-1}$$

and

$$mb = (4.68)(0.04) = 0.187$$

The efficiency is

$$\eta = \frac{\tanh mb}{mb} = \frac{\tanh 0.187}{0.187} = 0.989$$

Thus justifying the earlier assumption of $\eta \approx 1$.

The Total Surface: The total surface is

$$\begin{aligned} S &= S_b + n\eta S_f \\ &= 0.0786 + 52(0.989)(0.0144) \\ &= 0.819\,\mathrm{m}^2 \end{aligned}$$

It may be observed that for the high thermal efficiency attained by these fins, heat transfer from the fin tips should be included. In the practical case, such a correction would increase S by 2.5 percent.

The Heat Sink Resistance: Because, the total surface, $S = S_b + \eta S_f$, is to be used in the computation of the thermal resistance, the heat sink resistance is equal to the convection resistance. Thus, in accordance with Eq. (1.13), it will be

$$\begin{aligned} R_{cv} = R_{sk} &= \frac{1}{hS} \\ &= \frac{1}{(4.90)(0.819)} \\ &= 0.249\,\mathrm{K/W} \end{aligned}$$

which is less than the required

$$R_{cv} = R_{sk} = 0.286\,\mathrm{K/W}$$

and makes this design satisfactory.

Attention now turns to Chap. 2 which introduces a new paramerization to replace the fin efficiency. This new parametization, which will be referred to as the fin or array *input admittance*, depends on a set of linear transformations that relate and freely mix conditions of temperature and heat flow at the bases and tips of individual fins.

Chapter

2

LINEAR TRANSFORMATIONS

2.1 INTRODUCTION

There is a delicate interplay between conduction and convection in a single extended surface or fin and, as was shown in Sec. 1.3.4, an inescapable loss of performance due to the negative temperature gradient that occurs within the fin. This loss of performance is handled analytically through the use of a design parameter called the *fin efficiency*, which was defined in Sec. 1.3.4 as the ratio of the actual heat dissipated by the fin to the ideal heat dissipated if the entire fin were to operate at the fin base temperature

$$\eta = \frac{q}{q_{id}}$$

For example, for the longitudinal fin of rectangular profile shown in Fig. 2.1, the fin efficiency, as derived in the next section, is given by

$$\eta = \frac{\tanh mb}{mb}$$

A knowledge of the value of the fin efficiency permits the total surface in a finned configuration to be represented as the sum of the *base* or *prime* surface, S_b, and a fin surface, S_f, that is modified by the fin efficiency

$$S = S_b + \eta S_f$$

This relationship was used in Ex. 1.1 in Sec. 1.5.

If the total surface is to be considered as an entity so that the total surface, S_T, is merely the sum of the base surface and the finned surface, $S_T = S_b + S_f$, the use of an *overall surface efficiency*, η_o, is sometimes found preferable. This

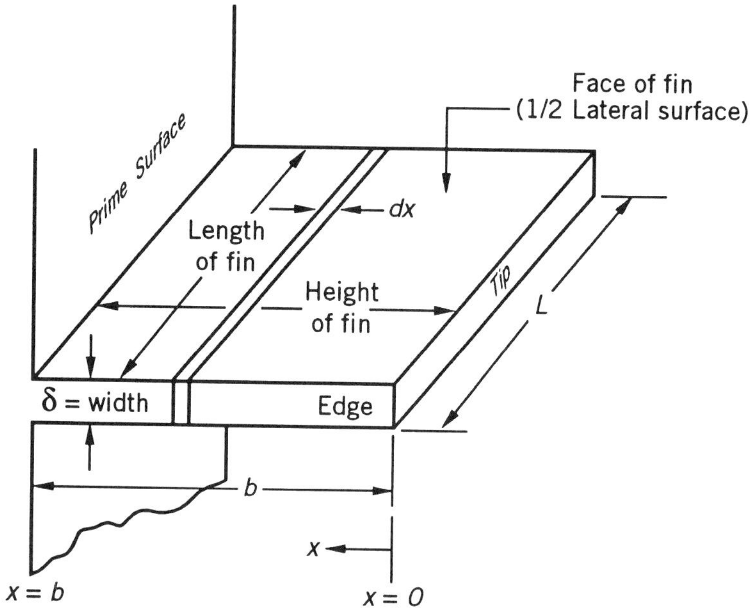

Figure 2.1 Terminology and coordinate system, longitudinal fin of rectangular profile.

parameter is based on the total surface, S_T, operating at η_o; the base surface, S_b, operating at an efficiency of 1.00; and the finned surface, S_f, operating at its efficiency, η,

$$\eta_o S_T = (1.00)S_b + \eta S_f$$

$$\eta_o = \frac{S_b}{S_T} + \eta\frac{S_f}{S_T}$$

But

$$S_b = S_T - S_f$$

so that

$$\eta_o = \frac{S_T - S_f}{S_T} + \eta\frac{S_f}{S_T}$$

$$\eta_o = 1 - \frac{S_f}{S_T} + \eta\frac{S_f}{S_T}$$

$$\eta_o = 1 - \frac{S_f}{S_T}\left[1 - \eta\right]$$

Use of the fin efficiency as a finned-surface adjustment factor has served heat transfer equipment designers well ever since it was originally proposed by Gardner (1945). However, while its use in the design and analysis of finned arrays, such as the electronic-component heat sink, can be useful, its determination or evaluation can require a lengthy procedure.

A study of an alternative parameterization called the *fin* or *array input admittance* begins in this chapter. This new parameter requires the use of any one of six linear transformations relating conditions of temperature and heat flow at the bases and tips of individual fins. These linear transformations are derived by solving the governing differential equations for the temperature excess profile of each fin as an *initial value problem*. This, of course, is a departure from the *textbook solutions* that solve the governing differential equation as a *boundary value problem*.

Thus, the first task for this chapter is to set the stage for fin analysis using the input admittance with a review of the boundary value solution for a common fin shape. The fin efficiency is then formally defined and shown to be of limited value in the analysis of finned arrays. This common fin shape is then revisited and an initial value solution is developed. This leads to the linear transformations that are needed to provide the fin or array input admittance.

2.2 THE LONGITUDINAL FIN OF RECTANGULAR PROFILE

In the longitudinal fin of rectangular fin profile, shown with its terminology and coordinate system in Fig. 2.1, the origin of the length coordinate, x, is taken at the fin tip, and its positive sense is in a direction towards the base of the fin. The differential equation for the temperature excess, θ, is formulated by considering a differential element, dx, having a cross-sectional area normal to the path of heat flow

$$A = \delta L$$

and a surface area

$$dS = 2(L + \delta)dx$$

The temperature excess is defined as the temperature difference between the fin and the constant-temperature surrounding fluid. At any point on the fin,

$$\theta = T - T_s$$

and

$$d\theta = dT$$

so that, at the base of the fin at $x = b$,

$$\theta_b = T_b - T_s$$

Conservation of energy dictates that the difference between the heat entering and leaving the element by conduction

$$kA\frac{d}{dx}\left[\frac{dT}{dx}\right]dx$$

must equal the heat leaving by convection

$$2h(L+\delta)(T-T_s)dx$$

Thus, with the cross-sectional area, $A=\delta L$ and the perimeter, $P=2(L+\delta)$,

$$k\delta L\frac{d}{dx}\left[\frac{dT}{dx}\right]dx=2h(L+\delta)(T-T_s)dx$$

or, with $\theta=T-T_s$ and $d\theta=dT$,

$$k\delta L\frac{d}{dx}\left[\frac{d\theta}{dx}\right]dx=2h(L+\delta)\theta dx$$

If $L>>\delta$, Murray-Gardner assumptions 5 and 9 permit writing the differential equation for temperature excess as

$$\frac{d^2\theta}{dx^2}-m^2\theta=0 \tag{2.1}$$

where m is called the *fin performance parameter*:

$$m=\left(\frac{2h}{k\delta}\right)^{1/2} \qquad [\mathrm{m}^{-1}]$$

The general solution to Eq. (2.1) is

$$\theta=C_1e^{mx}+C_2e^{-mx}$$

where the arbitrary constants, C_1 and C_2, are evaluated from the boundary conditions at $x=b$ where

$$\theta(x=b)=\theta_b$$

and at $x=0$, the adiabatic fin tip, where $k\neq 0$ and $A\neq 0$,

$$\left.\frac{d\theta}{dx}\right|_{x=0}=0$$

Use of these boundary conditions yields the particular solution for the point-to-point temperature excess

$$\theta(x)=\theta_b\frac{\cosh mx}{\cosh mb} \tag{2.2}$$

The heat flow through the base of the fin, q_b, is obtained from the derivative of Eq. (2.2), multiplied by $kA=k\delta L$ and evaluated at $x=b$

$$q_b=k\delta L\left.\frac{d\theta}{dx}\right|_{x=b}$$

or

$$q_b = k\delta L m\theta_b \tanh mb$$

and with $m = (2h/k\delta)^{1/2}$,

$$q_b = (2hk\delta)^{1/2} L\theta_b \tanh mb \tag{2.3}$$

The fin efficiency is defined as the ratio of the heat actually transferred from the fin to the heat that would be ideally transferred if the entire fin were at the temperature excess of the fin base, θ_b

$$\eta = \frac{q_b}{q_{id}}$$

For the longitudinal fin of rectangular profile, the actual heat flow is given by Eq. (2.3). Because the ideal heat flow is

$$q_{id} = 2hbL\theta_b$$

the efficiency becomes

$$\eta = \frac{k\delta L m\theta_b \tanh mb}{2hbL\theta_b}$$

and by noting that $k\delta/2h = 1/m^2$,

$$\eta = \frac{\tanh mb}{mb} \tag{2.4}$$

An example that concerns the analysis of an individual fin is now offerred to help cement the concepts just developed. It will further demonstrate the use of the fin efficiency as an analysis and design parameter.

Example 2.1: A longitudinal fin of rectangular profile is exposed to surroundings at a temperature of 50°C. The temperature at the fin base is 90°C. The fin is made of aluminum with a thermal conductivity of $k = 202$ W/m-K and is 101.6 mm (4 in) high, 9.525 mm (0.375 in) thick, and 250 mm long. Determine:

1. For a heat-transfer coefficient of $h = 50.5$ W/m^2-K, the
   (a) fin efficiency
   (b) temperature at the tip of the fin
   (c) dissipation of the fin
2. For a quintupled heat-transfer coefficient ($h = 252.5$ W/m^2-K), the
   (a) fin efficiency
   (b) temperature at the tip of the fin and
   (c) dissipation of the fin

SOLUTION

1. For the surroundings with $h = 50.5\,\mathrm{W/m^2\text{-}K}$[1]

   (a) the fin efficiency

$$\begin{aligned}
\theta_b &= 90 - 50 = 40^\circ\mathrm{C} \\
\delta &= 9.525/1000 = 9.525 \times 10^{-3}\,\mathrm{m} \\
m &= \left(\frac{2h}{k\delta}\right)^{1/2} = \left[\frac{2(50.5)}{202(9.525 \times 10^{-3})}\right]^{1/2} = 7.2452\,\mathrm{m}^{-1} \\
b &= 101.6/1000 = 0.1016\,\mathrm{m} \\
mb &= 7.2452(0.1016) = 0.7361
\end{aligned}$$

   From Eq. (2.4),

$$\begin{aligned}
\eta &= \frac{\tanh mb}{mb} \\
&= \frac{\tanh 0.7361}{0.7361} \\
&= \frac{0.6268}{0.7361} \\
&= 0.852
\end{aligned}$$

   (b) the tip temperature of the fin, determined from Eq. (2.2) at $x = 0$, where $\theta(x = 0) = \theta_a$:

$$\begin{aligned}
\theta_a &= \theta_b \frac{\cosh mx}{\cosh mb}\bigg|_{x=a=0} \\
&= 40 \frac{\cosh 0}{\cosh 0.7361} \\
&= \frac{40}{1.2834} \\
&= 31.2^\circ\mathrm{C}
\end{aligned}$$

   so that at $x = 0$ where $T(x = 0) = T_a$,

$$T_a = \theta_a + T_s = 31.2 + 50 = 81.2^\circ\mathrm{C}$$

[1] The computations are carried out to three or four significant figures, not because such accuracy is is really warranted but because this book relies heavily on matrix computation. This accuracy is dictated by the phenomenon in matrix computation called *catastrophic cancellation.*

(c) the heat dissipated by the fin, calculated from Eq. (2.3), which considers a unit length of fin (L =1 m),

$$\begin{aligned} q_b &= k\delta m L \theta_b \tanh mb \\ &= (202)(9.525 \times 10^{-3})(7.2452)(1.00)(40)(0.6268) \\ &= 349.50\,\text{W} \end{aligned}$$

and for $L = 250$ mm:

$$q_b = 0.25(349.50) = 87.38\,\text{W}$$

2. For the surroundings, with $h = 252.5\,\text{W/m}^2\text{-K}$,

(a) the fin efficiency

$$\begin{aligned} \theta_b &= 40^\circ\text{C} \\ \delta &= 9.525 \times 10^{-3}\,\text{m} \\ m &= \left(\frac{2h}{k\delta}\right)^{1/2} \\ &= \left[\frac{2(252.5)}{202(9.525 \times 10^{-3})}\right]^{1/2} \\ &= 16.2008\,\text{m}^{-1} \\ b &= 0.1016\,\text{m} \\ mb &= 16.2008(0.1016) = 1.6460 \end{aligned}$$

From Eq. (2.4)

$$\eta = \frac{\tanh mb}{mb} = \frac{\tanh 1.6460}{1.6460} = \frac{0.9283}{1.6460} = 0.564$$

(b) the tip temperature, determined from Eq. (2.2) at $x = 0$,

$$\begin{aligned} \theta_a &= \theta_b \frac{\cosh mx}{\cosh mb} \\ &= 40 \frac{\cosh 0}{\cosh 1.6460} \\ &= \frac{40(1)}{2.6895} = 14.9^\circ\text{C} \end{aligned}$$

so that

$$T_a = \theta_a + T_s = 14.9 + 50 = 64.9^\circ\text{C}$$

(c) the heat dissipated by the fin, calculated from Eq. (2.3), For $L = 1$ m,

$$\begin{aligned} q_b &= k\delta m\theta_b \tanh mb \\ &= (202)(9.525\times 10^{-3})(16.2008)(40)(0.9283) \\ &= 1157.46\,\text{W} \end{aligned}$$

and for $L = 250$ mm,

$$q_b = 0.25(1157.46) = 289.36\,\text{W}$$

The heat dissipated can also be obtained by multiplying the ideal heat dissipation by the fin efficiency:

$$\begin{aligned} \theta_b &= 2bLh\theta_b\eta \\ &= 2(0.1016)(0.25)(252.5)(40)(0.564) = 289.36\,\text{W} \end{aligned}$$

2.3 THE LIMITATIONS OF THE FIN EFFICIENCY

The general idea of an efficiency as a performance parameter is sound; it is a dimensionless ratio comparing performance with a certain standard. However, the particular way that efficiency has been defined for fins compares every fin with a different standard—the actual fin performance as compared with what a geometrically identical fin could do if conditions were perfect. Two fins of different dimensions in the same environment may have the same efficiency, but they may transmit different quantities of heat. A simple example, based on a real-world optimization, can show that one fin can transmit more heat than another under identical environmental conditions and operate at a lower value of fin efficiency. This is vividly demonstrated in the following paragraphs.

Using the terminology in Fig. 2.1, one may consider an aluminum longitudinal fin of rectangular profile (k =202 W/m-K) with a base temperature of 200°C dissipating to an environment at 100°C under natural convection conditions where h is taken as $h = 10\,\text{W/m}^2$-K. The fin dimensions are: height, $b = 10$ cm; thickness, $\delta = 0.2286$ cm (0.090 in); and length, $L = 25$ cm. Using Eq. (2.3),

$$q_b = (2hk\delta)^{1/2} L\theta_b \tanh mb \tag{2.3}$$

with $\theta_b = 200 - 100 = 100$°C, one may compute the heat dissipation as 43.85 W. Then, using Eq. (2.4),

$$\eta = \frac{\tanh mb}{mb} \tag{2.4}$$

the efficiency may be computed as $\eta = 0.877$.

In a desire to save mass, it is proposed that a magnesium fin (k =148 W/m-K), with the same length and thickness dimensions operating at otherwise identical thermal conditions, be employed. Using the dissipation of

43.85 W, one may solve Eq. (2.3) to obtain $b = 10.64$ cm, which shows that, because of its poorer conductivity, the height of the magnesium fin must be increased in order to accommodate an identical dissipation requirement. For the magnesium fin, the efficiency, computed from Eq. (2.4), is $\eta = 0.824$. Observe that the magnesium fin, which has a huge advantage in a weight optimization dissipates the *same* quantity of heat, but at a *lower* fin efficiency.

If the heat-transfer coefficient is reduced to, say, $h = 8\,\text{W/m}^2\text{-K}$, Eqs. (2.3) and (2.4) show that the magnesium fin (with the greater fin height) will dissipate 36.30 W at an efficiency of 0.853, while the aluminum fin will dissipate less heat, 35.94 W, at the higher efficiency of 0.899. The fins are no longer behaving identically, and the magnesium fin is clearly outperforming its aluminum *design-point twin* at a lower efficiency.

At this point, it is noted that the fault is probably not in the efficiency concept but in the efficiency definition. If the efficiency had originally been called *the surface utilization factor* (which, as demonstrated in Sec. 1.5, is exactly what it is), it might never have entered either the preceding discussion or the calculations.

The usefulness of the fin efficiency is debatable when a finned array is considered. Recall that the definition of the fin efficiency is also defined as the ratio of the heat dissipated, q_1, to the heat dissipated if the fin were entirely at the base temperature, $q_{id} = hS\theta_b$,

$$\eta = \frac{q_1}{q_{id}} \tag{2.5}$$

For a single fin without tip heat dissipation, $q_1 = q_b$ and Eq. (2.5), consistent with the development that is based on no heat flow from the fin tip, results. However, for a single fin with heat dissipation from the tip, $q_a = h_a A_a$, where, as Sparrow et al. (1978) and others have shown, h_a, the coefficient of heat transfer at the fin tip, does not necessarily equal the coefficient of heat transfer on the fin faces. In this case, A_a, the fin cross-sectional area at the fin tip, is equal to the fin surface area at the fin tip. But, to be sure, the entire convective dissipation (dissipation from the fin faces and the fin tip) passes through the base of the fin. Thus, with $q_1 = q_b$, the efficiency of the fin is also given by Eq. (2.5).

But, in a finned array, heat loss from an individual fin includes both convection to the surrounding environment and conduction from its tip to the contiguous fin or a cluster of fins located downstream. Consequently, the convective heat dissipation is $q_1 = q_b - q_a$, and the fin efficiency becomes

$$\eta = \frac{q_1}{q_{id}} = \frac{q_b - q_a}{q_{id}}$$

It is observed that the efficiency of a fin in a finned array depends not only on the heat-transfer condition, h; the fin thermal conductivity, k; and the fin dimensions, L, b, and δ; but also on where the fin is mounted in the array.

This, it is felt, is a major inadequacy of the concept of the fin efficiency and justifies the development of an alternative approach.

2.4 THE LONGITUDINAL FIN OF RECTANGULAR PROFILE REVISITED

Once again, consider Fig. 2.1, which displays the longitudinal fin of rectangular profile. The governing differential equation for the point-to-point temperature excess along the fin is Eq. (2.1):

$$\frac{d^2\theta}{dx^2} - m^2\theta = 0 \tag{2.1}$$

This equation was solved using *boundary value data*, that is, the temperature excess at the fin base:

$$\theta(x = b) = \theta_b$$

and the heat flow at the fin tip

$$q(x = 0) = 0$$

which led to

$$\left.\frac{d\theta}{dx}\right|_{x=0} = 0$$

The general differential equation may also be solved using *initial value data*, which involve conditions of temperature excess and heat flow at the same point on the fin. If this point is taken at the fin base, the initial conditions are

$$\theta(x = b) = \theta_b$$

and

$$q(x = b) = q_b$$

Use of these conditions with the general solution of Eq. (2.1),

$$\theta = C_1 e^{mx} + C_2 e^{-mx}$$

and its derivative

$$\frac{d\theta}{dx} = m[C_1 e^{mx} - C_2 e^{-mx}]$$

yields a pair of simultaneous, linear algebraic equations in the arbitrary constants C_1 and C_2, which can be written in matrix form as

$$\begin{bmatrix} e^{mb} & e^{-mb} \\ kAme^{mb} & -kAme^{-mb} \end{bmatrix} \begin{bmatrix} C_1 \\ C_2 \end{bmatrix} = \begin{bmatrix} \theta_b \\ q_b \end{bmatrix}$$

The solution for C_1 and C_2 is obtained expeditiously using a matrix inversion.[2] With the fin cross-sectional area, $A = \delta L$,

$$C_1 = \frac{k\delta L e^{-mb}\theta_b + e^{-mb} q_b}{2k\delta Lm}$$

and

$$C_2 = \frac{k\delta L e^{mb}\theta_b - e^{mb} q_b}{2k\delta Lm}$$

Substitution of these values into the general solution gives

$$\theta(x) = \left[\frac{e^{-mb}}{2}\theta_b + \frac{e^{-mb}}{2k\delta Lm} q_b\right] e^{mx} + \left[\frac{e^{mb}}{2}\theta_b - \frac{e^{mb}}{2k\delta Lm} q_b\right] e^{-mx}$$

or on rearrangement

$$\theta(x) = \left[\frac{e^{m(b-x)} + e^{-m(b-x)}}{2}\right]\theta_b - \frac{1}{k\delta Lm}\left[\frac{e^{m(b-x)} - e^{m(b-x)}}{2}\right] q_b$$

Observe that it has been possible to separate the effects of base temperature excess and base heat flow on the fin temperature excess at any point, x. Noting the importance of the fin parameter, $k\delta Lm = (2hk\delta)^{1/2}L$, it is convenient to define the *characteristic admittance of the fin* as

$$Y_o \equiv (2hk\delta)^{1/2}L \tag{2.6}$$

Thus, the temperature excess at any point on the fin may be written in terms of the hyperbolic functions

$$\theta(x) = [\cosh m(b-x)]\theta_b - \left[\frac{1}{Y_o}\sinh m(b-x)\right] q_b \tag{2.7}$$

and a simple differentiation yields the heat flow

$$q(x) = k\delta L\frac{d\theta(x)}{dx} = -[Y_o \sinh m(b-x)]\theta_b + [\cosh m(b-x)]q_b \tag{2.8}$$

Observe in the foregoing that at $x = b$, Eq. (2.7) gives, as it should, $\theta(x = b) = \theta_b$, and Eq. (2.8) reduces, as it should, to $q = q_b$.

The temperature excess and heat flow at the tip of the fin can be determined by merely letting $x = 0$ in Eqs. (2.7) and (2.8):

$$\theta(x=0) = \theta_a = [\cosh mb]\,\theta_b - \left[\frac{1}{Y_o}\sinh mb\right] q_b \tag{2.9}$$

[2]Recall that the inverse of a 2×2 matrix is easily obtained by swapping the elements of the principal diagonal, changing the sign of the off-diagonal elements, and then dividing all elements by the determinant of the matrix. Appendix A presents a brief introduction to matrices and some of the basic operations concerning matrices.

and

$$q(x=0) = q_a = -[Y_o \sinh mb]\,\theta_b + [\cosh mb]\,q_b \tag{2.10}$$

Equations (2.9) and (2.10) are very significant to an understanding of fin analysis because they lead to three important conclusions, one of which is intuitively physical and two of which are mathematical:

- Aside from the physical dimensions and the two important heat-transfer properties, h and k, conditions of temperature excess and heat flow at the fin tip can be said to be induced by conditions of temperature excess and heat flow at the fin base. This, of course, is intuitively true physically.
- Conditions of temperature excess and heat flow at the fin tip and, indeed, as shown by Eqs. (2.7) and (2.8) at any point on the fin, can be determined as a superposition of the effects of temperature excess and heat flow at the fin base. This is to be expected because the differential equation for temperature excess is a linear differential equation.
- There is a linear transformation relating conditions of temperature excess and heat flow at the fin tip and, indeed, at any point on the fin with conditions of temperature excess and heat flow at the fin base.

2.5 A LINEAR TRANSFORMATION

The linear transformation suggested by Eqs. (2.9) and (2.10) may be written in matrix form as

$$\begin{bmatrix} \theta_a \\ q_a \end{bmatrix} = \begin{bmatrix} \cosh mb & -Z_o \sinh mb \\ -Y_o \sinh mb & \cosh mb \end{bmatrix} \begin{bmatrix} \theta_b \\ q_b \end{bmatrix} \tag{2.11}$$

where, as defined by Eq. (2.6), $Y_o = (2hk\delta)^{1/2}L$. The *characteristic impedance*, Z_o, is the reciprocal of the characteristic admittance. Thus,

$$Z_o = \frac{1}{Y_o} = \frac{1}{(2hk\delta)^{1/2}L}$$

The matrix in Eq. (2.11) containing the hyperbolic functions is the linear transformation matrix that results from the solution of Eq. (2.1) as an initial value problem. In Eq. (2.11) the independent variables are θ_b and q_b, and the dependent variables are θ_a and q_a. Attention is often focused on a different set of dependent and independent variables. Thus, other linear transformations are useful, and they can be readily obtained.

2.6 OTHER LINEAR TRANSFORMATIONS

The general problem concerns the linear transformation from two variables to another two, and it is seen that four variables taken two at a time yields six

TABLE 2.1 Summary of Linear Transformations

Case	dep_1	dep_2	indep_1	indep_2	α_{11}	α_{12}	α_{21}	α_{22}	Matrix
1	θ_b	θ_a	q_b	q_a	z_{11}	z_{12}	z_{21}	z_{22}	**Z**
2	q_b	q_a	θ_b	θ_a	y_{11}	y_{12}	y_{21}	y_{22}	**Y**
3	θ_b	q_b	θ_a	q_a	τ_{11}	τ_{12}	τ_{21}	τ_{22}	**T**
4	θ_a	q_a	θ_b	q_b	γ_{11}	γ_{12}	γ_{21}	γ_{22}	**Γ**
5	θ_b	q_a	q_b	θ_a	h_{11}	h_{12}	h_{21}	h_{22}	**H**
6	q_b	θ_a	θ_b	q_a	g_{11}	g_{12}	g_{21}	g_{22}	**G**

combinations. Thus, six linear transformations are required to characterize completely *all* of the combinations of temperature excesses and heat flows at both fin tip and fin base. The general formulation involves the matrix representation

$$\begin{bmatrix} \text{dep}_1 \\ \text{dep}_2 \end{bmatrix} = \begin{bmatrix} \alpha_{11} & \alpha_{12} \\ \alpha_{21} & \alpha_{22} \end{bmatrix} \begin{bmatrix} \text{indep}_1 \\ \text{indep}_2 \end{bmatrix}$$

where *dep* and *indep* stand for dependent and independent variables, respectively.

Here, the α-matrix represents a linear combination of variables, as a simple matrix multiplication clearly shows

$$\text{dep}_1 = \alpha_{11}[\text{indep}_1] + \alpha_{12}[\text{indep}_2]$$

and

$$\text{dep}_2 = \alpha_{21}[\text{indep}_1] + \alpha_{22}[\text{indep}_2]$$

Here the α's may take different forms depending on which two of the variables are selected as independent variables

Table 2.1 presents a summary of the six possible combinations of dependent and independent variables. Observe that the choice of parameters shown in the table is clearly a function of the variables chosen as independent variables. This, in turn, may depend on which variables are easily measured or calculated and possibly on how individual fins may be connected with other fins to form a finned array.

2.6.1 The Z- and Y-Matrices

The z-parameters relate the dependent variables of temperature excess at fin base and fin tip, θ_b and θ_a, to the independent variables of base and tip heat flows. This is case 1 in Table 2.1, and the matrix representation is

$$\begin{bmatrix} \theta_b \\ \theta_a \end{bmatrix} = \mathbf{Z} \begin{bmatrix} q_b \\ q_a \end{bmatrix} = \begin{bmatrix} z_{11} & z_{12} \\ z_{21} & z_{22} \end{bmatrix} \begin{bmatrix} q_b \\ q_a \end{bmatrix} \tag{2.12}$$

The y-parameters, also shown in Table 2.1 (case 2), relate the base and tip heat flows, q_b and q_a, as dependent variables to the base and tip temperature excesses, θ_b and θ_a as independent variables. The matrix representation is

$$\begin{bmatrix} q_b \\ q_a \end{bmatrix} = \mathbf{Y}\begin{bmatrix} \theta_b \\ \theta_a \end{bmatrix} = \begin{bmatrix} y_{11} & y_{12} \\ y_{21} & y_{22} \end{bmatrix}\begin{bmatrix} \theta_b \\ \theta_a \end{bmatrix} \tag{2.13}$$

If Eq. (2.13) is put into Eq. (2.12), the result is

$$\begin{bmatrix} \theta_b \\ \theta_a \end{bmatrix} = \begin{bmatrix} z_{11} & z_{12} \\ z_{21} & z_{22} \end{bmatrix}\begin{bmatrix} y_{11} & y_{12} \\ y_{21} & y_{22} \end{bmatrix}\begin{bmatrix} \theta_b \\ \theta_a \end{bmatrix}$$

which shows that the z-parameter matrix is the inverse of the y-parameter matrix:

$$\mathbf{Z} = \mathbf{Y}^{-1} \tag{2.14a}$$

and

$$\mathbf{Y} = \mathbf{Z}^{-1} \tag{2.14b}$$

Equations (2.14) form the basis for the conversion from the z-parameters to the y-parameters, and vice versa. With the determinants of $\mathbf{H}$ and $\mathbf{G}$ designated, respectively, as Δ_H and Δ_G, this procedure yields

$$z_{11} = \frac{y_{22}}{\Delta_Y} \tag{2.15a}$$

$$z_{12} = -\frac{y_{12}}{\Delta_Y} \tag{2.15b}$$

$$z_{21} = -\frac{y_{21}}{\Delta_Y} \tag{2.15c}$$

and

$$z_{22} = \frac{y_{11}}{\Delta_Y} \tag{2.15d}$$

and a similar procedure gives the y's in terms of the z's:

$$y_{11} = \frac{z_{22}}{\Delta_Z} \tag{2.16a}$$

$$y_{12} = -\frac{z_{12}}{\Delta_Z} \tag{2.16b}$$

$$y_{21} = -\frac{z_{21}}{\Delta_Z} \tag{2.16c}$$

and

$$y_{22} = \frac{z_{11}}{\Delta_Z} \tag{2.16d}$$

2.6.2 The T- and Γ- Matrices

The τ-parameters and the γ-parameters differ from the z- and y-parameters in that they relate the temperature excess and heat flows at either the base or tip to the temperature excesses and heat flow at the tip or base. These are represented as cases 3 and 4 in Table 2.1, and their matrix representations are

$$\begin{bmatrix} \theta_b \\ q_b \end{bmatrix} = \mathbf{T} \begin{bmatrix} \theta_a \\ q_a \end{bmatrix} = \begin{bmatrix} \tau_{11} & \tau_{12} \\ \tau_{21} & \tau_{22} \end{bmatrix} \begin{bmatrix} \theta_a \\ q_a \end{bmatrix} \tag{2.17}$$

and

$$\begin{bmatrix} \theta_a \\ q_a \end{bmatrix} = \mathbf{\Gamma} \begin{bmatrix} \theta_b \\ q_b \end{bmatrix} = \begin{bmatrix} \gamma_{11} & \gamma_{12} \\ \gamma_{21} & \gamma_{22} \end{bmatrix} \begin{bmatrix} \theta_b \\ q_b \end{bmatrix} \tag{2.18}$$

Equation (2.17) is very significant, and it forms an important cornerstone for the combination of individual fins into finned arrays. It is the *matrix of the linear transformation* that maps fin tip conditions into fin base conditions.

If Eq. (2.18) is put into Eq. (2.17), the result is

$$\begin{bmatrix} \theta_b \\ q_b \end{bmatrix} = \begin{bmatrix} \tau_{11} & \tau_{12} \\ \tau_{21} & \tau_{22} \end{bmatrix} \begin{bmatrix} \gamma_{11} & \gamma_{12} \\ \gamma_{21} & \gamma_{22} \end{bmatrix} \begin{bmatrix} \theta_b \\ q_b \end{bmatrix}$$

which shows that the τ-parameter matrix is the inverse of the γ-parameter matrix:

$$\mathbf{T} = \mathbf{\Gamma}^{-1} \tag{2.19a}$$

and

$$\mathbf{\Gamma} = \mathbf{T}^{-1} \tag{2.19b}$$

Equations (2.19) form the basis for the conversion from the τ-parameters to the γ-parameters, and vice versa. With the determinants of $\mathbf{T}$ and $\mathbf{\Gamma}$ designated, respectively, as Δ_T and Δ_Γ,

$$\tau_{11} = \frac{\gamma_{22}}{\Delta_\Gamma} \tag{2.20a}$$

$$\tau_{12} = -\frac{\gamma_{12}}{\Delta_\Gamma} \tag{2.20b}$$

$$\tau_{21} = -\frac{\gamma_{21}}{\Delta_\Gamma} \tag{2.20c}$$

and

$$\tau_{22} = \frac{\gamma_{11}}{\Delta_\Gamma} \tag{2.20d}$$

and one may work in the opposite direction to obtain

$$\gamma_{11} = \frac{\tau_{22}}{\Delta_T} \tag{2.21a}$$

$$\gamma_{12} = -\frac{\tau_{12}}{\Delta_T} \tag{2.21b}$$

$$\gamma_{21} = -\frac{\tau_{21}}{\Delta_T} \tag{2.21c}$$

and

$$\gamma_{22} = \frac{\tau_{11}}{\Delta_T} \tag{2.21d}$$

2.6.3 The H- and G- Matrices

The h-parameters and g-parameters shown in Table 2.1 as cases 5 and 6 bear no resemblance to cases 1 through 4. They are called *hybrids* because they freely mix both heat flows, q_b and q_a, and both temperature excesses, θ_b and θ_a, without regard to any consistency at either fin base or fin tip. Their matrix representations are

$$\begin{bmatrix} \theta_b \\ q_a \end{bmatrix} = \mathbf{H} \begin{bmatrix} q_b \\ \theta_a \end{bmatrix} = \begin{bmatrix} h_{11} & h_{12} \\ h_{21} & h_{22} \end{bmatrix} \begin{bmatrix} q_b \\ \theta_a \end{bmatrix} \tag{2.22}$$

and

$$\begin{bmatrix} q_b \\ \theta_a \end{bmatrix} = \mathbf{G} \begin{bmatrix} \theta_b \\ q_a \end{bmatrix} = \begin{bmatrix} g_{11} & g_{12} \\ g_{21} & g_{22} \end{bmatrix} \begin{bmatrix} \theta_b \\ q_a \end{bmatrix} \tag{2.23}$$

If Eq. (2.23) is put into Eq. (2.22), the result is

$$\begin{bmatrix} \theta_b \\ q_a \end{bmatrix} = \begin{bmatrix} h_{11} & h_{12} \\ h_{21} & h_{22} \end{bmatrix} \begin{bmatrix} g_{11} & g_{12} \\ g_{21} & g_{22} \end{bmatrix} \begin{bmatrix} \theta_b \\ q_a \end{bmatrix}$$

which shows that the h-parameter matrix is the inverse of the g-parameter matrix:

$$\mathbf{H} = \mathbf{G}^{-1} \tag{2.24a}$$

and

$$\mathbf{G} = \mathbf{H}^{-1} \tag{2.24b}$$

Equations (2.24) form the basis for the conversion from the h-parameters to g-parameters, and vice versa. With the determinants of $\mathbf{H}$ and $\mathbf{G}$ designated, respectively, as Δ_H and Δ_G,

$$h_{11} = \frac{g_{22}}{\Delta_G} \tag{2.25a}$$

$$h_{12} = -\frac{g_{12}}{\Delta_G} \tag{2.25b}$$

$$h_{21} = -\frac{g_{21}}{\Delta_G} \tag{2.25c}$$

and

$$h_{22} = \frac{g_{11}}{\Delta_G} \tag{2.25d}$$

and, working in reverse,

$$g_{11} = \frac{h_{22}}{\Delta_H} \tag{2.26a}$$

$$g_{12} = -\frac{h_{12}}{\Delta_H} \tag{2.26b}$$

$$g_{21} = -\frac{h_{21}}{\Delta_H} \tag{2.26c}$$

and

$$g_{22} = \frac{h_{11}}{\Delta_H} \tag{2.26d}$$

2.6.4 Relationships between Parameters

Each of the six sets of parameters can be represented in terms of the other five. It is the intent of this subsection to show how all parameters may be represented in terms of the z-parameters and then to provide a tabulation of all possible conversions.

Begin with the defining equations for the z, τ, and h-parameters,

$$\begin{aligned} \theta_b &= z_{11}q_b + z_{12}q_a && (a) \\ \theta_a &= z_{21}q_b + z_{22}q_a && (b) \end{aligned}$$

$$\begin{aligned} \theta_b &= \tau_{11}\theta_a + \tau_{12}q_a && (c) \\ q_b &= \tau_{21}\theta_a + \tau_{22}q_a && (d) \end{aligned}$$

$$\begin{aligned} \theta_b &= h_{11}q_b + h_{12}\theta_a && (e) \\ q_a &= h_{21}q_b + h_{22}\theta_a && (f) \end{aligned}$$

The y-parameters in terms of the z-parameters have already been established by Eqs. (2.14). These equations may be summarized by an equality of two matrices:

$$\mathbf{Y} = \begin{bmatrix} y_{11} & y_{12} \\ y_{21} & y_{22} \end{bmatrix} = \begin{bmatrix} z_{22}/\Delta_Z & -z_{12}/\Delta_Z \\ -z_{21}/\Delta_Z & z_{11}/\Delta_Z \end{bmatrix} \tag{2.27}$$

To find the τ-parameters in terms of the z-parameters, begin with Eq. (b) and solve for q_b:

$$q_b = \frac{1}{z_{21}}\theta_a - \frac{z_{22}}{z_{21}}q_a \qquad (g)$$

Then make a term-by-term comparison between Eqs. (d) and (g) to obtain

$$\tau_{21} = \frac{1}{z_{21}}$$

and

$$\tau_{22} = -\frac{z_{22}}{z_{21}}$$

Next, put Eq. (*g*) into Eq. (*a*) to obtain

$$\theta_b = z_{11}\left[\frac{1}{z_{21}}\theta_a - \frac{z_{22}}{z_{21}}q_a\right] + z_{12}q_a$$

$$= \frac{z_{11}}{z_{21}}\theta_a + \left[z_{12} - \frac{z_{11}z_{22}}{z_{21}}\right]q_a$$

$$= \frac{z_{11}}{z_{21}}\theta_a - \frac{\Delta_Z}{z_{21}}q_a$$

where $\Delta_Z = z_{11}z_{22} - z_{12}z_{21}$. Then, a term-by-term comparison with Eq. (*c*) shows that

$$\tau_{11} = \frac{z_{11}}{z_{21}}$$

and

$$\tau_{12} = -\frac{\Delta_Z}{z_{21}}$$

These may be summarized by the matrix representation

$$\mathbf{T} = \begin{bmatrix} \tau_{11} & \tau_{12} \\ \tau_{21} & \tau_{22} \end{bmatrix} = \begin{bmatrix} z_{11}/z_{21} & -\Delta_Z/z_{21} \\ 1/z_{21} & -z_{22}/z_{11} \end{bmatrix} \tag{2.28}$$

The elements of $\mathbf{\Gamma}$ in terms of the z-parameters may be obtained by taking the inverse of Eq. (2.28). First obtain $\det \mathbf{T}$:

$$\det \mathbf{T} = \frac{1}{(z_{21})^2}\left[\Delta_Z - z_{11}z_{22}\right]$$

$$= \frac{1}{(z_{21})^2}\left[z_{11}z_{22} - z_{12}z_{21} - z_{11}z_{22}\right]$$

$$= -\frac{z_{12}z_{21}}{(z_{21})^2} = -\frac{z_{12}}{z_{21}}$$

Then

$$\mathbf{\Gamma} = -\frac{z_{21}}{z_{12}}\begin{bmatrix} -z_{22}/z_{21} & \Delta_Z/z_{21} \\ -1/z_{21} & z_{11}/z_{21} \end{bmatrix}$$

or

$$\mathbf{\Gamma} = \begin{bmatrix} \gamma_{11} & \gamma_{12} \\ \gamma_{21} & \gamma_{22} \end{bmatrix} = \begin{bmatrix} z_{22}/z_{12} & -\Delta_Z/z_{12} \\ 1/z_{12} & -z_{11}/z_{12} \end{bmatrix} \tag{2.29}$$

To find the h-parameters in terms of the z-parameters, begin with Eq. (*b*) and solve for q_a:

$$q_a = -\frac{z_{21}}{z_{22}}q_b + \frac{1}{z_{22}}\theta_a \qquad (h)$$

Then make a term-by-term comparison between Eqs. (*f*) and (*h*) to obtain

$$h_{21} = -\frac{z_{21}}{z_{22}}$$

and

$$h_{22} = \frac{1}{z_{22}}$$

Next, put Eq. (h) into Eq. (a) to obtain

$$\begin{aligned}\theta_b &= z_{11} q_b + z_{12} \left[-\frac{z_{21}}{z_{22}} q_b + \frac{1}{z_{22}} \theta_a \right] \\ &= \left[z_{11} - \frac{z_{12} z_{21}}{z_{22}} \right] q_b + \frac{z_{12}}{z_{22}} \theta_a \\ &= \frac{\Delta_Z}{z_{22}} q_b + \frac{z_{12}}{z_{22}} \theta_a\end{aligned}$$

where $\Delta_Z = z_{11} z_{22} - z_{12} z_{21}$. Then, a term-by-term comparison with Eq. (e) shows that

$$h_{11} = \frac{\Delta_Z}{z_{22}}$$

and

$$h_{12} = \frac{z_{12}}{z_{22}}$$

In summary,

$$\mathbf{H} = \begin{bmatrix} h_{11} & h_{12} \\ h_{21} & h_{22} \end{bmatrix} = \begin{bmatrix} \Delta_Z / z_{22} & z_{12} / z_{22} \\ -z_{21} / z_{22} & 1 / z_{22} \end{bmatrix} \tag{2.30}$$

The elements of **G** in terms of the z-parameters may be obtained by taking the inverse of Eq. (2.30). First obtain det **H**:

$$\begin{aligned}\det \mathbf{H} &= \frac{\Delta_Z + z_{12} z_{21}}{(z_{22})^2} \\ &= \frac{z_{11} z_{22} - z_{12} z_{21} + z_{12} z_{21}}{(z_{22})^2} \\ &= \frac{z_{11}}{z_{22}}\end{aligned}$$

Then

$$\mathbf{G} = \frac{z_{22}}{z_{11}} \begin{bmatrix} 1 / z_{22} & -z_{12} / z_{22} \\ z_{21} / z_{22} & \Delta_Z / z_{22} \end{bmatrix}$$

or

$$\mathbf{G} = \begin{bmatrix} g_{11} & g_{12} \\ g_{21} & g_{22} \end{bmatrix} = \begin{bmatrix} 1 / z_{11} & -z_{12} / z_{11} \\ z_{21} / z_{11} & \Delta_Z / z_{11} \end{bmatrix} \tag{2.31}$$

2.7 SUMMARY OF ALL CONVERSIONS

Machinations similar to those employed in the previous section can yield expressions for each of the parameters in terms of all of the others. A summary of these is provided (without derivation) in Table 2.2.

TABLE 2.2 Summary of Parameter Conversions

z_{11}	$=$	y_{22}/Δ_Y	τ_{11}/τ_{21}	γ_{22}/γ_{21}	Δ_H/h_{22}	$1/g_{11}$
z_{12}	$=$	$-y_{12}/\Delta_Y$	$-\Delta_T/\tau_{21}$	$1/\gamma_{21}$	h_{12}/h_{22}	$-g_{12}/g_{11}$
z_{21}	$=$	$-y_{21}/\Delta_Y$	$-\Delta_\Gamma/\tau_{21}$	$-\Gamma/\gamma_{21}$	$-h_{21}/h_{22}$	g_{21}/g_{11}
z_{22}	$=$	y_{11}/Δ_Y	$-\tau_{22}/\tau_{21}$	γ_{11}/γ_{21}	$1/h_{22}$	Δ_G/g_{11}
y_{11}	$=$	z_{22}/Δ_Z	τ_{22}/τ_{12}	$-\gamma_{11}/\gamma_{12}$	$1/h_{11}$	Δ_G/g_{22}
y_{12}	$=$	$-z_{12}/\Delta_Z$	$-\Delta_T/\tau_{12}$	$1/\gamma_{12}$	$-h_{12}/h_{11}$	g_{12}/g_{22}
y_{21}	$=$	$-z_{21}/\Delta_Z$	$1/\tau_{12}$	$-\Delta_\Gamma/\gamma_{12}$	h_{21}/h_{11}	$-g_{21}/g_{22}$
y_{22}	$=$	z_{11}/Δ_Z	$-\tau_{11}/\tau_{12}$	γ_{22}/γ_{12}	Δ_H/h_{11}	$1/g_{22}$
τ_{11}	$=$	z_{11}/z_{21}	$-y_{22}/y_{21}$	$\gamma_{22}/\Delta_\Gamma$	$-\Delta_H/h_{21}$	$1/g_{21}$
τ_{12}	$=$	$-\Delta_Z/z_{21}$	$1/y_{21}$	$-\gamma_{12}/\Delta_\Gamma$	h_{11}/h_{21}	$-g_{22}/g_{21}$
τ_{21}	$=$	$1/z_{21}$	$-\Delta_Y/y_{21}$	$-\gamma_{21}/\Delta_\Gamma$	$-h_{22}/h_{21}$	g_{11}/g_{21}
τ_{22}	$=$	$-z_{22}/z_{21}$	y_{11}/y_{21}	$\gamma_{11}/\Delta_\Gamma$	$1/h_{21}$	$-\Delta_G/g_{21}$
γ_{11}	$=$	z_{22}/z_{12}	$-y_{11}/y_{12}$	τ_{22}/Δ_T	$1/h_{12}$	$-\Delta_G/g_{21}$
γ_{12}	$=$	$-\Delta_Z/z_{12}$	$1/y_{12}$	$-\tau_{12}/\Delta_T$	$-h_{11}/h_{12}$	g_{22}/g_{12}
γ_{21}	$=$	$1/z_{12}$	$-\Delta_Y/y_{12}$	$-\tau_{21}/\Delta_T$	h_{22}/h_{12}	$-g_{11}/g_{12}$
γ_{22}	$=$	z_{11}/z_{12}	y_{22}/y_{12}	τ_{11}/Δ_T	$-\Delta_H/h_{12}$	$1/g_{12}$
h_{11}	$=$	Δ_Z/z_{22}	$1/y_{11}$	τ_{12}/τ_{22}	$-\gamma_{12}/\gamma_{11}$	g_{22}/Δ_G
h_{12}	$=$	z_{12}/z_{22}	$-y_{12}/y_{11}$	Δ_T/τ_{22}	$1/\gamma_{11}$	$-g_{12}/\Delta_G$
h_{21}	$=$	$-z_{21}/z_{22}$	y_{21}/y_{11}	$1/\tau_{22}$	$\Delta_\Gamma/\gamma_{11}$	$-g_{21}/\Delta_G$
h_{22}	$=$	$1/z_{22}$	Δ_Y/y_{11}	$-\tau_{21}/\tau_{22}$	γ_{21}/γ_{11}	g_{11}/Δ_G
g_{11}	$=$	$1/z_{11}$	Δ_Y/y_{22}	τ_{21}/τ_{11}	$-\gamma_{21}/\gamma_{22}$	h_{22}/Δ_H
g_{12}	$=$	$-z_{12}/z_{11}$	y_{12}/y_{22}	Δ_T/τ_{11}	$1/\gamma_{22}$	$-h_{12}/\Delta_H$
g_{21}	$=$	z_{21}/z_{11}	$-y_{21}/y_{22}$	$1/\tau_{11}$	$\Delta_\Gamma/\gamma_{22}$	$-h_{21}/\Delta_H$
g_{22}	$=$	Δ_Z/z_{11}	$1/y_{22}$	$-\tau_{12}/\tau_{11}$	γ_{12}/γ_{22}	h_{11}/Δ_H

2.8 FORMAL DEVELOPMENT OF THE LINEAR TRANSFORMATIONS

In order to develop the linear transformations for fins that are shaped irregularly and in which the effects of the slope of the fin faces is to be included, consider Fig. 2.2. Here too, x is the height coordinate, but with origin set arbitrarily at some point remote from the fin tip, which is located at $x = a$. For the specific cases where the origin of the height coordinate is set at the fin tip, $a = 0$. The fin cross-sectional area is $A(x)$, and the fin thickness, $\delta(x)$, is the difference between the two curves, $y_1(x)$ and $y_2(x)$. These curves are often referred to as the *profile curves* or *profile functions*.

With the Murray-Gardner assumptions still firmly in mind and with

$$\theta(x) = T(x) - T_s$$

and

$$A(x) = [y_1(x) - y_2(x)]L$$

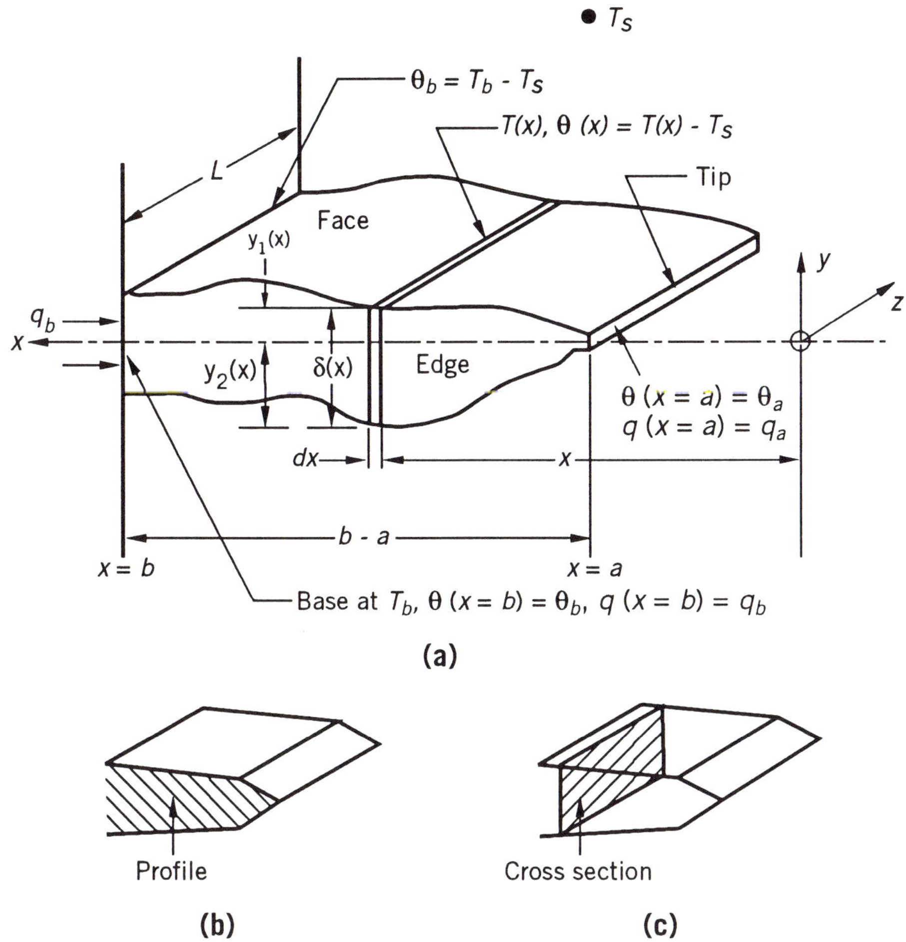

Figure 2.2 (*a*) Longitudinal fin of arbitrary profile and cross section and illustrations of the terminology, (*b*) profile and (*c*) cross section.

the difference between the heat entering and leaving the element dx by conduction will be

$$dq = \frac{d}{dx}\left\{k(x)L[y_1(x) - y_2(x)]\frac{d\theta(x)}{dx}\right\}$$

The heat dissipated by the fin by convection is based on the actual fin face surface area, which depends on the actual height of the fin faces. Recall from calculus that if the actual *fin profile* is $y(x)$, the differential arc length of the fin faces will be

$$ds(x) = \left[1 + \left(\frac{d\theta}{dx}\right)^2\right]^{1/2}$$

The actual fin surface is a function of x:

$$dS(x) = \{[1 + y_1'(x)^2]^{1/2} + [1 + y_2'(x)^2]^{1/2}\}L dx$$

where the primes designate the first derivative. Thus, with the edge area considered negligible, the convective dissipation is

$$dq = h(x)\theta(x)dS(x)$$

or[3]

$$dq = h(x)\theta(x)\{[1 + y_1'(x)^2]^{1/2} + [1 + y_2'(x)^2]^{1/2}\}L dx$$

Conservation of energy dictates that the difference between the heat entering and leaving the element dx by conduction must equal the heat leaving the element by convection. Hence

$$\frac{d}{dx}\left\{k(x)L[y_1(x) - y_2(x)]\frac{d\theta(x)}{dx}\right\}$$
$$= h(x)\theta(x)\{[1 + y_1'(x)^2]^{1/2} + [1 + y_2'(x)^2]^{1/2}\}L$$

and with the heat-transfer coefficient and the thermal conductivity constant

$$\frac{d}{dx}\left\{[y_1(x) - y_2(x)]\frac{d\theta(x)}{dx}\right\}$$
$$- \frac{h}{k}\theta(x)\{[1 + y_1'(x)^2]^{1/2} + [1 + y_2'(x)^2]^{1/2}\} = 0 \tag{2.32}$$

Equation (2.32) is a linear, homogeneous, second-order differential equation for $\theta(x)$. Its solution can be used to determine the heat flow in any longitudinal fin[4] that satisfies the Murray-Gardner assumptions. According to the theory of differential equations, Eq. (2.32) is *regular* at all points where

$$k(x)A(x) = k(x)L[y_1(x) - y_2(x)] \neq 0 \tag{2.33}$$

and *singular* when $k(x)A(x)$ vanishes. The latter situation can only occur physically when $y_1(x) = y_2(x)$, that is, when the fin has zero thickness. This happens at the tips of certain fins or spines such as the longitudinal fins of triangular and concave parabolic profiles and the conical and concave parabolic spines.

[3]If the height, b, rather than the actual arc length, s, is used to compute the surface area of the fin, the procedure is known as *making the length of arc assumption*.

[4]Or spine or indeed any radial fin as long as it is noted that the radial height coordinate has its positive sense from fin base to fin tip.

The theory also guarantees that Eq. (2.32) possesses two independent solutions: $\lambda_1(x)$ and $\lambda_2(x)$ satisfying the *initial conditions* (the conditions at $x = b$)

$$\lambda_1(b) = 1; \qquad \frac{d\lambda_1(b)}{dx} = 0 \tag{2.34a}$$

and

$$\lambda_2(b) = 0; \qquad \frac{d\lambda_2(b)}{dx} = \frac{1}{k(b)A(b)} \tag{2.34b}$$

subject to certain continuity conditions that must be imposed on k and A. If k and A happen to have discontinuities, one merely breaks the fin into shorter segments. The temperature excess and heat flow are then given in terms of the values at the base of the fin:

$$\theta(x) = \theta_b\lambda_1(x) + q_b\lambda_2(x) \tag{2.35}$$

and

$$q(x) = k(x)A(x)[\theta_b\lambda_1'(x) + q_b\lambda_2'(x)] \tag{2.36}$$

where the primes denote derivatives with respect to x,

The linear transformation that maps the temperature excess and heat flow at any point on the fin in terms of conditions at the fin base is

$$\begin{bmatrix} \theta(x) \\ q(x) \end{bmatrix} = \begin{bmatrix} \lambda_1(x) & \lambda_2(x) \\ k(x)A(x)\lambda_1'(x) & k(x)A(x)\lambda_2'(x) \end{bmatrix} \begin{bmatrix} \theta_b \\ q_b \end{bmatrix} \tag{2.37}$$

which is used in

$$\begin{bmatrix} \theta(x) \\ q(x) \end{bmatrix} = \mathbf{\Lambda} \begin{bmatrix} \theta_b \\ q_b \end{bmatrix} \tag{2.38}$$

where $\mathbf{\Lambda}$ may be written as

$$\mathbf{\Lambda} = \begin{bmatrix} 1 & 0 \\ 0 & k(x)A(x) \end{bmatrix} \begin{bmatrix} \lambda_1(x) & \lambda_2(x) \\ \lambda_1'(x) & \lambda_2'(x) \end{bmatrix} \tag{2.39}$$

The determinant of the second matrix is the familiar *Wronskian*

$$w\{\lambda_1, \lambda_2 : x\} = \det \begin{bmatrix} \lambda_1(x) & \lambda_2(x) \\ \lambda_1'(x) & \lambda_2'(x) \end{bmatrix}$$

The γ-matrix is generated when x is set equal to a (a may equal zero) in Eq. (2.39):

$$\mathbf{\Gamma} = \begin{bmatrix} \gamma_{11} & \gamma_{12} \\ \gamma_{21} & \gamma_{22} \end{bmatrix}$$

so that the actual linear transformation from base conditions to tip conditions is

$$\begin{bmatrix} \theta_a \\ q_a \end{bmatrix} = \mathbf{\Gamma} \begin{bmatrix} \theta_b \\ q_b \end{bmatrix} = \begin{bmatrix} \gamma_{11} & \gamma_{12} \\ \gamma_{21} & \gamma_{22} \end{bmatrix} \begin{bmatrix} \theta_b \\ q_b \end{bmatrix} \tag{2.40}$$

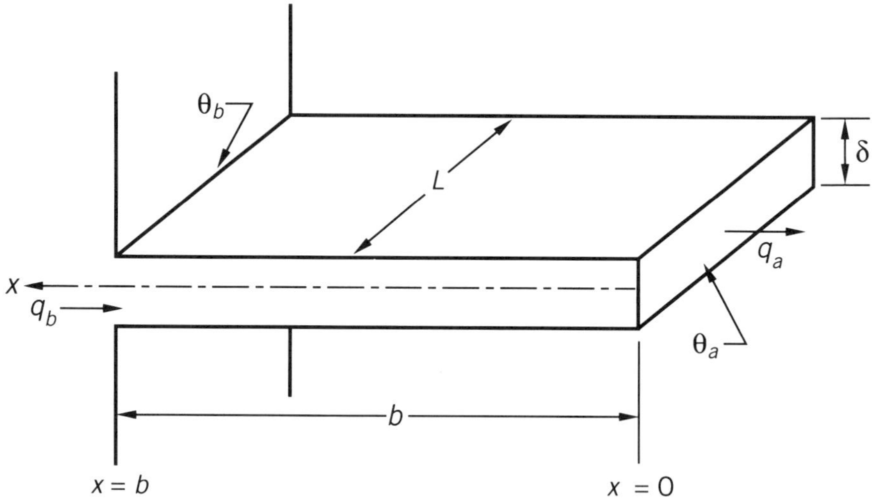

Figure 2.3 Longitudinal fin of rectangular profile.

Here

$$\gamma_{11} = \lambda_1(a) \tag{2.41a}$$

$$\gamma_{12} = \lambda_2(a) \tag{2.41b}$$

$$\gamma_{21} = k(a)A(a)\lambda_1'(a) \tag{2.41c}$$

and

$$\gamma_{22} = k(a)A(a)\lambda_2'(a) \tag{2.41d}$$

Observe that if $A(a) = 0, \gamma_{21} = \gamma_{22} = 0$ and

$$\mathbf{\Gamma} = \begin{bmatrix} \lambda_1(0) & \lambda_2(0) \\ 0 & 0 \end{bmatrix} = \begin{bmatrix} \gamma_{11} & \gamma_{12} \\ 0 & 0 \end{bmatrix}$$

This determinant of this matrix is equal to zero, and the matrix is termed *singular* because it has no inverse. This is why fins and spines that taper to a zero cross section are called *singular* and why a different approach (to be considered in Chap. 4) is necessary for their parameterization.

An example of the determination of the parameters now follows. It concerns a familiar configuration, the longitudinal fin of rectangular profile.

2.9 AN EXAMPLE OF FINDING THE PARAMETERS

For the longitudinal fin of rectangular profile shown in Fig. 2.3, the profile functions are $y_1(x) = \delta/2$ and $y_2(x) = -\delta/2$. With these in Eq. (2.32), one obtains the differential equation for temperature excess of Eq. (2.1):

$$\frac{d^2\theta}{dx^2} - m^2\theta = 0 \tag{2.1}$$

where

$$m = \left(\frac{2h}{k\delta}\right)^{1/2}$$

With the initial conditions given by

$$\theta(x = b) = \theta_b \tag{2.35a}$$

and

$$q(x = b) = q_b \tag{2.35b}$$

the development follows that employed in Sec. 2.4 to arrive at the temperature excess at any point on the fin:

$$\theta(x) = [\cosh m(b - x)]\theta_b - \frac{1}{Y_o}[\sinh m(b - x)]q_b \tag{2.7}$$

and the heat flow at any point in the fin:

$$q(x) = k\delta L\frac{d\theta(x)}{dx} = -[Y_o \sinh m(b - x)]\theta_b + [\cosh m(b - x)]q_b \tag{2.8}$$

where Y_o is the characteristic admittance of the fin defined by Eq. (2.6):

$$Y_o = (2hk\delta)^{1/2}L \tag{2.6}$$

Reference to Eqs. (2.35) shows that

$$\lambda_1(x) = \cosh\, m(b - x); \qquad \lambda_1(b) = 1 \tag{2.42a}$$

$$\lambda_2(x) = -\frac{1}{Y_o}\sinh\, m(b - x) \qquad \lambda_2(b) = 0 \tag{2.42b}$$

$$\lambda_1'(x) = -m \sinh\, m(b - x); \qquad \lambda_1'(b) = 0 \tag{2.42c}$$

and

$$\lambda_2'(x) = \frac{m}{Y_o}\cosh\, m(b - x); \qquad \lambda_2'(b) = \frac{m}{Y_o} = \frac{1}{k\delta L} \tag{2.42d}$$

and at $x = a = 0$, the γ-parameters are obtained from Eqs. (2.41):

$$\gamma_{11} = \lambda_1(0) = \cosh\, mb \tag{2.43a}$$

$$\gamma_{12} = \lambda_2(0) = -\frac{1}{Y_o}\sinh\, mb \tag{2.43b}$$

$$\gamma_{21} = k\delta L\lambda_1'(0) = -k\delta Lm \sinh\, mb = -Y_o \sinh\, mb \tag{2.43c}$$

and

$$\gamma_{22} = k\delta L\lambda_2'(0) = \frac{k\delta Lm}{Y_o}\cosh\, mb = \cosh\, mb \tag{2.43d}$$

2.10 THE INPUT ADMITTANCE AND THE THERMAL TRANSMISSION RATIO

The fin efficiency has been recognized for years as the key to the analysis and design of equipment containing extended surfaces. An attempt has been made here (in Sec. 2.3) to show that the fin efficiency has its limitations. Indeed it must be emphasized that *the fin efficiency is certainly not a design parameter that can be used in the design and analysis of the intricate heat sinks employed with electronic components* because, for complicated shapes, the fin efficiency of individual fins segments may be difficult to obtain.

Two completely new parameterizations were developed by Kraus et al. (1978). One of them, the *input admittance,* pertains to the so-called *regular fins'* The second, the *thermal transmission ratio,* involves the *singular fins.* Both parameters are based on the Murray-Gardner limiting assumptions.

2.10.1 The Input Admittance

The input admittance applies to an individual fin or a finned array and is defined by

$$Y_{in} \equiv \frac{q_b}{\theta_b} \tag{2.44}$$

It is the ratio of the base heat flow to the base temperature excess and it is related to the fin efficiency, η by

$$Y_{in} = \eta h S \tag{2.45}$$

where S is the total convecting surface in either the single fin under consideration or the finned array containing many fins. It should be emphasized that the use of the input admittance in conjunction with the base or prime surface is additive. The total input admittance, $Y_{in,T}$, is the sum of the input admittance of the base or prime surface, $Y_{in,b} = hS_b$ and the fin or array input admittance, Y_{in} so that

$$q_T = Y_{in,T}\theta_b = (Y_{in,b} + Y_{in})\theta_b$$

Recall that the fin or array input admittance is easily formulated from the linear transformation involving $\mathbf{T}$:

$$\begin{bmatrix} \theta_b \\ q_b \end{bmatrix} = \mathbf{T} \begin{bmatrix} \theta_a \\ q_a \end{bmatrix} = \begin{bmatrix} \tau_{11} & \tau_{12} \\ \tau_{21} & \tau_{22} \end{bmatrix} \begin{bmatrix} \theta_a \\ q_a \end{bmatrix} \tag{2.46}$$

A matrix multiplication yields

$$\begin{aligned} \theta_b &= \tau_{11}\theta_a + \tau_{12}q_a \\ q_b &= \tau_{21}\theta_a + \tau_{22}q_a \end{aligned}$$

and now the input admittance may be defined using its defininition Eq. (2.44),

$$Y_{in} = \frac{q_b}{\theta_b} = \frac{\tau_{21}\theta_a + \tau_{22}q_a}{\tau_{11}\theta_a + \tau_{12}q_a}$$

or

$$Y_{in} = \frac{q_b}{\theta_b} = \frac{\tau_{21} + \tau_{22}(q_a/\theta_a)}{\tau_{11} + \tau_{12}(q_a/\theta_a)} \tag{2.47}$$

Observe that the tip heat flow may be zero or negligible, in which case

$$Y_{in} = \frac{q_b}{\theta_b} = \frac{\tau_{21}}{\tau_{11}} \tag{2.48}$$

2.10.2 The Thermal Transmission Ratio

As indicated by Eq. (2.47), the input admittance is easily obtained from the elements of $\mathbf{T}$. However, an attempt to follow the same procedure for the singular fin is foolhardy because, as observed at the end of Sec. 2.8, $\mathbf{\Gamma}$ for the singular fin,

$$\mathbf{\Gamma} = \begin{bmatrix} \lambda_1(0) & \lambda_2(0) \\ 0 & 0 \end{bmatrix} = \begin{bmatrix} \gamma_{11} & \gamma_{12} \\ 0 & 0 \end{bmatrix}$$

possesses a zero determinant. Thus $\mathbf{T}$, which is the inverse of $\mathbf{\Gamma}$, does not exist.

The use of the input admittance for a regular fin or finned array suggests that, because no heat can ever leave the tip of a singular fin, it is sufficent to parameterize the singular fin with a single parameter

$$\mu = \frac{q_b}{\theta_b} \tag{2.49}$$

called the *thermal transmission ratio*, which is seen to be, like the input admittance, the ratio of the base heat flow to the base temperature excess. While both are ratios, the input admittance is a ratio reserved for regular fins or the base of an array of fins, and the thermal transmission ratio is reserved for use with singular fins. This is but a subtle difference, and the usage of the terminology will be quite apparent as the reader proceeds.

Attention now turns to Chap. 3 which deals with the development of the six linear transformations for several longitudinal fins and spines and the radial fin of rectangular profile. The thermal transmission ratios for four singular fins and spines are developed in Chap. 4.

Chapter

3

ELEMENTS OF THE LINEAR TRANSFORMATIONS

3.1 INTRODUCTION

It has been noted that those fins that possess linear transformations relating two of the four physical quantities of base and tip heat flow and base and tip temperature excess to the other two are termed *regular fins*. Several of the useful shapes are regular, and the following types will be treated in this chapter:

1. Longitudinal fin of rectangular profile
2. Longitudinal fin of rectangular profile with one face insulated
3. Longitudinal fin of trapezoidal profile
4. Longitudinal fin of trapezoidal profile with one face insulated
5. Longitudinal fin of truncated parabolic profile
6. Radial fin of rectangular profile
7. Spines of constant cross section
8. Truncated conical spine
9. Truncated concave parabolic spine

Moreover, in contrast to the regular fins, there are several useful fins that cannot be characterized by any linear transformations. Such shapes will be considered in Chap. 4.

Because interest is most often focused on fin base conditions (temperature excess and heat flow), the differential equations for the individual fins and spines treated in this chapter are solved using initial value data. This guarantees that the solutions of the equations are analytically well posed (as opposed to solutions obtained from boundary value data), easily computed numerically, and convenient to tabulate. The procedure here will involve the use of the generalized differential equation derived in Sec. 2.8 and the development of the $\boldsymbol{\Gamma}$-matrix, which maps base conditions of temperature excess and heat flow into similar conditions at the tip. Then the other five sets of parameters are developed through the use of Table 2.2.

It was shown in Sec. 2.8 that a generalized differential equation for the point-to-point temperature excess on a regular longitudinal fin could be developed. The result was Eq. (2.32). which is repeated here as Eq. (3.1):

$$\frac{d}{dx}\{[y_1(x) - y_2(x)]\frac{d\theta(x)}{dx}\} - \frac{h}{k}\theta(x)\{[1 + y_1'(x)^2]^{1/2} + [1 + y_2'(x)^2]^{1/2}\} = 0 \tag{3.1}$$

where the primes represent derivatives.

Equation (3.1) is a linear, homogeneous, second-order differential equation for $\theta(x)$, and its solution can be used to determine the heat flow in any longitudinal fin that satisfies the Murray-Gardner assumptions. Moreover, it was shown that the theory guarantees two solutions to Eq. (3.1), $\lambda_1(x)$ and $\lambda_2(x)$, that satisfy the *initial conditions*:

$$\lambda_1(b) = 1; \qquad \frac{d\lambda_1(b)}{dx} = 0 \tag{3.2a}$$

and

$$\lambda_2(b) = 0; \qquad \frac{d\lambda_2(b)}{dx} = \frac{1}{k(b)A(b)} \tag{3.2b}$$

subject to certain continuity conditions that must be imposed on k and A.

The solutions are

$$\theta(x) = \theta_b\lambda_1(x) + q_b\lambda_2(x) \tag{3.3}$$

and

$$q(x) = k(x)A(x)[\theta_b\lambda_1'(x) + q_b\lambda_2'(x)] \tag{3.4}$$

Equations (3.3) and (3.4) produce a linear transformation that maps the temperature excess and heat flow at the fin tip from conditions at the fin base. This is frequently referred to as the $\boldsymbol{\Gamma}$-matrix, which is generated when x is set equal to the tip location, a (a may equal zero)

$$\boldsymbol{\Gamma} = \begin{bmatrix} \gamma_{11} & \gamma_{12} \\ \gamma_{21} & \gamma_{22} \end{bmatrix}$$

so that the actual linear transformation from base conditions to tip conditions is

$$\begin{bmatrix} \theta_a \\ q_a \end{bmatrix} = \mathbf{\Gamma} \begin{bmatrix} \theta_b \\ q_b \end{bmatrix} = \begin{bmatrix} \gamma_{11} & \gamma_{12} \\ \gamma_{21} & \gamma_{22} \end{bmatrix} \begin{bmatrix} \theta_b \\ q_b \end{bmatrix} \tag{3.5}$$

Here, because the fin tip is located at $x = a$,

$$\gamma_{11} = \lambda_1(a) \tag{3.6a}$$

$$\gamma_{12} = \lambda_2(a) \tag{3.6b}$$

$$\gamma_{21} = k(a)A(a)\lambda_1'(a) \tag{3.6c}$$

and

$$\gamma_{22} = k(a)A(a)\lambda_2'(a) \tag{3.6d}$$

and if $a = 0$, these become

$$\gamma_{11} = \lambda_1(0) \tag{3.7a}$$

$$\gamma_{12} = \lambda_2(0) \tag{3.7b}$$

$$\gamma_{21} = k(0)A(0)\lambda_1'(0) \tag{3.7c}$$

and

$$\gamma_{22} = k(0)A(0)\lambda_2'(0) \tag{3.7d}$$

3.2 THE LONGITUDINAL FIN OF RECTANGULAR PROFILE

3.2.1 The Γ-matrix

The $\mathbf{\Gamma}$-matrix for the longitudinal fin of rectangular profile, displayed in Fig. 3.1, was developed in Sec. 2.4 and summarized in Sec. 2.5. The matrix representation is

$$\begin{bmatrix} \theta_a \\ q_a \end{bmatrix} = \mathbf{\Gamma} \begin{bmatrix} \theta_b \\ q_b \end{bmatrix} = \begin{bmatrix} \cosh mb & -Z_0 \sinh mb \\ -Y_0 \sinh mb & \cosh mb \end{bmatrix} \begin{bmatrix} \theta_b \\ q_b \end{bmatrix} \tag{3.8}$$

where

$$Y_0 = (2hk\delta)^{1/2} L$$

$$Z_0 = \frac{1}{Y_0} = \frac{1}{(2hk\delta)^{1/2} L}$$

and

$$m = \left(\frac{2h}{k\delta} \right)^{1/2}$$

The use of Table 2.2 provides the other five matrices

3.2.2 The $\mathbf{T}$-matrix

$$\begin{bmatrix} \theta_b \\ q_b \end{bmatrix} = \mathbf{T} \begin{bmatrix} \theta_a \\ q_a \end{bmatrix} = \begin{bmatrix} \cosh mb & Z_0 \sinh mb \\ Y_0 \sinh mb & \cosh mb \end{bmatrix} \begin{bmatrix} \theta_a \\ q_a \end{bmatrix} \tag{3.9}$$

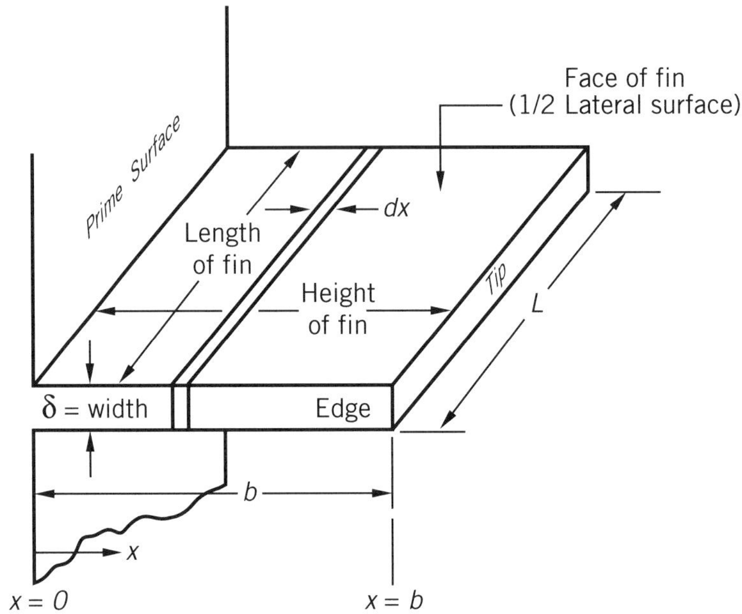

Figure 3.1 The longitudinal fin of rectangular profile.

3.2.3 The Z-matrix

$$\begin{bmatrix} \theta_b \\ \theta_a \end{bmatrix} = \mathbf{Z}\begin{bmatrix} q_b \\ q_a \end{bmatrix} = \begin{bmatrix} Z_0 \coth mb & -Z_0 \operatorname{csch} mb \\ Z_0 \operatorname{csch} mb & -Z_0 \coth mb \end{bmatrix}\begin{bmatrix} q_b \\ q_a \end{bmatrix} \tag{3.10}$$

3.2.4 The Y-matrix

$$\begin{bmatrix} q_b \\ q_a \end{bmatrix} = \mathbf{Y}\begin{bmatrix} \theta_b \\ \theta_a \end{bmatrix} = \begin{bmatrix} Y_0 \coth mb & -Y_0 \operatorname{csch} mb \\ Y_0 \operatorname{csch} mb & -Y_0 \coth mb \end{bmatrix}\begin{bmatrix} \theta_b \\ q_b \end{bmatrix} \tag{3.11}$$

3.2.5 The H-matrix

$$\begin{bmatrix} \theta_b \\ q_a \end{bmatrix} = \mathbf{H}\begin{bmatrix} q_b \\ \theta_a \end{bmatrix} = \begin{bmatrix} Z_0 \tanh mb & \operatorname{sech} mb \\ \operatorname{sech} mb & -Y_0 \tanh mb \end{bmatrix}\begin{bmatrix} q_b \\ \theta_a \end{bmatrix} \tag{3.12}$$

3.2.6 The G-matrix

$$\begin{bmatrix} q_b \\ \theta_a \end{bmatrix} = \mathbf{G}\begin{bmatrix} \theta_b \\ q_a \end{bmatrix} = \begin{bmatrix} Y_0 \tanh mb & \operatorname{sech} mb \\ \operatorname{sech} mb & -Z_0 \tanh mb \end{bmatrix}\begin{bmatrix} \theta_b \\ q_a \end{bmatrix} \tag{3.13}$$

Examples 3.1 and 3.2 now follow. Example 3.1 employs the **G**-matrix to make some rudimentary calculations involving a longitudinal fin of rectangular profile. It is substantially similar to Ex. 2.1 in Kern and Kraus (1972), which does not use the **G**-matrix. Thus, it is noted that the advent of the linear transformation matrices does not imply that the simple fin temperature excess, heat flow, and efficiency relationships cannot be used. It only means that there is now another alternative. Example 3.2 concerns the heat loss from the tip of a longitudinal fin of rectangular profile. It is similar to Ex. 3.1 in Kern and Kraus (1972) and involves the **T**-matrix. The solution contained here seems to be less involved than the solution in Kern and Kraus (1972). Nevertheless, the Kern and Kraus solution has validity.

The point to be made is that solution methods that do not involve the linear transformations developed here have validity, and the linear transformations do not replace methods that heretofore have been customary. But there will be cases where the linear transformations provide the only possible expeditious solutions, and these facts will be demonstrated as this study proceeds.

Example 3.1: A longitudinal fin of rectangular profile is exposed to surroundings with a temperature of 50°C and a heat-transfer coefficient of 50 W/m²-K. The temperature at the fin base is 100°C, and the fin is made of a material with thermal conductivity, $k = 160\,\mathrm{W/m\text{-}K}$. The fin is 10 cm high, 40 cm long, and 0.9525 cm thick. Determine, if the fin tip is insulated, the tip temperature and the base heat flow.

SOLUTION Here, $T_b = 100°\mathrm{C}$ is a given and $q_a = 0$ because the tip of the fin is stated as being insulated. The values of θ_a and q_b are to be found. This suggests the use of the **G**-matrix, and the necessary computations now follow using four significant figures:

$$\begin{aligned} m &= \left(\frac{2h}{k\delta}\right)^{1/2} \\ &= \left[\frac{2(50)}{160(0.009525)}\right]^{1/2} \\ &= 8.1004\,\mathrm{m}^{-1} \end{aligned}$$

$$mb = 8.1004(0.1000) = 0.8100$$

$$\begin{aligned} Y_0 &= (2hk\delta)^{1/2}L \\ &= [2(50)(160)(0.009525)]^{1/2}(0.40) \\ &= 4.9380\,\mathrm{W/°C} \end{aligned}$$

$$Z_0 = 1/Y_0 = 0.2025\,\mathrm{K/W}$$

and from Eqs. (3.13):

$$g_{11} = Y_0 \tanh mb$$

$$\begin{aligned}
&= 4.9380 \tanh 0.8100 \\
&= 4.9380(0.6696) = 3.3066\,\mathrm{W/K} \\
g_{12} &= \operatorname{sech} mb \\
&= \operatorname{sech} 0.8100 \\
&= 0.7427 \\
g_{21} &= \operatorname{sech} mb \\
&= \operatorname{sech} 0.8100 \\
&= 0.7427 \\
g_{22} &= -Z_0 \tanh mb \\
&= -0.2025 \tanh 0.8100 \\
&= -0.2025(0.6696) = -0.1356\,\mathrm{K/W}
\end{aligned}$$

The linear transformation may be written as

$$\begin{bmatrix} q_b \\ \theta_a \end{bmatrix} = \mathbf{G} \begin{bmatrix} \theta_b \\ q_a \end{bmatrix} = \begin{bmatrix} 3.3066 & 0.7427 \\ 0.7427 & -0.1356 \end{bmatrix} \begin{bmatrix} \theta_b \\ q_a \end{bmatrix}$$

and with

$$\theta_b = 100 - 50 = 50°\mathrm{C}$$

and

$$q_a = 0\,\mathrm{W}$$

then

$$\begin{bmatrix} q_b \\ \theta_a \end{bmatrix} = \mathbf{G} \begin{bmatrix} \theta_b \\ q_a \end{bmatrix} = \begin{bmatrix} 3.3066 & 0.7427 \\ 0.7427 & -0.1356 \end{bmatrix} \begin{bmatrix} 50 \\ q_a = 0 \end{bmatrix}$$

or to one decimal place

$$\begin{bmatrix} q_b \\ \theta_a \end{bmatrix} = \begin{bmatrix} 165.3\,\mathrm{W} \\ 37.1\,°\mathrm{C} \end{bmatrix}$$

Thus

$$\theta_b = 37.1 + 50 = 87.1°\mathrm{C}$$

and

$$q_b = 165.3\,\mathrm{W}$$

Example 3.2: Show that if the tip heat transfer coefficient is h_a, the heat flowing through the base of a longitudinal fin of rectangular profile is given by

$$q_b = \frac{k\delta L m \theta_b \left[\left(\dfrac{h_a}{mk} \right) + \tanh mb \right]}{1 + \left(\dfrac{h_a}{mk} \right) \tanh mb}$$

where all symbols are as previously defined.

SOLUTION The given expression contains q_b and θ_b, and the strategy is to obtain these base conditions in terms of the tip conditions q_a and θ_a. This leads to the use of the **T**-matrix with tip heat loss, $q_a = h_a \delta L \theta_a$,

$$\begin{bmatrix} \theta_b \\ q_b \end{bmatrix} = \mathbf{T} \begin{bmatrix} \theta_a \\ q_a \end{bmatrix} = \begin{bmatrix} \cosh mb & Z_0 \sinh mb \\ Y_0 \sinh mb & \cosh mb \end{bmatrix} \begin{bmatrix} \theta_a \\ h_a \delta L \theta_a \end{bmatrix}$$

After expansion,

$$\begin{aligned} \theta_b &= \cosh mb\, \theta_a + (h_a \delta L) Z_0 \sinh mb\, \theta_a \\ q_b &= Y_0 \sinh mb\, \theta_a + (h_a \delta L) \cosh mb\, \theta_a \end{aligned}$$

a rearrangement provides

$$\begin{aligned} \theta_b &= [\cosh mb + h_a \delta L Z_0 \sinh mb] \theta_a \\ q_b &= [Y_0 \sinh mb + h_a \delta L \cosh mb] \theta_a \end{aligned}$$

A division of the second by the first yields

$$\frac{q_b}{\theta_b} = \frac{Y_0 \sinh mb + h_a \delta L \cosh mb}{\cosh mb + h_a \delta L Z_0 \sinh mb}$$

The Y_0 in the numerator and the $\cosh mb$ in both the numerator and denominator may be factored. With $Z_0 = 1/Y_0$, this leads to

$$\frac{q_b}{\theta_b} = \frac{Y_0[\tanh mb + h_a \delta L Z_0]}{1 + h_a \delta L Z_0 \tanh mb}$$

and when it is noted that

$$Y_0 = (2hk\delta)^{1/2} L = k\delta L m$$

so that

$$h_a \delta L Z_0 = \frac{h_a \delta L}{k \delta L m} = \frac{h_a}{mk}$$

the result is obtained

$$q_b = \frac{k\delta L m \theta_b \left[\left(\dfrac{h_a}{mk} \right) + \tanh mb \right]}{1 + \left(\dfrac{h_a}{mk} \right) \tanh mb}$$

3.3 THE LONGITUDINAL FIN OF RECTANGULAR PROFILE WITH ONE FACE INSULATED

In the event that one face of the longitudinal fin of rectangular profile (see Fig. 3.1) is insulated, all of Eqs. (3.8) to (3.13) apply but with

$$Y_0 = (hk\delta)^{1/2} L$$

$$Z_0 = \frac{1}{Y_0} = \frac{1}{(hk\delta)^{-1/2}L}$$

and

$$m = \left(\frac{h}{k\delta}\right)^{1/2}$$

3.4 THE LONGITUDINAL FIN OF TRAPEZOIDAL PROFILE

For the longitudinal fin of trapezoidal profile shown in Fig. 3.2, the origin of the x-coordinate system is taken at the fictitious point of intersection of the two fin faces. This puts the tip of the fin at $x = a$. The fin profile is confined by the two smooth planes, $y_1(x)$ and $-y_2(x)$. The fin taper angle is defined by

$$\phi = \arctan \frac{\delta_b}{2b}$$

and a consideration of similar triangles provides a relationship between the base and tip coordinates and thicknesses

$$\frac{\delta_a}{\delta_b} = \frac{a}{b}$$

The differential equation for the temperature excess can be obtained from Eq. (3.1) as

$$2x \tan\phi \frac{d^2\theta}{dx^2} + 2\tan\phi \frac{d\theta}{dx} - \frac{2h}{k\cos\phi}\theta = 0 \tag{3.14}$$

The presence of the $\cos\phi$ in the last term accounts for the fact that the fin face surface element is not dx but $dx/\cos\phi$. This accounts for the taper at the fin faces, and a slight adjustment of Eq. (3.14) then gives

$$x\frac{d^2\theta}{dx^2} + \frac{d\theta}{dx} - n^2\theta = 0 \tag{3.15}$$

where

$$n = \left(\frac{2h}{2k\cos\phi\tan\phi}\right)^{1/2} = \left(\frac{h}{k\sin\phi}\right)^{1/2}$$

Equation (3.15) is a linear, second-order, homogeneous differential equation with variable coefficients that possesses a general solution in terms of the modified Bessel functions

$$\theta(x) = C_1 I_0(2nx^{1/2}) + C_2 K_0(2nx^{1/2}) \tag{3.16}$$

where C_1 and C_2 are to be evaluated from the initial conditions

$$\theta(x = b) = \theta_b \tag{3.17a}$$

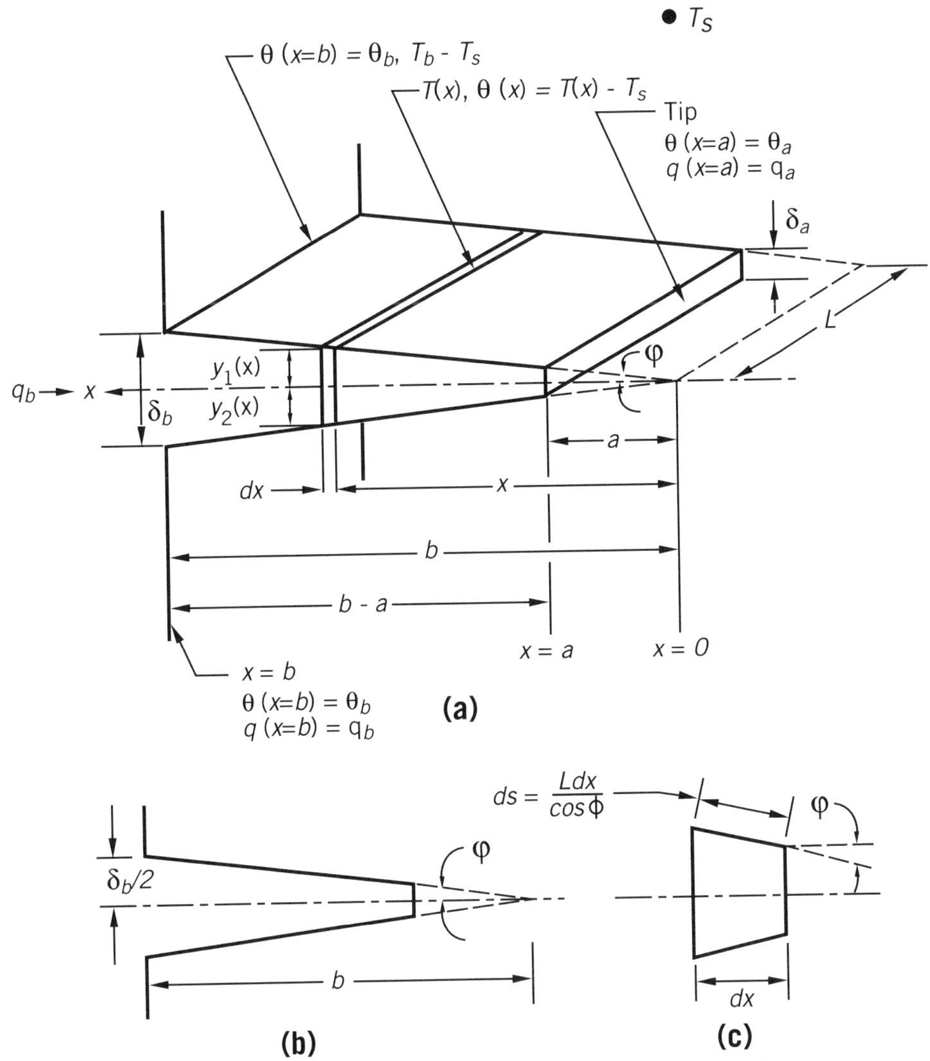

Figure 3.2 Coordinate system for the analysis of the longitudinal fin of trapezoidal profile: (*a*) entire fin, (*b*) detail showing the taper angle, and (*c*) detail showing a differential height element.

and

$$q(x = b) = q_b \tag{3.17b}$$

If a transformation is made

$$u = 2nx^{1/2}; \qquad x = u^2/4n^2 \tag{3.18}$$

so that at $x = b$

$$u_b = 2nb^{1/2}$$

and at $x = a$

$$u_a = 2na^{1/2}$$

then, Eq. (3.16) can be written as

$$\theta(u) = C_1 I_0(u) + C_2 K_0(u) \tag{3.19}$$

with a derivative

$$\frac{d\theta(x)}{dx} = \frac{d}{dx}[C_1 I_0(u) + C_2 K_0(u)]$$

With $x = u^2/4n^2$ [see Eq. (3.18)] so that $dx = (u/2n^2)du$, this becomes

$$\frac{d\theta(x)}{dx} = \frac{2n^2}{u}\frac{d}{du}[C_1 I_0(u) + C_2 K_0(u)]$$

or

$$\frac{d\theta(x)}{dx} = \frac{2n^2}{u}[C_1 I_1(u) - C_2 K_1(u)]$$

Now at $x = b$ where $u = u_b$,

$$\theta(x = b) = \theta_b = C_1 I_0(u_b) + C_2 K_0(u_b) \tag{3.20a}$$

and with $A(b) = \delta_b L$,

$$\begin{aligned} q(x = b) = q_b &= kA(b)\frac{d\theta(x)}{dx}\bigg|_{x=b} \\ &= \frac{2k\delta_b L n^2}{u_b}[C_1 I_1(u_b) - C_2 K_1(u_b)] \end{aligned} \tag{3.20b}$$

Equations (3.20), when solved simultaneously for C_1 and C_2, provide

$$C_1 = \frac{K_1(u_b)\theta_b + \left[\dfrac{u_b}{2k\delta_b L n^2}\right] K_0(u_b) q_b}{I_0(u_b)K_1(u_b) + I_1(u_b)K_0(u_b)}$$

and

$$C_2 = \frac{I_1(u_b)\theta_b - \left[\dfrac{u_b}{2k\delta_b L n^2}\right] I_0(u_b) q_b}{I_0(u_b)K_1(u_b) + I_1(u_b)K_0(u_b)}$$

Use of the Bessel function identity

$$I_0(z)K_1(z) + I_1(z)K_0(z) = \frac{1}{z}$$

permits the arbitrary constants to be written as

$$C_1 = u_b \left[K_1(u_b)\theta_b + \left(\frac{u_b}{2k\delta_b L n^2} \right) K_0(u_b) q_b \right]$$

and

$$C_2 = u_b \left[I_1(u_b)\theta_b - \left(\frac{u_b}{2k\delta_b L n^2} \right) I_0(u_b) q_b \right]$$

Now, with C_1 and C_2 inserted into Eq. (3.19),

$$\theta(u) = u_b[K_1(u_b)I_0(u) + I_1(u_b)K_0(u)]\theta_b + \frac{u_b^2}{2k\delta_b L n^2}[K_0(u_b)I_0(u) - I_0(u_b)K_0(u)]q_b \quad (3.21)$$

which shows that

$$\lambda_1(u) = u_b[K_1(u_b)I_0(u) + I_1(u_b)K_0(u)] \quad (3.22a)$$

and

$$\lambda_2(u) = \frac{u_b^2}{2k\delta_b L n^2}[K_0(u_b)I_0(u) - I_0(u_b)K_0(u)] \quad (3.22b)$$

Observe that $\lambda_1(u_b) = 1$ and $\lambda_2(u_b) = 0$ as required by Eq. (3.2*a*).

To determine $q(x)$, it is necessary to obtain expressions for $\lambda_1'(u)$ and $\lambda_2'(u)$. First

$$\lambda_1'(u) = \frac{d}{dx}\{u_b[K_1(u_b)I_0(u) + I_1(u_b)K_0(u)]\}$$

and with $dx = (u/2n^2)du$,

$$\lambda_1'(u) = 2n^2 \left(\frac{u_b}{u} \right) [K_1(u_b)I_1(u) - I_1(u_b)K_1(u)] \quad (3.22c)$$

Then

$$\lambda_2'(u) = \frac{d}{dx}\left\{ \frac{u_b^2}{2k\delta_b L n^2}[K_0(u_b)I_0(u) - I_0(u_b)K_0(u)] \right\}$$

and again with $dx = (u/2n^2)du$

$$\lambda_2'(u) = \frac{u_b}{k\delta_b L} \left(\frac{u_b}{u} \right) [K_0(u_b)I_1(u) + I_0(u_b)K_1(u)] \quad (3.22d)$$

Observe that at $x = b$ where $u = u_b$, $\lambda_1'(u_b) = 0$ and $\lambda_2'(u_b) = 1/k\delta_b L = 1/kA(b)$ both of which are required by Eq. (3.2*b*).

Then with

$$q(x) = k(x)A(x)[\theta_b\lambda_1'(x) + q_b\lambda_2'(x)] \tag{3.4}$$

the $\mathbf{\Gamma}$-matrix can now be formulated. Equations (3.22) can be evaluated at $x = a$ to form

$$\begin{bmatrix} \theta_a \\ q_a \end{bmatrix} = \mathbf{\Gamma}\begin{bmatrix} \theta_b \\ q_b \end{bmatrix} = \begin{bmatrix} \gamma_{11} & \gamma_{12} \\ \gamma_{21} & \gamma_{22} \end{bmatrix}\begin{bmatrix} \theta_b \\ q_b \end{bmatrix}$$

and, in accordance with Eqs. (3.6),

$$\gamma_{11} = \lambda_1(u_a) = u_b[I_0(u_a)K_1(u_b) + I_1(u_b)K_0(u_a)] \tag{3.23a}$$

$$\gamma_{12} = \lambda_2(u_a) = \frac{u_b^2}{2k\delta_b Ln^2}[I_0(u_a)K_0(u_b) - I_0(u_b)K_0(u_a)] \tag{3.23b}$$

$$\gamma_{21} = kA(a)\lambda_1'(u_a)$$

or

$$\gamma_{21} = 2k\delta_a Ln^2\left(\frac{u_b}{u_a}\right)[I_1(u_a)K_1(u_b) - I_1(u_b)K_1(u_a)]\} \tag{3.23c}$$

and with $\delta_a/\delta_b = a/b$,

$$\begin{aligned} \frac{\delta_a u_b^2}{\delta_b u_a} &= \frac{a(4n^2b)}{b(2na^{1/2})} \\ &= 2na^{1/2} \\ &= u_a \end{aligned}$$

and

$$\gamma_{22} = kA(a)\lambda_2'(u_a) = u_a[I_1(u_a)K_0(u_b) + I_0(u_b)K_1(u_a)] \tag{3.23d}$$

The elements of the other linear tranformations can now be formed using Table 2.2. However, before employing this procedure, it is necessary to obtain the value of the determinant of $\mathbf{\Gamma}$. If the γ's are written as

$$\begin{aligned} \gamma_{11} &= u_b\gamma_1 \\ \gamma_{12} &= \frac{u_b^2}{2k\delta_b Ln^2}\gamma_2 \\ \gamma_{21} &= 2k\delta_a Ln^2\left(\frac{u_b}{u_a}\right)\gamma_3 \\ \gamma_{22} &= u_a\gamma_4 \end{aligned}$$

where

$$\begin{aligned}\gamma_1 &= I_0(u_a)K_1(u_b) + I_1(u_b)K_0(u_a)\\ \gamma_2 &= I_0(u_a)K_0(u_b) - I_0(u_b)K_0(u_a)\\ \gamma_3 &= I_1(u_a)K_1(u_b) - I_1(u_b)K_1(u_a)\end{aligned}$$

and

$$\gamma_4 = [I_1(u_a)K_0(u_b) + I_0(u_b)K_1(u_a)$$

Then

$$\begin{aligned}\det\mathbf{\Gamma} &= \det\begin{bmatrix}\gamma_{11} & \gamma_{12}\\ \gamma_{21} & \gamma_{22}\end{bmatrix}\\ &= \gamma_{11}\gamma_{22} - \gamma_{12}\gamma_{21}\\ &= u_a u_b \gamma_1\gamma_4 - \left(\frac{\delta_a}{\delta_b}\right)\left(\frac{u_b^3}{u_a}\right)\gamma_2\gamma_3\\ &= u_a u_b[\gamma_1\gamma_4 - \gamma_2\gamma_3]\end{aligned}$$

With

$$\begin{aligned}\gamma_1\gamma_4 &= [I_0(u_a)K_1(u_b) + I_1(u_b)K_0(u_a)]\cdot\\ &\qquad [I_1(u_a)K_0(u_b) + I_0(u_b)K_1(u_a)]\\ &= I_0(u_a)I_1(u_a)K_0(u_b)K_1(u_b) + I_0(u_a)I_0(u_b)K_1(u_a)K_1(u_b)\\ &\quad + I_1(u_a)I_1(u_b)K_0(u_a)K_0(u_b) + I_0(u_b)I_1(u_b)K_0(u_a)K_1(u_a)\end{aligned}$$

and

$$\begin{aligned}\gamma_2\gamma_3 &= [I_0(u_a)K_0(u_b) - I_0(u_b)K_0(u_a)]\cdot\\ &\qquad [I_1(u_a)K_1(u_b) - I_1(u_b)K_1(u_a)]\\ &= I_0(u_a)I_1(u_a)K_0(u_b)K_1(u_b) - I_0(u_a)I_1(u_b)K_0(u_b)K_1(u_a)\\ &\quad - I_0(u_b)I_1(u_a)K_0(u_a)K_1(u_b)\\ &\quad + I_0(u_b)I_1(u_b)K_0(u_a)K_1(u_a)\end{aligned}$$

A subtraction provides

$$\begin{aligned}\gamma_1\gamma_4 - \gamma_2\gamma_4 &= I_0(u_a)I_0(u_b)K_1(u_a)K_1(u_b)\\ &\quad + I_0(u_a)I_1(u_b)K_0(u_b)K_1(u_a)\\ &\quad + I_1(u_a)I_1(u_b)K_0(u_a)K_0(u_b)\\ &\quad + I_0(u_b)I_1(u_a)K_0(u_a)K_1(u_b)\\ &= I_0(u_a)K_1(u_a)[I_0(u_b)K_1(u_b) + I_1(u_b)K_0(u_b)]\\ &\quad + I_1(u_a)K_0(u_a)[I_0(u_b)K_1(u_b) + I_1(u_b)K_0(u_b)]\\ &= [I_0(u_a)K_1(u_a) + I_1(u_a)K_0(u_a)]\cdot\end{aligned}$$

$$= \frac{[I_0(u_b)K_1(u_b) + I_1(u_b)K_0(u_b)]}{} \left(\frac{1}{u_a}\right)\left(\frac{1}{u_b}\right)$$

so that

$$\det \mathbf{\Gamma} = u_a u_b \left(\frac{1}{u_a}\right)\left(\frac{1}{u_b}\right) = 1 \tag{3.24}$$

3.4.1 The T-matrix

$$\tau_{11} = u_a[I_1(u_a)K_0(u_b) + I_0(u_b)K_1(u_a)] \tag{3.25a}$$

$$\tau_{12} = \frac{u_b^2}{2k\delta_b L n^2}[I_0(u_b)K_0(u_a) - I_0(u_a)K_0(u_b)] \tag{3.25b}$$

$$\tau_{21} = 2k\delta_a L n^2 \left(\frac{u_b}{u_a}\right)[I_1(u_b)K_1(u_a) - I_1(u_a)K_1(u_b)] \tag{3.25c}$$

and

$$\tau_{22} = u_b[I_0(u_a)K_1(u_b) + I_1(u_b)K_0(u_a)] \tag{3.25d}$$

3.4.2 The Z-matrix

$$z_{11} = \frac{u_b}{2k\delta_b L n^2}\left[\frac{I_1(u_a)K_0(u_b) + I_0(u_b)K_1(u_a)}{I_1(u_b)K_1(u_a) - I_1(u_a)K_1(u_b)}\right] \tag{3.26a}$$

$$z_{12} = \frac{\dfrac{1}{2k\delta_a L n^2}\left(\dfrac{u_a}{u_b}\right)}{I_1(u_a)K_1(u_b) - I_1(u_b)K_1(u_a)} \tag{3.26b}$$

$$z_{21} = \frac{\dfrac{1}{2k\delta_a L n^2}\left(\dfrac{u_a}{u_b}\right)}{I_1(u_b)K_1(u_a) - I_1(u_a)K_1(u_b)} \tag{3.26c}$$

and

$$z_{22} = \frac{u_a}{2k\delta_a L n^2}\left[\frac{I_0(u_a)K_1(u_b) + I_1(u_b)K_0(u_a)}{I_1(u_a)K_1(u_b) - I_1(u_b)K_1(u_a)}\right] \tag{3.26d}$$

3.4.3 The Y-matrix

$$y_{11} = \frac{2k\delta_b L n^2}{u_b}\left[\frac{I_0(u_a)K_1(u_b) + I_1(u_b)K_0(u_a)}{I_0(u_b)K_0(u_a) - I_0(u_a)K_0(u_b)}\right] \tag{3.27a}$$

$$y_{12} = \frac{\dfrac{2k\delta_b L n^2}{u_b^2}}{I_0(u_a)K_0(u_b) - I_0(u_b)K_0(u_a)} \tag{3.27b}$$

$$y_{21} = \frac{\dfrac{2k\delta_b L n^2}{u_b^2}}{I_0(u_b)K_0(u_a) - I_0(u_a)K_0(u_b)} \tag{3.27c}$$

and

$$y_{22} = \frac{2k\delta_a L n^2}{u_a}\left[\frac{I_1(u_a)K_0(u_b) + I_0(u_b)K_1(u_a)}{I_0(u_a)K_0(u_b) - I_0(u_b)K_0(u_a)}\right] \tag{3.27d}$$

3.4.4 The H-matrix

$$h_{11} = \frac{u_b}{2k\delta_b L n^2}\left[\frac{I_0(u_b)K_0(u_a) - I_0(u_a)K_0(u_b)}{I_0(u_a)K_1(u_b) + I_1(u_b)K_0(u_a)}\right] \tag{3.28a}$$

$$h_{12} = \frac{1}{u_b[I_0(u_a)K_1(u_b) + I_1(u_b)K_0(u_a)]} \tag{3.28b}$$

$$h_{21} = \frac{1}{u_b[I_0(u_a)K_1(u_b) + I_1(u_b)K_0(u_a)]} \tag{3.28c}$$

and

$$h_{22} = \frac{2k\delta_a L n^2}{u_a}\left[\frac{I_1(u_a)K_1(u_b) - I_1(u_b)K_1(u_a)}{I_0(u_a)K_1(u_b) + I_1(u_b)K_0(u_a)}\right] \tag{3.28d}$$

3.4.5 The G-matrix

$$g_{11} = \frac{2k\delta_b L n^2}{u_b}\left[\frac{I_1(u_b)K_1(u_a) - I_1(u_a)K_1(u_b)}{I_1(u_a)K_0(u_b) + I_0(u_b)K_1(u_a)}\right] \tag{3.29a}$$

$$g_{12} = \frac{1}{u_a[I_1(u_a)K_0(u_b) + I_0(u_b)K_1(u_a)]} \tag{3.29b}$$

$$g_{21} = \frac{1}{u_a[I_1(u_a)K_0(u_b) + I_0(u_b)K_1(u_a)]} \tag{3.29c}$$

and

$$g_{22} = \frac{u_a}{2k\delta_a L n^2}\left[\frac{I_0(u_a)K_0(u_b) - I_0(u_b)K_0(u_a)}{I_1(u_a)K_0(u_b) + I_0(u_b)K_1(u_a)}\right] \tag{3.29d}$$

Example 3.3: A longitudinal fin of trapezoidal profile with a thermal conductivity of 217 W/m-K is dissipating heat in a convective environment with a coefficient of heat transfer of 100 W/m^2-K. The fin is 7.62 cm high, 25 cm long, and has base and tip thicknesses of 0.9525 cm (0.375 in) and 0.2381 cm (0.0938 in), respectively. Precise measurements indicate a tip temperature excess of $\theta_a = 27.65°$C and a base heat flow of $q_b = 138.75$ W. Evaluate the base temperature excess and the tip heat flow.

SOLUTION Here, θ_a and q_b are given and θ_b and q_a are to be found. This suggests the use of the **H**-matrix. Observe that $b - a = 7.62$ cm and that by similar triangles, $a/b = \delta_a/\delta_b = 0.2381/0.9525 = 1/4$. Use of these two relationships shows that

$$b = 10.16\,\text{cm} \qquad \text{and} \qquad a = 2.54\,\text{cm}$$

The taper angle is

$$\phi = \arctan\frac{\delta_b}{2b} = \arctan\frac{0.009525}{2(0.1016)} = 2.68^\circ$$

so that

$$\sin\phi = \sin 2.68^\circ = 0.0468$$

Then

$$n = \left(\frac{h}{k\sin\phi}\right)^{1/2} = \left[\frac{100}{217(0.0468)}\right]^{1/2} = 3.1372\,\text{m}^{-1/2}$$

and

$$\begin{aligned} u_a &= 2na^{1/2} = 2(3.1372)(0.0254)^{1/2} = 1.0000 \\ u_b &= 2nb^{1/2} = 2(3.1372)(0.1016)^{1/2} = 2.0000 \end{aligned}$$

The result will come from the use of the **H**-matrix. The elements of this matrix are given in Eqs. (3.28), and the values of eight modified Bessel functions will be needed. A table of Bessel functions or a software routine provides for $u_b = 2.0000$ and $u_a = 1.0000$

$$\begin{aligned} I_0(u_b) &= 2.27959 & I_0(u_a) &= 1.26607 \\ I_1(u_b) &= 1.59064 & I_1(u_a) &= 0.56516 \\ K_0(u_b) &= 0.11389 & K_0(u_a) &= 0.42102 \\ K_1(u_b) &= 0.13987 & K_1(u_a) &= 0.60191 \end{aligned}$$

From Eqs. (3.28),

$$h_{11} = \frac{u_b}{2k\delta_b L n^2}\left[\frac{I_0(u_b)K_0(u_a) - I_0(u_a)K_0(u_b)}{I_0(u_a)K_1(u_b) + I_1(u_b)K_0(u_a)}\right]$$

and with

$$\frac{u_b}{2k\delta_b L n^2} = \frac{2.0000}{2(217)(.009525)(0.25)(3.1372)^2} = 0.1966$$

this becomes

$$h_{11} = 0.1966\left[\frac{2.27959(0.42102) - 1.26607(0.11389)}{1.26607(0.13987) + 1.59064(0.42102)}\right]$$

or

$$h_{11} = 0.1894\,\mathrm{K/W}$$

$$h_{12} = h_{21} = \frac{1}{u_b[I_0(u_a)K_1(u_b) + I_1(u_b)K_0(u_a)}$$

$$= \frac{1}{2.0000[1.26607(0.13987) + 1.59064(0.42102)]} = 0.5905$$

and

$$h_{22} = \frac{2k\delta_a L n^2}{u_a}\left[\frac{I_1(u_a)K_1(u_b) - I_1(u_b)K_1(u_a)}{I_0(u_a)K_1(u_b) + I_1(u_b)K_0(u_a)}\right]$$

With

$$\frac{2k\delta_a L n^2}{u_a} = \frac{2(217)(.002381)(0.25)(3.1372)^2}{1.0000} = 2.5428$$

this results in

$$h_{22} = 2.5428\left[\frac{0.56516(0.13987) - 1.59064(0.60191)}{1.26607(0.13987) + 1.59064(0.42102)}\right]$$

or

$$h_{22} = -2.6377\,\mathrm{W/K}$$

The conclusion of the example comes from a straightforward matrix multiplication:

$$\begin{aligned}
\begin{bmatrix} \theta_a \\ q_b \end{bmatrix} &= \mathbf{H}\begin{bmatrix} q_b \\ \theta_a \end{bmatrix} \\
&= \begin{bmatrix} 0.1894 & 0.5905 \\ 0.5905 & -2.6377 \end{bmatrix}\begin{bmatrix} q_b = 138.75\,\mathrm{W} \\ \theta_a = 27.65^\circ\mathrm{C} \end{bmatrix} \\
&= \begin{bmatrix} 42.6^\circ\mathrm{C} \\ 9.0\,\mathrm{W} \end{bmatrix}
\end{aligned}$$

Observe that a tip heat flow of 9.0 W when the base heat flow is 138.75 W is *not* negligible.

3.5 THE LONGITUDINAL FIN OF HALF TRAPEZOIDAL PROFILE

The configuration for the longitudinal fin of half trapezoidal profile is shown in Fig. 3.3. It may be observed that this fin is just half of the longitudinal fin of trapezoidal profile of Fig. 3.2, with a cut taken along the x-coordinate axis in Fig. 3.2. In both fins, the taper angle, ϕ, and the profile function, $y_1(x)$, are given by identical expressions:

$$\phi = \arctan\frac{\delta_b}{2b}$$

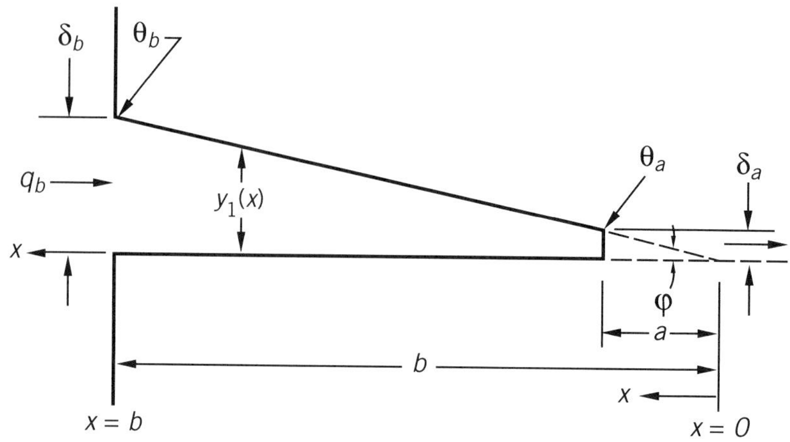

Figure 3.3 Longitudinal fin of half trapezoidal profile.

and

$$y_1(x) = x \tan\phi$$

However, in Fig. 3.3, the profile function, $y_2(x) = 0$, so that

$$y_1'(x) = \tan\phi$$

and

$$y_2'(x) = 0$$

Insertion of these into Eq. (3.1) yields

$$x \tan\phi \frac{d^2\theta}{dx^2} + \tan\phi \frac{d\theta}{dx} - \frac{h}{k}(\sec\phi + 1)\theta = 0$$

and a further simplification gives

$$x\frac{d^2\theta}{dx^2} + \frac{d\theta}{dx} - n^2\theta = 0 \tag{3.30}$$

where n for this case is given by

$$n = \left[\frac{h(\sec\phi + 1)}{k \tan\phi}\right]^{1/2}$$

Equations (3.14) and (3.30) are identical in form and therefore must have identical solutions. As long as it is recognized that the value of n differs in the two cases, Eqs. (3.23) and (3.25) through (3.29) may be used to evaluate the elements of all of the linear transformations for the longitudinal fin of half trapezoidal profile.

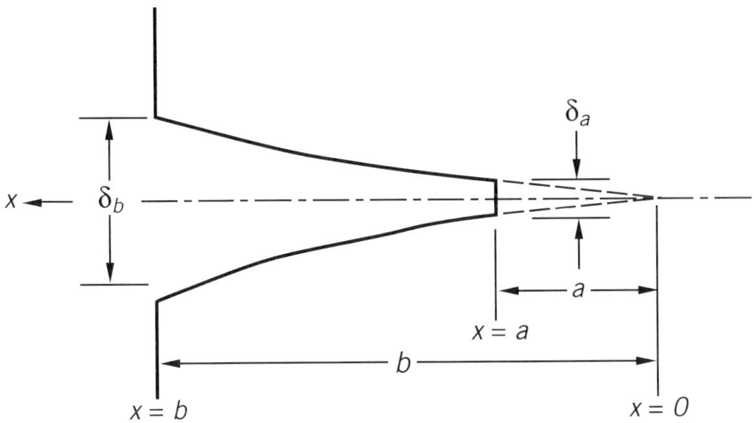

Figure 3.4 Longitudinal fin of truncated parabolic profile.

3.6 THE LONGITUDINAL FIN OF TRUNCATED CONCAVE PARABOLIC PROFILE

The configuration and coordinate system for the longitudinal fin of truncated concave parabolic profile is displayed in Fig. 3.4. It may be noted that the origin of the x-coordinate axis is *not* at the fin tip and that the fin *does not* taper to zero thickness at the tip. The profile function for this fin is given by

$$y_1(x) = -y_2(x) = \frac{\delta_b}{2}\left(\frac{x}{b}\right)^2$$

and with this in Eq. (3.1), and with no provision made for the length of arc of the fin faces, the differential equation for the temperature excess becomes

$$x^2\frac{d^2\theta}{dx^2} + 2x\frac{d\theta}{dx} - m^2b^2\theta = 0 \tag{3.31}$$

where again

$$m = \left(\frac{2h}{k\delta_b}\right)^{1/2}$$

Equation (3.31) is an Euler equation, and by appropriate transformation, its general solution can be shown to be

$$\theta(x) = C_1x^\alpha + C_2x^\beta \tag{3.32}$$

where

$$\alpha, \beta = -\frac{1}{2} \pm \frac{(1+4m^2b^2)^{1/2}}{2}$$

Application of the initial conditions of Eqs. (3.17) leads to the temperature excess and heat flow at the fin base in terms of the arbitrary constants:

$$\begin{aligned} \theta(b) = \theta_b &= C_1 b^\alpha + C_2 b^\beta \\ q(b) = q_b &= k\delta_b L(\alpha C_1 b^{\alpha-1} + \beta C_2 b^{\beta-1}) \end{aligned}$$

because

$$A(b) = 2y_1(b)L = 2\left(\frac{\delta_b}{2}\right)\left(\frac{b}{b}\right)^2 L = \delta_b L$$

Simultaneous solution for the arbitrary constants, C_1 and C_2, yields

$$C_1 = \frac{\beta b^{\beta-1}\theta_b - \dfrac{1}{k\delta_b L} b^\beta q_b}{(\beta-\alpha)b^{\alpha+\beta-1}}$$

and

$$C_2 = \frac{\dfrac{1}{k\delta_b L} b^\alpha q_b - \alpha b^{\alpha-1}\theta_b}{(\beta-\alpha)b^{\alpha+\beta-1}}$$

Consequently

$$\theta(x) = \frac{\alpha b^{\alpha-1}x^\beta - \beta b^{\beta-1}x^\alpha}{(\alpha-\beta)b^{\alpha+\beta-1}}\theta_b + \frac{b^\beta x^\alpha - b^\alpha x^\beta}{k\delta_b L(\alpha-\beta)b^{\alpha+\beta-1}} q_b \tag{3.33a}$$

and

$$\begin{aligned} q(x) = {} & \frac{k\delta_b L\alpha\beta[b^{\alpha-1}x^{\beta-1} - b^{\beta-1}x^{\alpha-1}]}{(\alpha-\beta)b^{\alpha+\beta-1}}\theta_b \\ & + \frac{\alpha b^\beta x^{\alpha-1} - \beta b^\alpha x^{\beta-1}}{(\alpha-\beta)b^{\alpha+\beta-1}} q_b \end{aligned} \tag{3.33b}$$

Then

$$\begin{aligned} \lambda_1(x) &= \frac{\alpha b^{\alpha-1}x^\beta - \beta b^{\beta-1}x^\alpha}{(\alpha-\beta)b^{\alpha+\beta-1}} \\ &= \frac{1}{\alpha-\beta}\left[\alpha\left(\frac{x}{b}\right)^\beta - \beta\left(\frac{x}{b}\right)^\alpha\right] \end{aligned} \tag{3.34a}$$

and

$$\begin{aligned} \lambda_2(x) &= \frac{b^\beta x^\alpha - b^\alpha x^\beta}{k\delta_b L(\alpha-\beta)b^{\alpha+\beta-1}} \\ &= \frac{b}{k\delta_b L(\alpha-\beta)}\left[\left(\frac{x}{b}\right)^\alpha - \left(\frac{x}{b}\right)^\beta\right] \end{aligned} \tag{3.34b}$$

and that, as required by Eq. (3.2*a*), $\lambda_1(b) = 1$ and $\lambda_2(b) = 0$. The derivatives of Eqs. (3.34) are

$$\lambda_1'(x) = \frac{\alpha\beta[b^{\alpha-1}x^{\beta-1} - b^{\beta-1}x^{\alpha-1}]}{(\alpha-\beta)b^{\alpha+\beta-1}} \tag{3.35a}$$

and

$$\lambda_2'(x) = \frac{\alpha b^{\beta}x^{\alpha-1} - \beta b^{\alpha}x^{\beta-1}}{k\delta_b L(\alpha-\beta)b^{\alpha+\beta-1}} \tag{3.35b}$$

and, as required by Eq. (3.2*b*), $\lambda_1'(b) = 0$ and $\lambda_2'(b) = 1/k\delta_b L$.

The elements of the $\mathbf{\Gamma}$-matrix can now be set down from an inspection of Eqs. (3.34) and (3.35) using $x = a$ at the tip of the fin:

$$\gamma_{11} = \lambda_1(a) = \frac{1}{\alpha-\beta}\left[\alpha\left(\frac{a}{b}\right)^{\beta} - \beta\left(\frac{a}{b}\right)^{\alpha}\right] \tag{3.36a}$$

$$\gamma_{12} = \lambda_2(a) = \frac{b}{k\delta_b L(\alpha-\beta)}\left[\left(\frac{a}{b}\right)^{\alpha} - \left(\frac{a}{b}\right)^{\beta}\right] \tag{3.36b}$$

and with $A(a) = a^2\delta_b L/b^2$

$$\gamma_{21} = kA(a)\lambda_1'(a) = \frac{k\delta_b L\alpha\beta}{b(\alpha-\beta)}\left[\left(\frac{a}{b}\right)^{\beta+1} - \left(\frac{a}{b}\right)^{\alpha+1}\right] \tag{3.36c}$$

and

$$\gamma_{22} = kA(a)\lambda_2'(a) = \frac{1}{\alpha-\beta}\left[\alpha\left(\frac{a}{b}\right)^{\alpha+1} - \beta\left(\frac{a}{b}\right)^{\beta+1}\right] \tag{3.36d}$$

The determinant of $\mathbf{\Gamma}$ is

$$\begin{aligned}\det\begin{bmatrix}\gamma_{11} & \gamma_{12}\\ \gamma_{21} & \gamma_{22}\end{bmatrix} &= \gamma_{11}\gamma_{22} - \gamma_{12}\gamma_{21}\\ &= \frac{1}{(\alpha-\beta)^2}\left[\alpha\left(\frac{a}{b}\right)^{\beta} - \beta\left(\frac{a}{b}\right)^{\alpha}\right]\cdot\\ &\qquad\left[\alpha\left(\frac{a}{b}\right)^{\alpha+1} - \beta\left(\frac{a}{b}\right)^{\beta+1}\right]\\ &\qquad - \frac{\alpha\beta}{(\alpha-\beta)^2}\left[\left(\frac{a}{b}\right)^{\alpha} - \left(\frac{a}{b}\right)^{\beta}\right]\cdot\\ &\qquad\left[\left(\frac{a}{b}\right)^{\beta+1} - \left(\frac{a}{b}\right)^{\alpha+1}\right]\end{aligned}$$

The algebra is really not that horrendous:

$$\det\mathbf{\Gamma} = \frac{1}{(\alpha-\beta)^2}\left[\alpha^2\left(\frac{a}{b}\right)^{\alpha+\beta+1} - \alpha\beta\left(\frac{a}{b}\right)^{2\alpha+1}\right.$$

$$
\begin{aligned}
& \qquad - \alpha\beta \left(\frac{a}{b}\right)^{2\beta+1} + \beta^2 \left(\frac{a}{b}\right)^{\alpha+\beta+1}\Bigg] \\
& \qquad - \frac{1}{(\alpha-\beta)^2}\left[-\alpha\beta\left(\frac{a}{b}\right)^{\alpha+\beta+1} + \alpha\beta\left(\frac{a}{b}\right)^{2\alpha+1}\right. \\
& \qquad \left. + \alpha\beta\left(\frac{a}{b}\right)^{2\beta+1} - \alpha\beta\left(\frac{a}{b}\right)^{\alpha+\beta+1}\right] \\
= & \frac{1}{(\alpha-\beta)^2}\left[\alpha^2 - 2\alpha\beta + \beta^2\right]\left(\frac{a}{b}\right)^{\alpha+\beta+1} \\
= & \frac{(\alpha-\beta)^2}{(\alpha-\beta)^2}\left(\frac{a}{b}\right)^{\alpha+\beta+1} \\
= & \left(\frac{a}{b}\right)^{\alpha+\beta+1}
\end{aligned}
$$

and because

$$
\alpha + \beta + 1 = \left[-\frac{1}{2} + \frac{(1+4m^2b^2)^{1/2}}{2}\right] + \left[-\frac{1}{2} - \frac{(1+4m^2b^2)^{1/2}}{2}\right] + 1 = 0
$$

the determinant reduces to unity

$$
\Delta = \left(\frac{a}{b}\right)^{\alpha+\beta+1} = 1
$$

The other five linear transformations can be determined through the use of Table 2.2. The elements of these matrices now follow.

3.6.1 The T-matrix

$$
\tau_{11} = \frac{1}{\alpha-\beta}\left[\alpha\left(\frac{a}{b}\right)^{\alpha+1} - \beta\left(\frac{a}{b}\right)^{\beta+1}\right] \tag{3.37a}
$$

$$
\tau_{12} = \frac{b}{k\delta_b L(\alpha-\beta)}\left[\left(\frac{a}{b}\right)^{\beta} - \left(\frac{a}{b}\right)^{\alpha}\right] \tag{3.37b}
$$

$$
\tau_{21} = \frac{k\delta_b L\alpha\beta}{b(\alpha-\beta)}\left[\left(\frac{a}{b}\right)^{\alpha+1} - \left(\frac{a}{b}\right)^{\beta+1}\right] \tag{3.37c}
$$

and

$$
\tau_{22} = \frac{1}{\alpha-\beta}\left[\alpha\left(\frac{a}{b}\right)^{\beta} - \beta\left(\frac{a}{b}\right)^{\alpha}\right] \tag{3.37d}
$$

3.6.2 The Z-matrix

$$z_{11} = \frac{b}{k\delta_b L\alpha\beta}\left[\frac{\beta\left(\frac{a}{b}\right)^{\beta+1} - \alpha\left(\frac{a}{b}\right)^{\alpha+1}}{\left(\frac{a}{b}\right)^{\beta+1} - \left(\frac{a}{b}\right)^{\alpha+1}}\right] \tag{3.38a}$$

$$z_{12} = \frac{b(\alpha-\beta)}{k\delta_b L\alpha\beta}\left[\left(\frac{a}{b}\right)^{\beta+1} - \left(\frac{a}{b}\right)^{\alpha+1}\right]^{-1} \tag{3.38b}$$

$$z_{21} = \frac{b(\beta-\alpha)}{k\delta_b L\alpha\beta}\left[\left(\frac{a}{b}\right)^{\beta+1} - \left(\frac{a}{b}\right)^{\alpha+1}\right]^{-1} \tag{3.38c}$$

and

$$z_{22} = \frac{b}{k\delta_b L\alpha\beta}\left[\frac{\alpha\left(\frac{a}{b}\right)^{\beta} - \beta\left(\frac{a}{b}\right)^{\alpha}}{\left(\frac{a}{b}\right)^{\beta+1} - \left(\frac{a}{b}\right)^{\alpha+1}}\right] \tag{3.38d}$$

3.6.3 The Y-matrix

$$y_{11} = \frac{k\delta_b L}{b}\left[\frac{\beta\left(\frac{a}{b}\right)^{\alpha} - \alpha\left(\frac{a}{b}\right)^{\beta}}{\left(\frac{a}{b}\right)^{\alpha} - \left(\frac{a}{b}\right)^{\beta}}\right] \tag{3.39a}$$

$$y_{12} = \frac{k\delta_b L(\alpha-\beta)}{b}\left[\left(\frac{a}{b}\right)^{\alpha} - \left(\frac{a}{b}\right)^{\beta}\right]^{-1} \tag{3.39b}$$

$$y_{21} = \frac{k\delta_b L(\beta-\alpha)}{b}\left[\left(\frac{a}{b}\right)^{\alpha} - \left(\frac{a}{b}\right)^{\beta}\right]^{-1} \tag{3.39c}$$

and

$$y_{22} = \frac{k\delta_b L}{b}\left[\frac{\alpha\left(\frac{a}{b}\right)^{\alpha+1} - \beta\left(\frac{a}{b}\right)^{\beta+1}}{\left(\frac{a}{b}\right)^{\alpha} - \left(\frac{a}{b}\right)^{\beta}}\right] \tag{3.39d}$$

3.6.4 The H-matrix

$$h_{11} = \frac{b}{k\delta_b L}\left[\frac{\left(\frac{a}{b}\right)^{\beta} - \left(\frac{a}{b}\right)^{\alpha}}{\alpha\left(\frac{a}{b}\right)^{\beta} - \beta\left(\frac{a}{b}\right)^{\alpha}}\right] \tag{3.40a}$$

$$h_{12} = (\alpha-\beta)\left[\alpha\left(\frac{a}{b}\right)^{\beta} - \beta\left(\frac{a}{b}\right)^{\alpha}\right]^{-1} \tag{3.40b}$$

$$h_{21} = (\alpha-\beta)\left[\alpha\left(\frac{a}{b}\right)^{\beta} - \beta\left(\frac{a}{b}\right)^{\alpha}\right]^{-1} \tag{3.40c}$$

and

$$h_{22} = \frac{k\delta_b L\alpha\beta}{b}\left[\frac{\left(\frac{a}{b}\right)^{\beta+1} - \left(\frac{a}{b}\right)^{\alpha+1}}{\alpha\left(\frac{a}{b}\right)^{\beta} - \beta\left(\frac{a}{b}\right)^{\alpha}}\right] \tag{3.40d}$$

3.6.5 The G-matrix

$$g_{11} = \frac{k\delta_b L\alpha\beta}{b}\left[\frac{\left(\frac{a}{b}\right)^{\alpha+1} - \left(\frac{a}{b}\right)^{\beta+1}}{\alpha\left(\frac{a}{b}\right)^{\alpha+1} - \beta\left(\frac{a}{b}\right)^{\beta+1}}\right] \tag{3.41a}$$

$$g_{12} = (\alpha - \beta)\left[\alpha\left(\frac{a}{b}\right)^{\alpha+1} - \beta\left(\frac{a}{b}\right)^{\beta+1}\right]^{-1} \tag{3.41b}$$

$$g_{21} = (\alpha - \beta)\left[\alpha\left(\frac{a}{b}\right)^{\alpha+1} - \beta\left(\frac{a}{b}\right)^{\beta+1}\right]^{-1} \tag{3.41c}$$

and

$$g_{22} = \frac{b}{k\delta_b L}\left[\frac{\left(\frac{a}{b}\right)^{\alpha} - \left(\frac{a}{b}\right)^{\beta}}{\alpha\left(\frac{a}{b}\right)^{\alpha+1} - \beta\left(\frac{a}{b}\right)^{\beta+1}}\right] \tag{3.41d}$$

3.7 THE RADIAL FIN OF RECTANGULAR PROFILE

The radial fin of rectangular profile is the only radial fin to be treated in this work. Thus, there is no need to develop a general differential equation for radial fins.

For the radial fin of rectangular profile displayed in Fig. 3.5*a*, observe that the origin of the height coordinate system is taken at the center of curvature of the fin base and fin tip. The height coordinate, r, has a positive orientation from fin base to fin tip, and it is to be noted that $r_a > r_b$.

Refer to Fig. 3.5*b* and consider the differential element, dr. The difference in heat conducted into the differential element at dr and that leaving at $r + dr$ is

$$dq = k\frac{d}{dr}\left[(2\pi r)\delta\frac{d\theta}{dr}\right]dr$$

This can be equated to the heat leaving the element, dr, by convection to the environment with a heat-transfer coefficient, h,

$$dq = 2h(2\pi r dr)\theta$$

so that the heat balance is

$$2k\pi\delta\frac{d}{dr}\left[r\frac{d\theta}{dr}\right]dr = 4h\pi\theta r dr$$

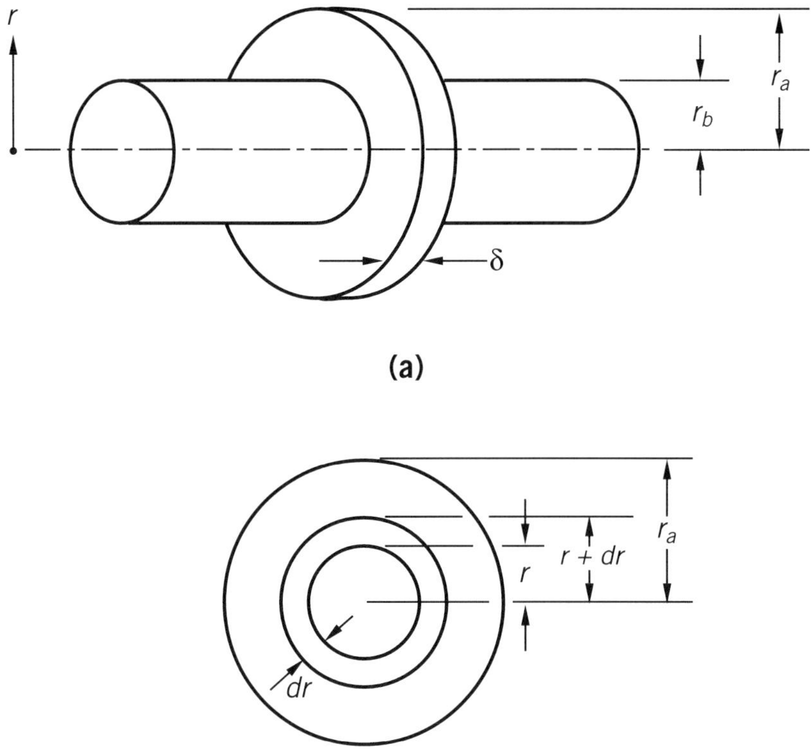

Figure 3.5 (a) Radial fin of rectangular profile and (b) a view of the face showing a differential element, dr. In (b) the inner radius, r_b, has been omitted.

This leads to the differential equation for temperature excess

$$r^2 \frac{d^2\theta}{dr^2} + r\frac{d\theta}{dr} - m^2 r^2 \theta = 0 \tag{3.42}$$

where

$$m = \left(\frac{2h}{k\delta}\right)^{1/2}$$

Equation (3.42) is Bessel's modified equation of zero-th order, which has a general solution:

$$\theta(r) = C_1 I_0(mr) + K_0(mr) \tag{3.43}$$

where the arbitrary constants, C_1 and C_2, are to be evaluated using the initial value data

$$\theta(r = r_b) = \theta_b \tag{3.44a}$$

and

$$q(r = r_b) = q_b \tag{3.44b}$$

Thus

$$\theta(r_b) = \theta_b = C_1 I_0(mr_b) + C_2 K_0(mr_b) \tag{3.44c}$$

and

$$q(r_b) = q_b = -2k\pi r_b \delta \theta_b m[C_1 I_1(mr_b) - C_2 K_1(mr_b)] \tag{3.44d}$$

A simultaneous solution for C_1 and C_2 then yields

$$C_1 = \frac{K_1(mr_b)\theta_b - \dfrac{1}{2k\pi r_b \delta m} K_0(mr_b) q_b}{I_0(mr_b)K_1(mr_b) + I_1(mr_b)K_0(mr_b)}$$

and

$$C_2 = \frac{I_1(mr_b)\theta_b + \dfrac{1}{2k\pi r_b \delta m} I_0(mr_b) q_b}{I_0(mr_b)K_1(mr_b) + I_1(mr_b)K_0(mr_b)}$$

and with the Bessel function identity

$$I_0(mr_b)K_1(mr_b) + I_1(mr_b)K_0(mr_b) = \frac{1}{mr_b}$$

the arbitrary constants can be expressed as

$$C_1 = mr_b \left[K_1(mr_b)\theta_b - \frac{1}{2k\pi r_b \delta m} K_0(mr_b) q_b \right]$$

$$C_2 = mr_b \left[I_1(mr_b)\theta_b + \frac{1}{2k\pi r_b \delta m} I_0(mr_b) q_b \right]$$

These make

$$\theta(r) = mr_b[K_1(mr_b)I_0(mr) + I_1(mr_b)K_0(mr)]\theta_b + \frac{1}{2k\pi\delta}\left[I_0(mr_b)K_0(mr) - K_0(mr_b)I_0(mr)\right] q_b \tag{3.45}$$

which indicates that

$$\lambda_1(r) = mr_b[K_1(mr_b)I_0(mr) + I_1(mr_b)K_0(mr)] \tag{3.46a}$$

and

$$\lambda_2(r) = \frac{1}{2k\pi\delta}\left[I_0(mr_b)K_0(mr) - K_0(mr_b)I_0(mr)\right] \tag{3.46b}$$

and that $\lambda_1(r_b) = 1$ and $\lambda_2(r_b) = 0$, as required by Eqs. (3.2*a*).

The heat flow at any point, r, is $q(r) = -2k\pi\delta r\theta'(r)$, and after a differentiation of Eq. (3.45), one obtains

$$q(r) = 2\pi k\delta m^2 r_b r[I_1(mr_b)K_1(mr) - K_1(mr_b)I_1(mr)]\theta_b$$

$$+mr[I_0(mr_b)K_1(mr) + K_0(mr_b)I_1(mr)]q_b$$

and this makes

$$\lambda_1'(r) = m^2 r_b[I_1(mr_b)K_1(mr) - K_1(mr_b)I_1(mr)] \tag{3.47a}$$

and

$$\lambda_2'(r) = \frac{m}{2k\pi\delta}[I_0(mr_b)K_1(mr) + K_0(mr_b)I_1(mr)] \tag{3.47b}$$

Here, it is to be noted that $\lambda_1'(r_b) = 0$ and $\lambda_2'(r_b) = -1/2k\pi\delta r_b = -1/kA(r_b)$ as required by Eqs. (3.2*b*) where the minus sign appears because the heat flow is in the direction of the positive sense of the coordinate system.

Let a radius ratio be proposed:

$$\rho \equiv \frac{r_b}{r_a} \qquad (\rho < 1) \tag{3.48}$$

Moreover, consider the profile area

$$A_p = \delta(r_a - r_b) \tag{3.49}$$

and note that

$$\delta = \frac{A_p}{r_a - r_b}$$

so that

$$m = \left(\frac{2h}{k\delta}\right)^{1/2} = (r_a - r_b)^{1/2}\left(\frac{2h}{kA_p}\right)^{1/2} \tag{3.50}$$

Then

$$mr_b = r_b(r_a - r_b)^{1/2}\left(\frac{2h}{kA_p}\right)^{1/2}$$

and if both numerator and denominator are multiplied by $(r_a - r_b)$,

$$mr_b = r_b\frac{(r_a - r_b)^{3/2}}{r_a - r_b}\left(\frac{2h}{kA_p}\right)^{1/2} \tag{3.51}$$

An identical procedure yields

$$mr_a = r_a\frac{(r_a - r_b)^{3/2}}{r_a - r_b}\left(\frac{2h}{kA_p}\right)^{1/2} \tag{3.52}$$

Now define

$$\phi \equiv \frac{(r_a - r_b)^{3/2}}{r_a - r_b}\left(\frac{2h}{kA_p}\right)^{1/2} \tag{3.53}$$

and write

$$mr_b = \frac{r_b\phi}{r_a - r_b} = \frac{\rho\phi}{1 - \phi} = \alpha \tag{3.54a}$$

and

$$mr_a = \frac{r_a \phi}{r_a - r_b} = \frac{\phi}{1 - \phi} = \beta \tag{3.54b}$$

This allows the Bessel function arguments to be written in terms of the very usable radius ratio, ρ.

With Eqs. (3.46) and (3.47) in hand, the $\mathbf{\Gamma}$-matrix can be set up in accordance with Eqs. (3.6) using α and β as defined by Eqs. (3.54):

$$\gamma_{11} = \lambda_1(r_a) = \alpha[I_0(\beta)K_1(\alpha) + I_1(\alpha)K_0(\beta) \tag{3.55a}$$

$$\gamma_{12} = \lambda_2(r_a) = \frac{1}{2k\pi\delta}[I_0(\alpha)K_0(\beta) - I_0(\beta)K_0(\alpha)] \tag{3.55b}$$

$$\gamma_{21} = kA(r_a)\lambda_1'(r_a) = 2k\pi\delta m r_a \alpha[I_1(\alpha)K_1(\beta) - I_1(\beta)K_1(\alpha)]$$

or with $mr_a = \beta$

$$\gamma_{21} = 2k\pi\delta\alpha\beta[I_1(\alpha)K_1(\beta) - I_1(\beta)K_1(\alpha)] \tag{3.55c}$$

and

$$\gamma_{22} = kA(r_a)\lambda_2'(r_a) = 2k\pi\delta r_a \frac{m}{2k\pi\delta}[I_0(\alpha)K_1(\beta) + I_1(\beta)K_0(\alpha)]$$

and again with $mr_a = \beta$,

$$\gamma_{22} = \beta[I_0(\alpha)K_1(\beta) + I_1(\beta)K_0(\alpha)] \tag{3.55d}$$

The determinant of the $\mathbf{\Gamma}$-matrix is

$$\det \mathbf{\Gamma} = \gamma_{11}\gamma_{22} - \gamma_{12}\gamma_{21}$$

and using Eqs. (3.55), it can be shown that

$$\det \mathbf{\Gamma} = 1$$

so that the elements of the other linear transformations can be developed from Table 2.2 using the $\mathbf{\Gamma}$-matrix.

3.7.1 The T-matrix

$$\tau_{11} = \beta[I_0(\alpha)K_1(\beta) + I_1(\beta)K_0(\alpha)] \tag{3.56a}$$

$$\tau_{12} = \frac{1}{2k\pi\delta}[I_0(\beta)K_0(\alpha) - I_0(\alpha)K_0(\beta)] \tag{3.56b}$$

$$\tau_{21} = 2k\pi\delta\alpha\beta[I_1(\beta)K_1(\alpha) - I_1(\alpha)K_1(\beta)] \tag{3.56c}$$

and

$$\tau_{22} = \alpha[I_0(\beta)K_1(\alpha) + I_1(\alpha)K_0(\beta)] \tag{3.56d}$$

3.7.2 The Z-matrix

$$z_{11} = \frac{1}{2k\pi\delta\alpha}\left[\frac{I_0(\alpha)K_1(\beta) + I_1(\beta)K_0(\alpha)}{I_1(\beta)K_1(\alpha) - I_1(\alpha)K_1(\beta)}\right] \tag{3.57a}$$

$$z_{12} = \frac{1}{2k\pi\delta\alpha\beta}\left[\frac{1}{I_1(\alpha)K_1(\beta) - I_1(\beta)K_1(\alpha)}\right] \tag{3.57b}$$

$$z_{21} = \frac{1}{2k\pi\delta\alpha\beta}\left[\frac{1}{I_1(\beta)K_1(\alpha) - I_1(\alpha)K_1(\beta)}\right] \tag{3.57c}$$

and

$$z_{22} = \frac{1}{2k\pi\delta\beta}\left[\frac{I_0(\beta)K_1(\alpha) + I_1(\alpha)K_0(\beta)}{I_1(\alpha)K_1(\beta) - I_1(\beta)K_1(\alpha)}\right] \tag{3.57d}$$

3.7.3 The Y-matrix

$$y_{11} = 2k\pi\delta\alpha\left[\frac{I_0(\beta)K_1(\alpha) + I_1(\alpha)K_0(\beta)}{I_0(\beta)K_0(\alpha) - I_0(\alpha)K_0(\beta)}\right] \tag{3.58a}$$

$$y_{12} = \frac{2k\pi\delta}{I_0(\alpha)K_0(\beta) - I_0(\beta)K_0(\alpha)} \tag{3.58b}$$

$$y_{21} = \frac{2k\pi\delta}{I_0(\beta)K_0(\alpha) - I_0(\alpha)K_0(\beta)} \tag{3.58c}$$

and

$$y_{22} = 2k\pi\delta\beta\left[\frac{I_0(\alpha)K_1(\beta) + I_1(\beta)K_0(\alpha)}{I_0(\alpha)K_0(\beta) - I_0(\beta)K_0(\alpha)}\right] \tag{3.58d}$$

3.7.4 The H-matrix

$$h_{11} = \frac{1}{2k\pi\delta}\left[\frac{I_0(\beta)K_0(\alpha) - I_0(\alpha)K_0(\beta)}{I_0(\beta)K_1(\alpha) + I_1(\alpha)K_0(\beta)}\right] \tag{3.59a}$$

$$h_{12} = \frac{1}{\alpha[I_0(\beta)K_1(\alpha) + I_1(\alpha)K_0(\beta)]} \tag{3.59b}$$

$$h_{21} = \frac{1}{\alpha[I_0(\beta)K_1(\alpha) + I_1(\alpha)K_0(\beta)]} \tag{3.59c}$$

and

$$h_{22} = 2k\pi\delta\beta\left[\frac{I_1(\alpha)K_1(\beta) - I_1(\beta)K_1(\alpha)}{I_0(\beta)K_1(\alpha) + I_1(\alpha)K_0(\beta)}\right] \tag{3.59d}$$

3.7.5 The G-matrix

$$g_{11} = 2k\pi\delta\beta \left[\frac{I_1(\beta)K_1(\alpha) - I_1(\alpha)K_1(\beta)}{I_0(\alpha)K_1(\beta) + I_1(\beta)K_0(\alpha)} \right] \tag{3.60a}$$

$$g_{12} = \frac{1}{\beta[I_0(\alpha)K_1(\beta) + I_1(\beta)K_0(\alpha)]} \tag{3.60b}$$

$$g_{21} = \frac{1}{\beta[I_0(\alpha)K_1(\beta) + I_1(\beta)K_0(\alpha)]} \tag{3.60c}$$

and

$$g_{22} = \frac{1}{2k\pi\delta\beta} \left[\frac{I_0(\alpha)K_0(\beta) - I_0(\beta)K_0(\alpha)}{I_0(\alpha)K_1(\beta) + I_1(\beta)K_0(\alpha)} \right] \tag{3.60d}$$

Example 3.4: A precise temperature measurement at the tip of a radial fin of rectangular profile indicates that the temperature excess is 18.4°C. The fin is fabricated from a material with a thermal conductivity of k = 328 W/m-K, and the inner and outer radii are 2 and 4 cm, respectively. The fin is 0.3175 cm (0.125 in) thick, and the heat-transfer coefficient between the fin faces and the surrounding environment is known to be $h = 133.3\,\mathrm{W/m^2\text{-}K}$. What are θ_b and q_b if the tip heat-transfer coefficient is $h_a = 52\,\mathrm{W/m^2\text{-}K}$?

SOLUTION The **T**-matrix can be employed after the value of q_a is computed. Here

$$S_a = 2\pi r_a \delta = 2\pi(0.04)(0.003175) = 7.980 \times 10^{-4}\,\mathrm{m^2}$$

so that

$$q_a = h_a S_a \theta_a = 52(7.980 \times 10^{-4})(18.4) = 0.763\,\mathrm{W}$$

Then

$$m = \left(\frac{2h}{k\delta}\right)^{1/2} = \left[\frac{2(133.3)}{328(.003175)}\right]^{1/2} = 16.000\,\mathrm{m^{-1}}$$

so that

$$\alpha = mr_b = 16(0.02) = 0.320$$

and

$$\beta = mr_a = 16(0.04) = 0.640$$

The values of eight modified Bessel functions will be needed

$\beta = mr_a = 0.640$	$\alpha = mr_b = 0.320$
$I_0(\beta) = 1.10505$	$I_0(\alpha) = 1.02576$
$I_1(\beta) = 0.33667$	$I_1(\alpha) = 0.16206$
$K_0(\beta) = 0.72767$	$K_0(\alpha) = 1.31356$
$K_1(\beta) = 1.19227$	$K_1(\alpha) = 2.83899$

By Eqs. (3.56),

$$\begin{aligned}\tau_{11} &= \beta[I_0(\alpha)K_1(\beta) + I_1(\beta)K_0(\alpha)] \\ &= 0.64[1.02576(1.19227) + 0.33667(1.31356)] \\ &= 0.64[1.2230 + 0.4422] = 1.0657 \\ \tau_{12} &= (1/2k\pi\delta)\,[I_0(\beta)K_0(\alpha) - I_0(\alpha)K_0(\beta)]\end{aligned}$$

With

$$\frac{1}{2k\pi\delta} = \frac{1}{2(328)\pi(0.003175)} = 0.1528$$

$$\begin{aligned}\tau_{12} &= 0.1528[1.10505(1.31356) - 1.02576(0.72767)] \\ &= 0.1528[1.4515 - 0.7464] = 0.1078\,\mathrm{K/W} \\ \tau_{21} &= 2k\pi\delta\alpha\beta[I_1(\beta)K_1(\alpha) - I_1(\alpha)K_1(\beta)]\end{aligned}$$

and with

$$2k\pi\delta\alpha\beta = 2(328)\pi(0.003175)(0.32)(0.64) = 1.3401$$

$$\begin{aligned}\tau_{21} &= 1.3401[0.33667(2.83899) - 0.16206(1.19227)] \\ &= 1.3401[0.9558 - 0.1932] = 1.0219\,\mathrm{W/K}\end{aligned}$$

and

$$\begin{aligned}\tau_{22} &= \alpha[I_0(\beta)K_1(\alpha) + I_1(\alpha)K_0(\beta)] \\ &= 0.32[1.10505(2.83899) - 0.16206(0.72767)] \\ &= 0.32[3.1372 - 0.1179] = 0.9662\end{aligned}$$

Now with $\theta_a = 18.4°\mathrm{C}$ and $q_a = 0.763\,W$

$$\begin{bmatrix}\theta_b \\ q_b\end{bmatrix} = \mathbf{T}\begin{bmatrix}\theta_a \\ q_a\end{bmatrix} = \begin{bmatrix}1.0657 & 0.1078 \\ 1.0219 & 0.9662\end{bmatrix}\begin{bmatrix}18.4 \\ 0.763\end{bmatrix} = \begin{bmatrix}19.7°\mathrm{C} \\ 19.54\,\mathrm{W}\end{bmatrix}$$

3.8 A GENERALIZED DIFFERENTIAL EQUATION FOR SPINES

The first consideration concerns spines of circular cross section, and Fig. 3.6 presents the terminology and coordinate system. Observe that the x-axis is along the axis of symmetry and its positive sense is directed from the spine tip at $x = a$ to the spine base at $x = b$. The coordinate, r, denotes the radius of the circular cross section and is a function of x but, in accordance with the Murray-Gardner assumptions, the steady-state temperature excess, $\theta = T - T_s$, is a function only of x.

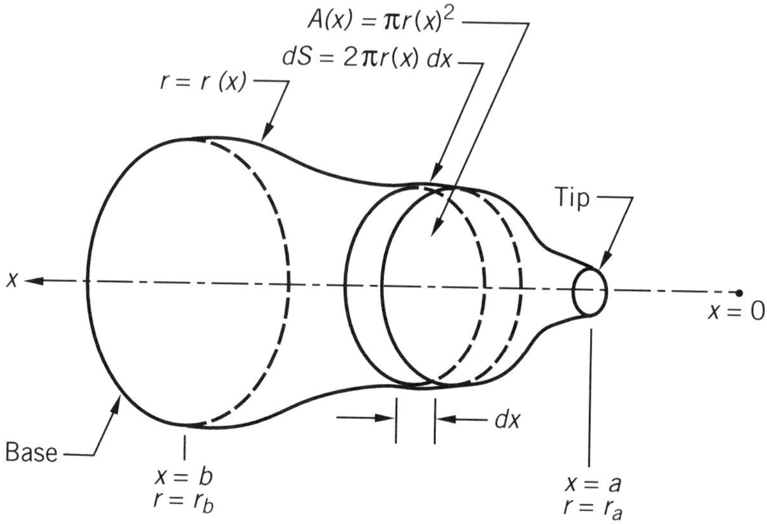

Figure 3.6 Coordinate system for the generalized spine of circular cross section.

The heat flow, q, is taken positive from spine base to spine tip, and the difference in heat conducted into the differential element at x, and that leaving at $x + dx$, is

$$dq(x) = \frac{d}{dx}\left[k(x)(\pi r^2)\frac{d\theta(x)}{dx}\right] dx$$

or, for the case of constant thermal conductivity,

$$dq(x) = k\pi\frac{d}{dx}\left[r^2\frac{d\theta(x)}{dx}\right] dx \tag{3.61a}$$

The heat dissipated through the differential element of spine surface is given by

$$dq(x) = h(x)dS(x)\theta(x)$$

and with the spine surface area given by

$$dS(x) = 2\pi r\left[1 + \left(\frac{dr}{dx}\right)^2\right]^{1/2} dx$$

and with $h(x) = h$ (a constant),[1] one obtains

$$dq(x) = 2h\pi r\left[1 + \left(\frac{dr}{dx}\right)^2\right]^{1/2} \theta(x)dx \tag{3.61b}$$

[1]Cases where h is not a constant are treated in Chap. 6.

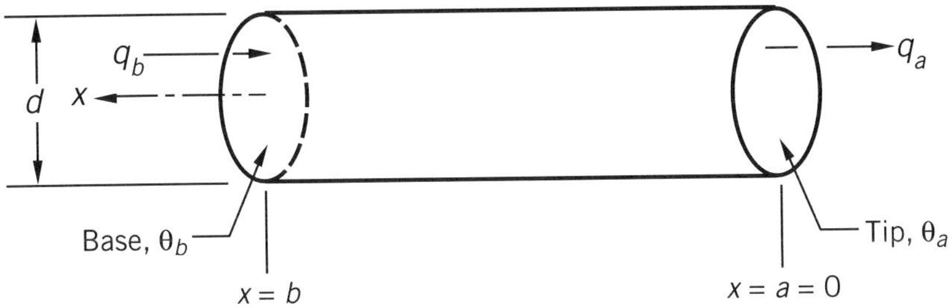

Figure 3.7 The cylindrical spine.

When Eqs. (3.61) are equated, the result is

$$k\pi \frac{d}{dx}\left[r^2 \frac{d\theta(x)}{dx}\right] dx = 2h\pi r \left[1 + \left(\frac{dr}{dx}\right)^2\right]^{1/2}$$

and simple adjustment yields the general equation for the temperature excess of the spine of circular cross section:

$$r\frac{d^2\theta}{dx^2} + 2\frac{dr}{dx}\frac{d\theta}{dx} - \frac{2h}{k}\left[1 + \left(\frac{dr}{dx}\right)^2\right]^{1/2} \theta = 0 \tag{3.62}$$

3.9 THE CYLINDRICAL SPINE

In Fig. 3.7, the cylindrical spine of diameter, d, has its origin of the height coordinate, x, taken at the fin tip ($x = a = 0$). Here, $r(x) = d/2, dr/dx = 0$, and Eq. (3.62) reduces to

$$\frac{d^2\theta}{dx^2} - m^2\theta = 0 \tag{3.63}$$

where

$$m = \left(\frac{4h}{kd}\right)^{1/2}$$

Equation (3.63) is identical in form to Eq. (2.1) (although the definition of m is different), with the solution for the temperature excess given by Eq. (2.2):

$$\theta = \theta_b \frac{\cosh mx}{\cosh mb}$$

The heat flow through the base of the spine is given by

$$q_b = kA\frac{d\theta}{dx}\bigg|_{x=b}$$

and with $A = \pi d^2/4$

$$q_b = \frac{\pi}{4} k d^2 m \frac{\sinh mx}{\cosh mb}\bigg|_{x=b}$$

or with $m = (4h/kd)^{1/2}$

$$q_b = \frac{1}{2}(\pi^2 hkd^3)^{1/2} \tanh mb$$

which makes

$$Y_0 = \frac{(\pi^2 hkd^3)^{1/2}}{2}$$

and

$$Z_0 = \frac{2}{(\pi^2 hkd^3)^{1/2}}$$

The linear transformations of Eqs. (3.8) through (3.13) apply as long as the proper values of m, Y_0, and Z_0 are employed. These are repeated here as Eqs. (3.64) through (3.69) for convenience.

$$\begin{bmatrix} \theta_a \\ q_a \end{bmatrix} = \mathbf{\Gamma} \begin{bmatrix} \theta_b \\ q_b \end{bmatrix} = \begin{bmatrix} \cosh mb & -Z_0 \sinh mb \\ -Y_0 \sinh mb & \cosh mb \end{bmatrix} \begin{bmatrix} \theta_b \\ q_b \end{bmatrix} \tag{3.64}$$

$$\begin{bmatrix} \theta_b \\ q_b \end{bmatrix} = \mathbf{T} \begin{bmatrix} \theta_a \\ q_a \end{bmatrix} = \begin{bmatrix} \cosh mb & Z_0 \sinh mb \\ Y_0 \sinh mb & \cosh mb \end{bmatrix} \begin{bmatrix} \theta_a \\ q_a \end{bmatrix} \tag{3.65}$$

$$\begin{bmatrix} \theta_b \\ \theta_a \end{bmatrix} = \mathbf{Z} \begin{bmatrix} q_b \\ q_a \end{bmatrix} = \begin{bmatrix} Z_0 \coth mb & -Z_0 \operatorname{csch} mb \\ Z_0 \operatorname{csch} mb & -Z_0 \coth mb \end{bmatrix} \begin{bmatrix} q_b \\ q_a \end{bmatrix} \tag{3.66}$$

$$\begin{bmatrix} q_b \\ q_a \end{bmatrix} = \mathbf{Y} \begin{bmatrix} \theta_b \\ \theta_a \end{bmatrix} = \begin{bmatrix} Y_0 \coth mb & -Y_0 \operatorname{csch} mb \\ Y_0 \operatorname{csch} mb & -Y_0 \coth mb \end{bmatrix} \begin{bmatrix} \theta_b \\ q_b \end{bmatrix} \tag{3.67}$$

$$\begin{bmatrix} \theta_b \\ q_a \end{bmatrix} = \mathbf{H} \begin{bmatrix} q_b \\ \theta_a \end{bmatrix} = \begin{bmatrix} Z_0 \tanh mb & \operatorname{sech} mb \\ \operatorname{sech} mb & -Y_0 \tanh mb \end{bmatrix} \begin{bmatrix} q_b \\ \theta_a \end{bmatrix} \tag{3.68}$$

$$\begin{bmatrix} q_b \\ \theta_a \end{bmatrix} = \mathbf{G} \begin{bmatrix} \theta_b \\ q_a \end{bmatrix} = \begin{bmatrix} Y_0 \tanh mb & \operatorname{sech} mb \\ \operatorname{sech} mb & -Z_0 \tanh mb \end{bmatrix} \begin{bmatrix} \theta_b \\ q_a \end{bmatrix} \tag{3.69}$$

3.10 THE SPINE OF RECTANGULAR CROSS SECTION

A spine of rectangular cross section is shown in Fig. 3.8[2] where it may be noted that the sides are designated by δ_1 and δ_2 and where, again, the origin of the x-coordinate is taken at the spine tip where $x = a = 0$. Of course, there is no radius function because the spine is rectangular, but the cross-sectional area is constant, $A = \delta_1 \delta_2$, and there is no spine taper.

[2]Be aware that if one of the δ's is very much greater than the other, this spine of rectangular cross section becomes a longitudinal fin of rectangular profile.

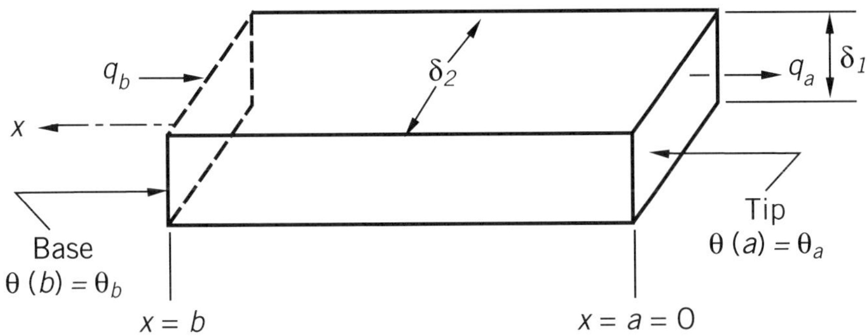

Figure 3.8 The spine of rectangular cross section. The spine of square cross section results when $\delta_1 = \delta_2$.

For an element of spine with surface area $dS = 2(\delta_1 + \delta_2)dx$, the theory leading to Eq. (3.62) yields Eq. (3.63):

$$\frac{d^2\theta}{dx^2} - m^2\theta = 0 \tag{3.63}$$

as the differential equation for temperature excess but with

$$m = \left[\frac{2h(\delta_1 + \delta_2)}{k\delta_1\delta_2}\right]^{1/2}$$

$$Y_0 = [2hk\delta_1\delta_2(\delta_1 + \delta_2)]^{1/2}$$

and

$$Z_0 = [2hk\delta_1\delta_2(\delta_1 + \delta_2)]^{-1/2}$$

As long as these values of m, Y_0, and Z_0 are used, Eqs. (3.64) through (3.69) provide the 24 elements of the linear transformations.

In the event that $\delta_1 = \delta_2 = \delta$, the spine is square, and Eqs. (3.64) through (3.69) may be used, but with

$$m = \left(\frac{4h}{k\delta}\right)^{1/2}$$

$$Y_0 = (4hk\delta^3)^{1/2}$$

and

$$Z_0 = (4hk\delta^3)^{-1/2}$$

3.11 THE ELLIPTICAL SPINE

The spine of elliptical cross section is displayed in Fig. 3.9. The semimajor axis is designated by δ_1, and the semiminor axis is designated by δ_2. The origin

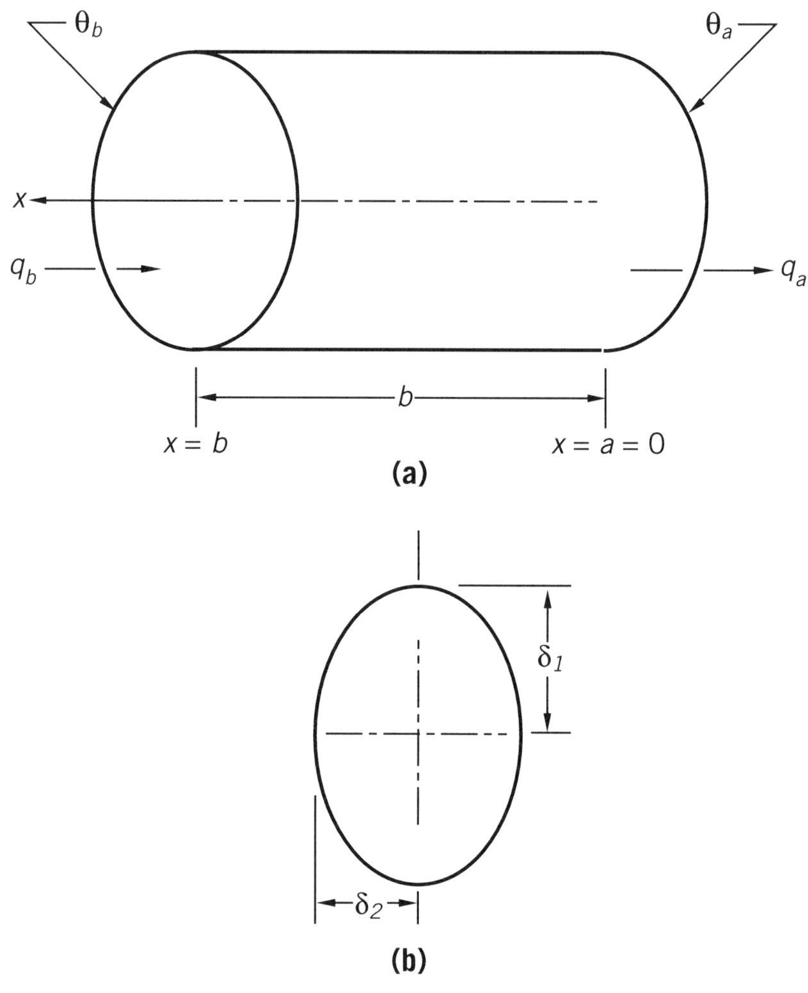

Figure 3.9 The elliptical spine.

of the height coordinate is taken at the tip of the spine where $x = a = 0$, and the positive sense of the coordinate system is toward the spine base at $x = b$. The cross-sectional area is constant and is given by

$$A = \pi\delta_1\delta_2$$

The differential surface area is given by $dS = Pdx$, with P, to three-decimal accuracy, given by

$$P = \pi(\delta_1 + \delta_2)\left[1 + \frac{p^2}{4} + \frac{p^4}{64} + \frac{p^6}{256} + \cdots\right]$$

where

$$p = \frac{\delta_1 - \delta_2}{\delta_1 + \delta_2}$$

Here too, the differential equation for temperature excess is Eq. (3.63):

$$\frac{d^2\theta}{dx^2} - m^2\theta = 0 \tag{3.63}$$

because the cross-sectional area is constant. However, for the fin of elliptic cross section:

$$\begin{aligned} m &= \left(\frac{hP}{kA}\right)^{1/2} \\ &= \left\{\frac{h(\delta_1+\delta_2)}{k\delta_1\delta_2}\left[1+\frac{p^2}{4}+\frac{p^4}{64}+\frac{p^6}{256}+\cdots\right]\right\}^{1/2} \end{aligned}$$

and

$$\begin{aligned} Y_0 &= kAm \\ &= k\pi\delta_1\delta_2\left\{\frac{h(\delta_1+\delta_2)}{k\delta_1\delta_2}\left[1+\frac{p^2}{4}+\frac{p^4}{64}+\frac{p^6}{256}+\cdots\right]\right\}^{1/2} \\ &= \left\{hk\pi^2\delta_1\delta_2(\delta_1+\delta_2)\left[1+\frac{p^2}{4}+\frac{p^4}{64}+\frac{p^6}{256}+\cdots\right]\right\}^{1/2} \end{aligned}$$

so that

$$\begin{aligned} Z_0 &= \frac{1}{Y_0} \\ &= \left\{hk\pi^2\delta_1\delta_2(\delta_1+\delta_2)\left[1+\frac{p^2}{4}+\frac{p^4}{64}+\frac{p^6}{256}+\cdots\right]\right\}^{-1/2} \end{aligned}$$

As long as these values of m, Y_0, and Z_0 are used, Eqs. (3.64) through (3.69) provide the 24 elements of the linear transformations.

3.12 THE TRUNCATED CONICAL SPINE

The x-coordinate in the truncated conical spine, shown in Fig. 3.10, has its origin located at a point remote from the fin tip. The fin tip is located at $x = a$, and the x-coordinate is positive in a direction from tip to base. The profile function is

$$r(x) = r_b\left(\frac{x}{b}\right)$$

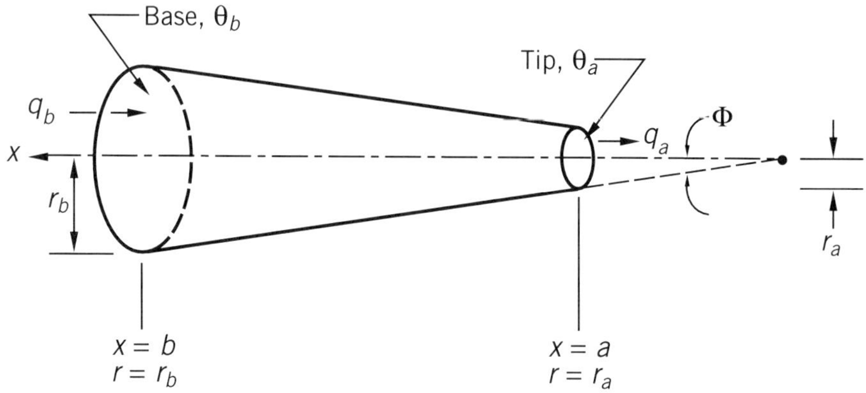

Figure 3.10 The truncated conical spine.

with derivative

$$\frac{dr}{dx} = \frac{r_b}{b}$$

Putting these in Eq. (3.62) gives

$$\frac{r_b}{b}x\frac{d^2\theta}{dx^2} + 2\frac{r_b}{b}\frac{d\theta}{dx} - \frac{2h}{k}\left[1 + \left(\frac{r_b}{b}\right)^2\right]^{1/2}\theta = 0$$

or

$$x\frac{d^2\theta}{dx^2} + 2\frac{d\theta}{dx} - \frac{2hb}{kr_b}\left[1 + \left(\frac{r_b}{b}\right)^2\right]^{1/2}\theta = 0 \tag{3.70}$$

With

$$n = \left[\frac{4h^2}{r_b^2k^2}(b^2 + r_b^2)\right]^{1/4}$$

Eq. (3.70) becomes

$$x^2\frac{d^2\theta}{dx^2} + 2x\frac{d\theta}{dx} - n^2x\theta = 0 \tag{3.71}$$

which has a solution

$$\theta(x) = x^{-1/2}[C_1I_1(2nx^{1/2}) + C_2K_1(2nx^{1/2})] \tag{3.72}$$

where C_1 and C_2 are to be evaluated from the initial conditions

$$\theta(x = b) = \theta_b$$

and

$$q(x = b) = q_b$$

If a transformation is made,

$$w = 2nx^{1/2}$$

so that

$$x = \frac{w^2}{4n^2}$$

and

$$dx = \frac{w}{2n^2} dw$$

then at $x = b$

$$w_b = 2nb^{1/2}$$

and at $x = a$

$$w_a = 2na^{1/2}$$

Equation (3.72) can then be written as

$$\theta(x) = \frac{2n}{w}[C_1 I_1(w) + C_2 K_1(w)] \tag{3.73}$$

and this possesses a derivative

$$\frac{d\theta(x)}{dx} = \frac{4n^3}{w^2}\frac{d}{dw}\left[C_1 \frac{I_1(w)}{w} + C_2 \frac{K_1(w)}{w}\right] \tag{3.74}$$

The indicated differentiations require the use of the relationships[3]

$$\frac{d}{dw}\left[\frac{I_1(w)}{w}\right] = \frac{I_2(w)}{w}$$

and

$$\frac{d}{dw}\left[\frac{K_1(w)}{w}\right] = -\frac{K_2(w)}{w}$$

and with these in Eq. (3.74) one obtains

$$\frac{d\theta(x)}{dx} = \frac{4n^3}{w^2}\left[C_1 \frac{I_2(w)}{w} - C_2 \frac{K_2(w)}{w}\right]$$

and with

$$A(x) = \pi r_b^2 \left(\frac{x}{b}\right)^2 = \frac{\pi r_b^2 w^4}{16 b^2 n^4}$$

the heat flow at any point will be

$$q(x) = \frac{k\pi r_b^2 w^2}{4b^2 n}[C_1 I_2(w) - C_2 K_2(w)] \tag{3.75}$$

Now, at $x = b$ where $w = w_b$

$$\theta_b = \frac{2n}{w_b}[C_1 I_1(w_b) + C_2 K_1(w_b)]$$

[3] A. L. Rabenstein, *Introduction to Ordinary Differential Equations*, Academic Press, New York (1966).

and

$$q_b = \frac{k\pi r_b^2 w_b^2}{4b^2 n}[C_1 I_2(w_b) - C_2 K_2(w_b)]$$

Solution of this set yields

$$C_1 = \frac{\dfrac{w_b}{2n}K_2(w_b)\theta_b + \dfrac{4nb^2}{k\pi r_b^2 w_b^2}K_1(w_b)q_b}{I_1(w_b)K_2(w_b) + I_2(w_b)K_1(w_b)}$$

and

$$C_2 = \frac{\dfrac{w_b}{2n}I_2(w_b)\theta_b - \dfrac{4nb^2}{k\pi r_b^2 w_b^2}I_1(w_b)q_b}{I_1(w_b)K_2(w_b) + I_2(w_b)K_1(w_b)}$$

With the Bessel function identity

$$I_1(w_b)K_2(w_b) + I_2(w_b)K_1(w_b) = \frac{1}{w_b}$$

these relationships for C_1 and C_2 may be inserted into Eq. (3.75) to provide

$$\theta(x) = \frac{w_b^2}{w}[K_2(w_b)I_1(w) + I_2(w_b)K_1(w)]\theta_b + \frac{8n^2b^2}{k\pi r_b^2 w_b w}[K_1(w_b)I_1(w) - I_1(w_b)K_1(w)]q_b \tag{3.76}$$

which shows that

$$\lambda_1(w) = \frac{w_b^2}{w}[K_2(w_b)I_1(w) + I_2(w_b)K_1(w)] \tag{3.77a}$$

and

$$\lambda_2(w) = \frac{8n^2b^2}{k\pi r_b^2 w_b w}[K_1(w_b)I_1(w) - I_1(w_b)K_1(w)] \tag{3.77b}$$

which, at $w = w_b$, reduce to $\lambda_1(w_b) = 1$ and $\lambda_2(w_b) = 0$ as required by Eqs. (3.2*a*).

Expressions for $\lambda_1'(w)$ and $\lambda_2'(w)$ can be obtained by differentiating Eqs. (3.77*a*) and (3.77*b*). Adopting the latter idea

$$\lambda_1'(w) = \frac{2n^2}{w^2}w_b^2[K_2(w_b)I_2(w) - I_2(w_b)K_2(w)] \tag{3.77c}$$

and

$$\lambda_2'(w) = \frac{16n^4b^2}{k\pi r_b^2 w_b w^2}[K_1(w_b)I_2(w) + I_1(w_b)K_2(w)] \tag{3.77d}$$

which, at $w = w_b$, reduce to

$$\lambda_1'(w_b) = 0 \qquad \text{and} \qquad \lambda_2'(w_b) = 16n^4b^2/k\pi r_b^2 w_b^4$$

By noting that $16n^4b^2 = w_b^4$, it is seen that

$$\lambda_2'(w_b) = 1/k\pi r_b^2 = 1/kA(b)$$

Both of these results are required by Eqs. (3.2*b*).

A listing of the $\mathbf{\Gamma}$-elements now follows. Their formulation requires the use of Eqs. (3.77) evaluated at $x = a$, where $w = w_a$, and the multiplication of $\lambda_1'(w)$ and $\lambda_2'(w)$ by $kA(a) = k\pi r_b^2(a/b)^2 = k\pi r_b^2(w_a/w_b)^4$:

$$\gamma_{11} = \lambda_1(w_a) = \frac{w_b^2}{w_a}[I_1(w_a)K_2(w_b) + I_2(w_b)K_1(w_a)] \tag{3.78a}$$

$$\gamma_{12} = \lambda_2(w_a) = \frac{8n^2b^2}{k\pi r_b^2 w_b w_a}[I_1(w_a)K_1(w_b) - I_1(w_b)K_1(w_a)] \tag{3.78b}$$

$$\begin{aligned}\gamma_{21} &= kA(a)\lambda_1'(w_a) \\ &= 2k\pi n^2 r_b^2\left(\frac{w_a}{w_b}\right)^2 [I_2(w_a)K_2(w_b) - I_2(w_b)K_2(w_a)]\}\end{aligned} \tag{3.78c}$$

and

$$\gamma_{22} = kA(a)\lambda_2'(w_a) = \frac{16n^4b^2w_a^2}{w_b^5}[I_2(w_a)K_1(w_b) + I_1(w_b)K_2(w_a)]$$

With $16n^4b^2 = w_b^4$, this becomes

$$\gamma_{22} = \frac{w_a^2}{w_b}[I_2(w_a)K_1(w_b) + I_1(w_b)K_2(w_a)] \tag{3.78d}$$

It can be shown that[4]

$$\det \mathbf{\Gamma} = 1$$

and use of this fact and Eqs. (3.78) in conjunction with Table 2.2 will provide the elements of all of the other linear transformations.

3.12.1 The T-matrix

$$\tau_{11} = \frac{w_a^2}{w_b}[I_2(w_a)K_1(w_b) + I_1(w_b)K_2(w_a)] \tag{3.79a}$$

$$\tau_{12} = \frac{8n^2b^2}{k\pi r_b^2 w_b w_a}[I_1(w_b)K_1(w_a) - I_1(w_a)K_1(w_b)] \tag{3.79b}$$

$$\tau_{21} = 2k\pi n^2 r_b^2\left(\frac{w_a}{w_b}\right)^2 [I_2(w_b)K_2(w_a) - I_2(w_a)K_2(w_b)] \tag{3.79c}$$

and

$$\tau_{22} = \frac{w_b^2}{w_a}[I_1(w_a)K_2(w_b) + I_2(w_b)K_1(w_a)] \tag{3.79d}$$

[4]In a manner that is substantially similar to the method used to show the same result for the longitudinal fins of trapezoidal and truncated concave parabolic profile.

3.12.2 The Z-matrix

$$z_{11} = \frac{w_b}{2k\pi n^2 r_b^2}\left[\frac{I_2(w_a)K_1(w_b) + I_1(w_b)K_2(w_a)}{I_2(w_b)K_2(w_a) - I_2(w_a)K_2(w_b)}\right] \tag{3.80a}$$

$$z_{12} = \frac{b/a}{2k\pi n^2 r_b^2[I_2(w_a)K_2(w_b) - I_2(w_b)K_2(w_a)]} \tag{3.80b}$$

$$z_{21} = \frac{b/a}{2k\pi n^2 r_b^2[I_2(w_b)K_2(w_a) - I_2(w_a)K_2(w_b)]} \tag{3.80c}$$

and

$$z_{22} = \frac{w_a}{2k\pi n^2 r_a^2}\left[\frac{I_1(w_a)K_2(w_b) + I_2(w_b)K_1(w_a)}{I_2(w_a)K_2(w_b) - I_2(w_b)K_2(w_a)}\right] \tag{3.80d}$$

3.12.3 The Y-matrix

$$y_{11} = \frac{k\pi r_b w_a}{2b}\left[\frac{I_1(w_a)K_2(w_b) + I_2(w_b)K_1(w_a)}{I_1(w_b)K_1(w_a) - I_1(w_a)K_1(w_b)}\right] \tag{3.81a}$$

$$y_{12} = \frac{k\pi r_b^2 w_a w_b}{8n^2 b^2[I_1(w_a)K_1(w_b) - I_1(w_b)K_1(w_a)} \tag{3.81b}$$

$$y_{21} = \frac{k\pi r_b^2 w_a w_b}{8n^2 b^2[I_1(w_b)K_1(w_a) - I_1(w_a)K_1(w_b)]} \tag{3.81c}$$

and

$$y_{22} = \frac{k\pi r_a^2 w_a}{2a}\left[\frac{I_2(w_a)K_1(w_b) + I_1(w_b)K_2(w_a)}{I_1(w_a)K_1(w_b) - I_1(w_b)K_1(w_a)}\right] \tag{3.81d}$$

3.12.4 The H-matrix

$$h_{11} = \frac{2b}{k\pi r_b^2 w_b}\left[\frac{I_1(w_b)K_1(w_a) - I_1(w_a)K_1(w_b)}{I_1(w_a)K_2(w_b) + I_2(w_b)K_1(w_a)}\right] \tag{3.82a}$$

$$h_{12} = \frac{w_a/w_b^2}{I_1(w_a)K_2(w_b) + I_2(w_b)K_1(w_a)} \tag{3.82b}$$

$$h_{21} = \frac{w_a/w_b^2}{I_1(w_a)K_2(w_b) + I_2(w_b)K_1(w_a)} \tag{3.82c}$$

and

$$h_{22} = \frac{2k\pi n^2 r_a^2}{w_a}\left[\frac{I_2(w_a)K_2(w_b) - I_2(w_b)K_2(w_a)}{I_1(w_a)K_2(w_b) + I_2(w_b)K_1(w_a)}\right] \tag{3.82d}$$

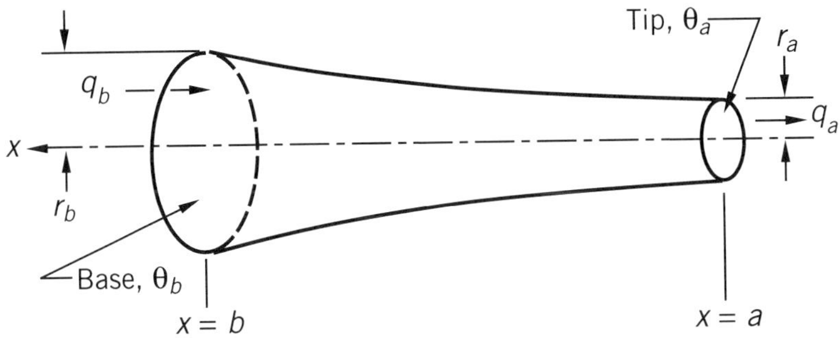

Figure 3.11 The truncated concave parabolic spine.

3.12.5 The G-matrix

$$g_{11} = \frac{2k\pi n^2 r_b^2}{w_b}\left[\frac{I_2(w_b)K_2(w_a) - I_2(w_a)K_2(w_b)}{I_2(w_a)K_1(w_b) + I_1(w_b)K_2(w_a)}\right] \tag{3.83a}$$

$$g_{12} = \frac{w_b/w_a^2}{I_2(w_a)K_1(w_b) + I_1(w_b)K_2(w_a)} \tag{3.83b}$$

$$g_{21} = \frac{w_b/w_a^2}{I_2(w_a)K_1(w_b) + I_1(w_b)K_2(w_a)} \tag{3.83c}$$

and

$$g_{22} = \frac{2a}{k\pi r_a^2 w_a}\left[\frac{I_1(w_a)K_1(w_b) - I_1(w_b)K_1(w_a)}{I_2(w_a)K_1(w_b) + I_1(w_b)K_2(w_a)}\right] \tag{3.83d}$$

3.13 THE TRUNCATED CONCAVE PARABOLIC SPINE

For the truncated concave parabolic spine displayed in Fig. 3.11, the profile function is

$$r(x) = r_b\left(\frac{x}{b}\right)^2$$

with derivative

$$\frac{dr(x)}{dx} = \frac{2r_b}{b^2}x$$

Insertion of these into Eq. (3.62) gives

$$r_b\left(\frac{x}{b}\right)^2\frac{d^2\theta}{dx^2} + \frac{4r_b}{b^2}x\frac{d\theta}{dx} - \frac{2h}{k}\left[1 + \frac{4r_b^2}{b^4}x^2\right]^{1/2}\theta = 0$$

This complicated differential equation can be simplified significantly if one recognizes that spines are usually long and thin so that the ratio $4r_b^2/b^2$ is small ($4r_b^2/b^2 << 1$). In this case, it reduces to

$$r_b\left(\frac{x}{b}\right)^2\frac{d^2\theta}{dx^2} + \frac{4r_b}{b^2}x\frac{d\theta}{dx} - \frac{2h}{k}\theta = 0$$

or

$$x^2 \frac{d^2\theta}{dx^2} + 4x \frac{d\theta}{dx} - m^2 b^2 \theta = 0 \tag{3.84}$$

where

$$m = \left(\frac{2h}{kr_b} \right)^{1/2}$$

Equation (3.84) is recognized as an Euler equation, and its general solution is

$$\theta = C_1 x^\alpha + C_2 x^\beta \tag{3.85}$$

where

$$\alpha, \beta = -\frac{3}{2} \pm \frac{(9 + 4m^2 b^2)^{1/2}}{2}$$

The arbitrary constants, C_1 and C_2, must be evaluated from the initial conditions

$$\theta(x = b) = \theta_b$$

and

$$q(x = b) = q_b$$

or from a solution of the pair of simultaneous equations

$$\theta_b = C_1 b^\alpha + C_2 b^\beta$$

and

$$q_b = k\pi r_b^2 [\alpha C_1 b^{\alpha-1} + \beta C_2 b^{\beta-1}]$$

Let

$$\psi = (9 + 4m^2 b^2)^{1/2}$$

so that the solutions for C_1 and C_2 can be written as

$$C_1 = \frac{\dfrac{b^\beta}{k\pi r_b^2} q_b - \beta b^{\beta-1} \theta_b}{\psi b^{\alpha+\beta-1}}$$

and

$$C_2 = \frac{\alpha b^{\alpha-1} \theta_b - \dfrac{b^\alpha}{k\pi r_b^2} q_b}{\psi b^{\alpha+\beta-1}}$$

With C_1 and C_2 inserted into Eq. (3.85)

$$\theta(x) = \frac{1}{\psi} \left[\alpha \left(\frac{x}{b} \right)^\beta - \beta \left(\frac{x}{b} \right)^\alpha \right] \theta_b + \frac{b}{k\pi r_b^2 \psi} \left[\left(\frac{x}{b} \right)^\alpha - \left(\frac{x}{b} \right)^\beta \right] q_b \tag{3.86}$$

and one can observe immediately that

$$\lambda_1(x) = \frac{1}{\psi}\left[\alpha\left(\frac{x}{b}\right)^{\beta} - \beta\left(\frac{x}{b}\right)^{\alpha}\right]$$

and

$$\lambda_2(x) = \frac{b}{k\pi r_b^2\psi}\left[\left(\frac{x}{b}\right)^{\alpha} - \left(\frac{x}{b}\right)^{\beta}\right]$$

When $x = b$, it is noted that $\lambda_1(b) = 1$ because $\alpha - \beta = \psi$, and that $\lambda_2(b) = 0$, both of which are required by Eqs. (3.2*a*).

Further, with $\alpha\beta = -(mb)^2$,

$$\lambda_1'(x) = \frac{2hb}{kr_b\psi}\left[\left(\frac{x}{b}\right)^{\alpha-1} - \left(\frac{x}{b}\right)^{\beta-1}\right]$$

and

$$\lambda_2'(x) = \frac{1}{k\pi r_b^2\psi}\left[\alpha\left(\frac{x}{b}\right)^{\alpha-1} - \beta\left(\frac{x}{b}\right)^{\beta-1}\right]$$

and one can observe that at $x = b$, $\lambda_1'(b) = 0$, and that

$$\lambda_2'(b) = (\alpha - \beta)/k\pi r_b^2\psi = 1/k\pi r_b^2 = 1/kA(b)$$

which also satisfies the requirements of Eqs. (3.2*b*).

The elements of the $\mathbf{\Gamma}$-matrix may be set down from the foregoing, evaluated at $x = a$, and with appropriate multiplication of $\lambda_1'(a)$ and $\lambda_2'(a)$ by $kA(a) = k\pi r_a^2$.

$$\gamma_{11} = \lambda_1(a) = \frac{1}{\psi}\left[\alpha\left(\frac{a}{b}\right)^{\beta} - \beta\left(\frac{a}{b}\right)^{\alpha}\right] \tag{3.87a}$$

$$\gamma_{12} = \lambda_2(a) = \frac{b}{k\pi r_b^2\psi}\left[\left(\frac{a}{b}\right)^{\alpha} - \left(\frac{a}{b}\right)^{\beta}\right] \tag{3.87b}$$

$$\gamma_{21} = kA(a)\lambda_1'(a) = \frac{2\pi r_b ah}{\psi}\left[\left(\frac{a}{b}\right)^{\alpha+2} - \left(\frac{a}{b}\right)^{\beta+2}\right] \tag{3.87c}$$

and

$$\gamma_{22} = kA(a)\lambda_2'(a) = \frac{1}{\psi}\left[\alpha\left(\frac{a}{b}\right)^{\alpha+3} - \beta\left(\frac{a}{b}\right)^{\beta+3}\right] \tag{3.87d}$$

It can be shown that

$$\det\mathbf{\Gamma} = 1$$

and use of this fact and Eqs. (3.87) in conjunction with Table 2.2 will generate the elements of all of the other linear transformations.

3.13.1 The T-matrix

$$\tau_{11} = \frac{1}{\psi}\left[\alpha\left(\frac{a}{b}\right)^{\alpha+3} - \beta\left(\frac{a}{b}\right)^{\beta+3}\right] \tag{3.88a}$$

$$\tau_{12} = \frac{b}{k\pi r_b^2\psi}\left[\left(\frac{a}{b}\right)^{\beta} - \left(\frac{a}{b}\right)^{\alpha}\right] \tag{3.88b}$$

$$\tau_{21} = \frac{2\pi r_b a h}{\psi}\left[\left(\frac{a}{b}\right)^{\beta+2} - \left(\frac{a}{b}\right)^{\alpha+2}\right] \tag{3.88c}$$

and

$$\tau_{22} = \frac{1}{\psi}\left[\alpha\left(\frac{a}{b}\right)^{\beta} - \beta\left(\frac{a}{b}\right)^{\alpha}\right] \tag{3.88d}$$

3.13.2 The Z-matrix

$$z_{11} = \frac{1}{2\pi r_b a h}\left[\frac{\beta\left(\frac{a}{b}\right)^{\beta+3} - \alpha\left(\frac{a}{b}\right)^{\alpha+3}}{\left(\frac{a}{b}\right)^{\alpha+2} - \left(\frac{a}{b}\right)^{\beta+2}}\right] \tag{3.89a}$$

$$z_{12} = \frac{\psi}{2\pi r_b a h\left[\left(\frac{a}{b}\right)^{\alpha+2} - \left(\frac{a}{b}\right)^{\beta+2}\right]} \tag{3.89b}$$

$$z_{21} = \frac{\psi}{2\pi r_b a h\left[\left(\frac{a}{b}\right)^{\beta+2} - \left(\frac{a}{b}\right)^{\alpha+2}\right]} \tag{3.89c}$$

and

$$z_{22} = \frac{1}{2\pi r_b a h}\left[\frac{\alpha\left(\frac{a}{b}\right)^{\beta} - \beta\left(\frac{a}{b}\right)^{\alpha}}{\left(\frac{a}{b}\right)^{\alpha+2} - \left(\frac{a}{b}\right)^{\beta+2}}\right] \tag{3.89d}$$

3.13.3 The Y-matrix

$$y_{11} = \frac{k\pi r_b^2}{b}\left[\frac{\beta\left(\frac{a}{b}\right)^{\alpha} - \alpha\left(\frac{a}{b}\right)^{\beta}}{\left(\frac{a}{b}\right)^{\alpha} - \left(\frac{a}{b}\right)^{\beta}}\right] \tag{3.90a}$$

$$y_{12} = \frac{k\pi r_b^2\psi}{b\left[\left(\frac{a}{b}\right)^{\alpha} - \left(\frac{a}{b}\right)^{\beta}\right]} \tag{3.90b}$$

$$y_{21} = \frac{k\pi r_b^2\psi}{b\left[\left(\frac{a}{b}\right)^{\beta} - \left(\frac{a}{b}\right)^{\alpha}\right]} \tag{3.90c}$$

and

$$y_{22} = \frac{k\pi r_b^2}{b}\left[\frac{\alpha\left(\dfrac{a}{b}\right)^{\alpha+3} - \beta\left(\dfrac{a}{b}\right)^{\beta+3}}{\left(\dfrac{a}{b}\right)^{\alpha} - \left(\dfrac{a}{b}\right)^{\beta}}\right] \tag{3.90d}$$

3.13.4 The H-matrix

$$h_{11} = \frac{b}{k\pi r_b^2}\left[\frac{\left(\dfrac{a}{b}\right)^{\beta} - \left(\dfrac{a}{b}\right)^{\alpha}}{\alpha\left(\dfrac{a}{b}\right)^{\beta} - \beta\left(\dfrac{a}{b}\right)^{\alpha}}\right] \tag{3.91a}$$

$$h_{12} = \frac{\psi}{\left[\alpha\left(\dfrac{a}{b}\right)^{\beta} - \beta\left(\dfrac{a}{b}\right)^{\alpha}\right]} \tag{3.91b}$$

$$h_{21} = \frac{\psi}{\left[\alpha\left(\dfrac{a}{b}\right)^{\beta} - \beta\left(\dfrac{a}{b}\right)^{\alpha}\right]} \tag{3.91c}$$

and

$$h_{22} = 2\pi r_b a h\left[\frac{\left(\dfrac{a}{b}\right)^{\alpha+2} - \left(\dfrac{a}{b}\right)^{\beta+2}}{\alpha\left(\dfrac{a}{b}\right)^{\alpha} - \beta\left(\dfrac{a}{b}\right)^{\beta}}\right] \tag{3.91d}$$

3.13.5 The G-matrix

$$g_{11} = 2\pi r_b a h\left[\frac{\left(\dfrac{a}{b}\right)^{\beta+2} - \left(\dfrac{a}{b}\right)^{\alpha+2}}{\alpha\left(\dfrac{a}{b}\right)^{\alpha+3} - \beta\left(\dfrac{a}{b}\right)^{\beta+3}}\right] \tag{3.92a}$$

$$g_{12} = \frac{\psi}{\alpha\left(\dfrac{a}{b}\right)^{\alpha+3} - \beta\left(\dfrac{a}{b}\right)^{\beta+3}} \tag{3.92b}$$

$$g_{21} = \frac{\psi}{\alpha\left(\dfrac{a}{b}\right)^{\alpha+3} - \beta\left(\dfrac{a}{b}\right)^{\beta+3}} \tag{3.92c}$$

and

$$g_{22} = \frac{b}{k\pi r_b^2}\left[\frac{\left(\dfrac{a}{b}\right)^{\alpha} - \left(\dfrac{a}{b}\right)^{\beta}}{\alpha\left(\dfrac{a}{b}\right)^{\alpha+3} - \beta\left(\dfrac{a}{b}\right)^{\beta+3}}\right] \tag{3.92d}$$

3.14 SUMMARY OF MATRIX ELEMENTS

Table 3.1 provides a directory to the listing of the elements of the linear transformations developed in this chapter.

TABLE 3.1 Summary of Linear Transformations for Regular Fins and Spines

Longitudinal Fin of Rectangular Profile

$$m = \left(\frac{2h}{k\delta}\right)^{1/2}$$

$$Y_0 = (2hk\delta)^{1/2}L \qquad Z_0 = \frac{1}{Y_0} = \frac{1}{(2hk\delta)^{1/2}L}$$

Γ-matrix	(3.8)
T-matrix	(3.9)
Z-matrix	(3.10)
Y-matrix	(3.11)
H-matrix	(3.12)
G-matrix	(3.13)

Longitudinal Fin of Rectangular Profile With One Face Insulated

$$m = \left(\frac{h}{k\delta}\right)^{1/2}$$

$$Y_0 = (hk\delta)^{1/2}L \qquad Z_0 = \frac{1}{(hk\delta)^{1/2}L}$$

Γ-matrix	(3.8)
T-matrix	(3.9)
Z-matrix	(3.10)
Y-matrix	(3.11)
H-matrix	(3.12)
G-matrix	(3.13)

TABLE 3.1 (cont'd): Summary of Linear Transformations for Regular Fins and Spines

Longitudinal Fin of Trapezoidal Profile

$$n = \left(\frac{h}{k \sin \phi} \right)^{1/2}$$

$$u_b = 2nb^{1/2} \qquad u_a = 2na^{1/2}$$

$$Y_0 = \frac{2k\delta_b L n^2}{u_b} \qquad Z_0 = \frac{u_b}{2k\delta_b L n^2}$$

Γ-matrix	(3.23)
T-matrix	(3.25)
Z-matrix	(3.26)
Y-matrix	(3.27)
H-matrix	(3.28)
G-matrix	(3.29)

Longitudinal Fin of Half Trapezoidal Profile

$$n = \left[\frac{h(\sec \phi + 1)}{k \tan \phi} \right]^{1/2}$$

$$u_b = 2nb^{1/2} \qquad u_a = 2na^{1/2}$$

$$Y_0 = \frac{2k\delta_b L n^2}{u_b} \qquad Z_0 = \frac{u_b}{2k\delta_b L n^2}$$

Γ-matrix	(3.23)
T-matrix	(3.25)
Z-matrix	(3.26)
Y-matrix	(3.27)
H-matrix	(3.28)
G-matrix	(3.29)

TABLE 3.1 (cont'd): Summary of Linear Transformations for Regular Fins and Spines

Longitudinal Fin of Truncated Concave Parabolic Profile

$$m = \left(\frac{2h}{k\delta_b} \right)^{1/2}$$

$$Y_0 = \frac{k\delta_b L}{b}$$

$$\alpha, \beta = -\frac{1}{2} \pm \frac{(1 + 4m^2b^2)^{1/2}}{2}$$

Γ-matrix	(3.36)
T-matrix	(3.37)
Z-matrix	(3.38)
Y-matrix	(3.39)
H-matrix	(3.40)
G-matrix	(3.41)

Radial Fin of Rectangular Profile

$$m = \left(\frac{2h}{k\delta} \right)^{1/2}$$

$$Y_0 = 2hk\delta \qquad Z_0 = \frac{1}{2hk\delta}$$

Γ-matrix	(3.55)
T-matrix	(3.56)
Z-matrix	(3.57)
Y-matrix	(3.58)
H-matrix	(3.59)
G-matrix	(3.60)

TABLE 3.1 (cont'd): Summary of Linear Transformations for Regular Fins and Spines.

Cylindrical Spine

$$m = \left(\frac{4h}{kd}\right)^{1/2}$$

$$Y_0 = \frac{(\pi^2 hkd^3)^{1/2}}{2} \qquad Z_0 = \frac{2}{(\pi^2 hkd^3)^{1/2}}$$

Γ-matrix	(3.64)
T-matrix	(3.65)
Z-matrix	(3.66)
Y-matrix	(3.67)
H-matrix	(3.68)
G-matrix	(3.69)

Rectangular Spine

$$n = \left(\frac{2h(\delta_1 + \delta_2}{k\delta_1\delta_2}\right)^{1/2}$$

$$Y_0 = [2hk\delta_1\delta_2(\delta_1 + \delta_2)]^{1/2}$$

$$Z_0 = [2hk\delta_1\delta_2(\delta_1 + \delta_2)]^{-1/2}$$

Γ-matrix	(3.64)
T-matrix	(3.65)
Z-matrix	(3.66)
Y-matrix	(3.67)
H-matrix	(3.68)
G-matrix	(3.69)

TABLE 3.1 (cont'd): Summary of Linear Transformations for Regular Fins and Spines.

Elliptical Spine

$$m = \left\{ \frac{h(\delta_1 + \delta_2)}{k\delta_1\delta_2} \left[1 + \frac{p^2}{4} + \frac{p^4}{64} + \frac{p^6}{256} + \cdots \right] \right\}^{1/2}$$

$$Y_0 = \left\{ hk\pi^2\delta_1\delta_2(\delta_1 + \delta_2) \left[1 + \frac{p^2}{4} + \frac{p^4}{64} + \frac{p^6}{256} + \cdots \right] \right\}^{1/2}$$

$$Z_0 = \left\{ hk\pi^2\delta_1\delta_2(\delta_1 + \delta_2) \left[1 + \frac{p^2}{4} + \frac{p^4}{64} + \frac{p^6}{256} + \cdots \right] \right\}^{-1/2}$$

Γ-matrix	(3.64)
T-matrix	(3.65)
Z-matrix	(3.66)
Y-matrix	(3.67)
H-matrix	(3.68)
G-matrix	(3.69)

Truncated Conical Spine

$$n = \left[\frac{4h^2}{r_b^2 k^2}(b^2 + r_b^2) \right]^{1/2}$$

$$w_b = 2nb^{1/2} \qquad w_a = 2na^{1/2}$$

$$Y_0 = \frac{k\pi r_b}{n^2 b} \qquad Z_0 = \frac{n^2 b}{k\pi r_b}$$

Γ-matrix	(3.78)
T-matrix	(3.79)
Z-matrix	(3.80)
Y-matrix	(3.81)
H-matrix	(3.82)
G-matrix	(3.83)

TABLE 3.1 (cont'd): Summary of Linear Transformations for Regular Fins and Spines.

Truncated Concave Parabolic Spine

$$m = \left(\frac{2h}{kr_b}\right)^{1/2}$$

$$\alpha, \beta = -\frac{3}{2} \pm \frac{(9 + 4m^2b^2)^{1/2}}{2}$$

$$Y_0 = \frac{k\pi r_b^2}{b} \qquad Z_0 = \frac{b}{k\pi r_b^2}$$

Γ-matrix	(3.87)
T-matrix	(3.88)
Z-matrix	(3.89)
Y-matrix	(3.90)
H-matrix	(3.91)
G-matrix	(3.92)

3.15 CLOSURE

This completes the development of the elements of the linear transformations for the regular fins to be considered in this book. The development of the thermal transmission ratios for four singular fins follows in Chap. 4. The design and analysis of finned arrays with no loops in the graph theoretical sense, begins with the algorithms for the combinations of fins in Chap. 5.

Chapter

4

SINGULAR FINS AND SPINES AND SINGLE ELEMENTS

4.1 INTRODUCTION

In contrast to the *regular* fins and spines discussed in Chap. 2, there are several useful shapes that cannot be characterized by the $\mathbf{\Gamma}$-matrix. These are termed *singular* fins and will be considered in this chapter. Treated here are:

1. Longitudinal fin of triangular profile
2. Longitudinal fin of concave parabolic profile
3. Conical spine
4. Concave parabolic spine

In addition, it is often useful to represent a bond resistance and a fin or spine tip heat dissipation as a linear transformation. The linear transformations in these two cases can be represented by a $\mathbf{T}$-matrix, and these will also be developed in this chapter.

As shown in Chap. 3, for any fin or spine, the temperature excess and heat flow at any point may be given as a superposition of base temperature excess and base heat flow.

$$\theta(x) = \theta_b \lambda_1(x) + q_b \lambda_2(x) \tag{4.1}$$

and

$$q(x) = k(x)A(x)[\theta_b \lambda_1'(x) + q_b \lambda_2'(x)] \tag{4.2}$$

These equations produce a linear transformation that maps the temperature excess and heat flow at the fin tip where $x = a$ (a may equal zero), from similar conditions at the fin base where $x = b$. This linear transformation is referred to as the $\mathbf{\Gamma}$-matrix, so that the linear transformation is represented by

$$\begin{bmatrix} \theta_a \\ q_a \end{bmatrix} = \mathbf{\Gamma} \begin{bmatrix} \theta_b \\ q_b \end{bmatrix} = \begin{bmatrix} \gamma_{11} & \gamma_{12} \\ \gamma_{21} & \gamma_{22} \end{bmatrix} \begin{bmatrix} \theta_b \\ q_b \end{bmatrix} \tag{4.3}$$

where, because the fin tip is located at $x = a$,

$$\gamma_{11} = \lambda_1(a) \tag{4.4a}$$

$$\gamma_{12} = \lambda_2(a) \tag{4.4b}$$

$$\gamma_{21} = k(a)A(a)\lambda_1'(a) \tag{4.4c}$$

and

$$\gamma_{22} = k(a)A(a)\lambda_2'(a) \tag{4.4d}$$

and if $a = 0$, these become

$$\gamma_{11} = \lambda_1(0) \tag{4.5a}$$

$$\gamma_{12} = \lambda_2(0) \tag{4.5b}$$

$$\gamma_{21} = k(0)A(0)\lambda_1'(0) \tag{4.5c}$$

and

$$\gamma_{22} = k(0)A(0)\lambda_2'(0) \tag{4.5d}$$

If the fin or spine tapers to a point at $x = 0$, where $A(0) = 0, \mathbf{\Gamma}$ becomes singular[1]:

$$\begin{bmatrix} \gamma_{11} & \gamma_{12} \\ \gamma_{21} & \gamma_{22} \end{bmatrix} = \begin{bmatrix} \lambda_1(0) & \lambda_2(0) \\ 0 & 0 \end{bmatrix} \tag{4.6}$$

It must be stressed that the $\mathbf{\Gamma}$-matrix, when regular as shown by Eq. (4.3) or when singular as indicated by Eq. (4.6), depends only on the fin or spine geometry and the heat transfer parameters of h and k. Observe that $\mathbf{\Gamma}$ is singular when the product $k(a)A(a)$ or $k(0)A(0)$ vanishes and that this condition does indeed occur for the longitudinal fins of triangular and concave parabolic profile and for the conical and concave parabolic spines. These four shapes require special interpretation through the use of a *thermal transmission ratio* defined on the basis of the heat flow and temperature excess at the fin or spine base,

$$\mu \equiv \frac{q_b}{\theta_b} \tag{4.7}$$

which will permit the inclusion of these fins and spines at remote locations in arrays of extended surface, which are commonly referred to as *heat sinks*.

[1] Recall that a singular matrix is a matrix whose determinant is equal to zero. This is the origin of the term *singular fin*.

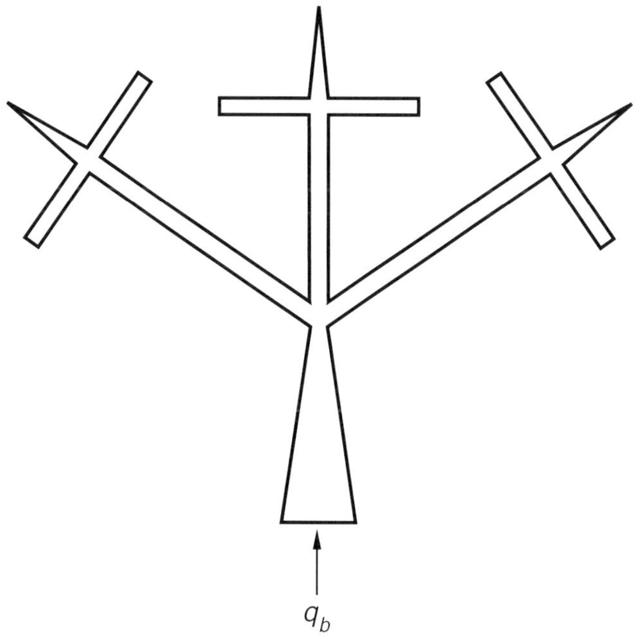

Figure 4.1 A finned array containing 13 fins. Three of the remote fins are longitudinal fins of triangular profile, and six of the remote fins are longitudinal fins of rectangular profile. The fin closest to the base of the array is a longitudinal fin of trapezoidal profile.

Figure 4.1 shows a sketch of a finned array containing 13 elements (fins). It may be observed that three of the remote fins are *singular* longitudinal fins of triangular profile. Moreover, six of the remote fins are *regular* longitudinal fins of rectangular profile, and the fin containing the base of the array is the *regular* longitudinal fin of trapezoidal profile.

4.2 THE LONGITUDINAL FIN OF TRIANGULAR PROFILE

The longitudinal fin of triangular profile, shown with its coordinate system in Fig. 4.2, is seen to be but a specific case of the longitudinal fin of trapezoidal profile (Fig. 3.2) with the origin located at $x = a = 0$ and with tip thickness, $\delta_a = 0$. Observe in Fig. 4.2 that the origin of the height coordinate is at the tip of the fin, with positive orientation toward the fin base, and that the fin height is b.

In this case, the differential equation for the temperature excess is identical to the differential equation for the temperature excess of the longitudinal fin

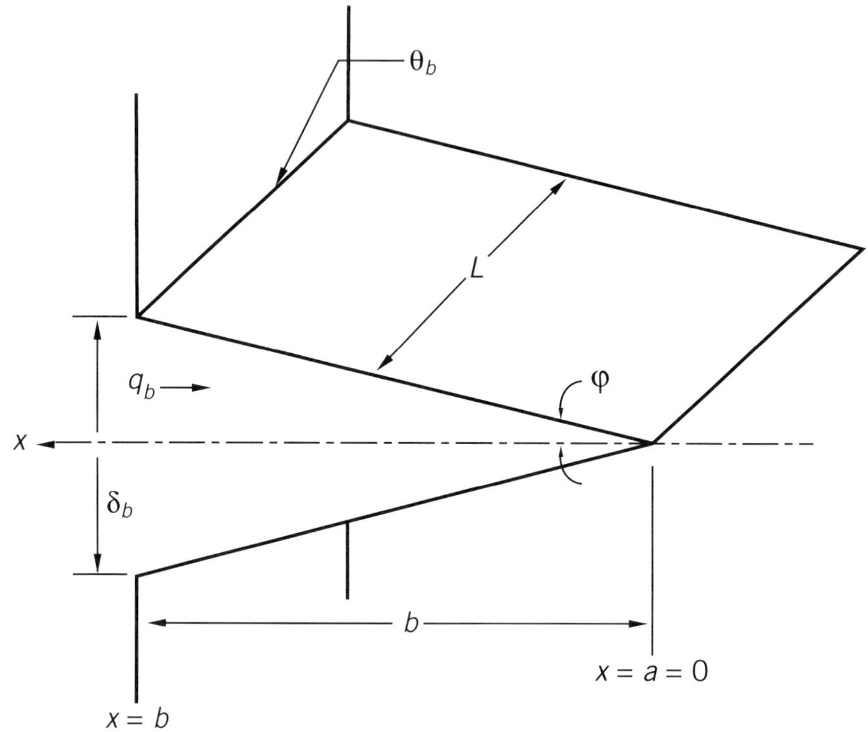

Figure 4.2 Longitudinal fin of triangular profile.

of trapezoidal profile

$$x\frac{d^2\theta}{dx^2} + \frac{d\theta}{dx} - n^2\theta = 0 \tag{3.15}$$

with

$$n = \left(\frac{h}{k \sin\phi}\right)^{1/2}$$

where ϕ is the taper angle

$$\phi = \arctan\frac{\delta_b}{2b}$$

With the initial conditions given by

$$\theta(x = b) = \theta_b$$

and

$$q(x = b) = q_b$$

the particular solution for the temperature excess in terms of a transformed variable

$$u = 2nx^{1/2}$$

is

$$\theta(u) = u_b[K_1(u_b)I_0(u) + I_1(u_b)K_0(u)]\theta_b + \frac{u_b^2}{2k\delta_b Ln^2}\left[K_0(u_b)I_0(u) - I_0(u_b)K_0(u)\right]q_b \tag{4.8}$$

where

$$u_b = 2nb^{1/2}$$

However, as $x \rightarrow 0, u \rightarrow 0$ and $\theta(0) \rightarrow \theta_a$. This temperature difference must be finite. With $I_0(u \rightarrow 0) = 1$, Eq. (4.8) becomes

$$\theta(u \rightarrow 0) = \theta_a = u_b[K_1(u_b) + I_1(u_b)K_0(u \rightarrow 0)]\theta_b + \frac{u_b^2}{2k\delta_b Ln^2}\left[K_0(u_b) \quad I_0(u_b)K_0(u \rightarrow 0)\right]q_b \tag{4.9}$$

Because $K_0(u)$ becomes unbounded as $u \rightarrow 0$, the terms involving $K_0(u)$ in Eq. (4.9) must cancel one another. This implies a relationship between θ_b and q_b:

$$u_b I_1(u_b)K_0(u \rightarrow 0)]\theta_b = \frac{u_b^2}{2k\delta_b Ln^2} I_0(u_b)K_0(u \rightarrow 0)q_b$$

so that after cancellation of the common $K_0(u \rightarrow 0)$ terms, one can obtain

$$q_b = \frac{2k\delta_b Ln^2}{u_b}\frac{I_1(u_b)}{I_0(u_b)}\theta_b$$

and this leads to the thermal transmission ratio, μ,

$$\mu = \frac{q_b}{\theta_b} = \frac{2k\delta_b Ln^2}{u_b}\frac{I_1(u_b)}{I_0(u_b)} \qquad \text{[W/K]} \tag{4.10}$$

which is all that is needed to characterize the longitudinal fin of triangular profile. This rather exceptional behavior is due to the fact that the fin tapers to zero thickness at its tip. It is consistent with the development of the fin resistance in Chap. 1 that for the longitudinal fin of triangular profile, the thermal resistance of the fin is the reciprocal of the thermal transmission ratio.

$$R_f = \frac{1}{\mu} = \frac{\theta_b}{q_b} = \frac{u_b}{2k\delta_b Ln^2}\frac{I_0(u_b)}{I_1(u_b)} \qquad \text{[K/W]}$$

Recognize that μ *does not* depend on the the base conditions of θ_b and q_b, but that μ is fixed by the physical dimensions, b, L, and δ_b; the physical property, k; and the operating condition, h, of the fin.

Example 4.1: A longitudinal fin of triangular profile, fabricated from a metal with a thermal conductivity of 217 W/m-K, has a base thickness of 0.9525 cm (0.375 in). Its height is 10.16 cm (4 in), and its length is 12 cm. It dissipates heat with a base temperature excess of 100°C to the surroundings via a heat-transfer coefficient of 100 W/m^2-K. What is its heat dissipation?

SOLUTION The taper angle is

$$\phi = \arctan \frac{\delta_b}{2b} = \arctan \frac{0.009525}{2(0.1016)} = 2.68^\circ$$

so that

$$\sin \phi = 0.0468$$

Then

$$n = \left(\frac{h}{k \sin \phi} \right)^{1/2} = \left[\frac{100}{217(0.0468)} \right]^{1/2} = 3.1372\,\text{m}^{-1}$$

and

$$u_b = 2nb^{1/2} = 2(3.1372)(0.1016)^{1/2} = 2.0000$$

With

$$I_0(u_b) = I_0(2.0000) = 2.27959$$

and

$$I_1(u_b) = I_1(2.0000) = 1.59064$$

by Eq. (4.10),

$$\begin{aligned} \mu &= \frac{q_b}{\theta_b} = \frac{2k\delta_b L n^2}{u_b} \left[\frac{I_1(u_b)}{I_0(u_b)} \right] \\ &= \frac{2(217)(0.009525)(0.12)(3.1372)^2}{2.0000} \left[\frac{1.59064}{2.27959} \right] \\ &= 2.4411(0.6978) = 1.7033\,\text{W/K} \end{aligned}$$

and, if $\theta_b = 100^\circ$C

$$q_b = 100(1.7033) = 170.33\,\text{W}$$

Observe that the thermal resistance is

$$R_f = \frac{\theta_b}{q_b} = \frac{1}{\mu} = \frac{1}{1.7033} = 0.5871\,\text{K/W}$$

4.3 LONGITUDINAL FIN OF CONCAVE PARABOLIC PROFILE

The coordinate system for the longitudinal fin of concave parabolic profile is shown in Fig. 4.3. Notice that the height coordinate is taken with origin at the fin tip where $x = a = 0$ and $\delta_a = 0$ and that the fin height is b. The height coordinate has a positive sense from tip to base.

Here too, the differential equation for temperature excess is identical to the differential equation for temperature excess when the fin is truncated and $\delta_a \neq 0$. This was shown in Chap. 3 to be

$$x^2 \frac{d^2\theta}{dx^2} + 2x \frac{d\theta}{dx} - m^2 b^2 \theta = 0 \tag{3.31}$$

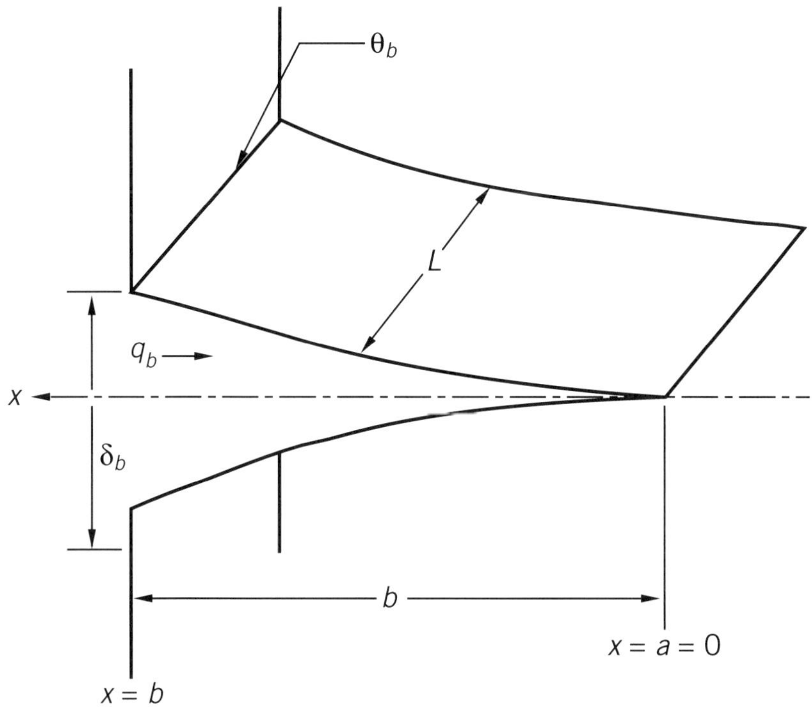

Figure 4.3 Longitudinal fin of concave parabolic profile.

where

$$m = \left(\frac{2h}{k\delta_b}\right)^{1/2}$$

After an appropriate transformation, the particular solution for $\theta(x)$ is obtained using the initial value data

$$\theta(x = b) = \theta_b$$

and

$$q(x = b) = q_b$$

and is

$$\theta(x) = \frac{\theta_b}{\alpha - \beta}\left[\alpha\left(\frac{x}{b}\right)^{\beta} - \beta\left(\frac{x}{b}\right)^{\alpha}\right] + \frac{q_b b}{k\delta_b L(\alpha - \beta)}\left[\left(\frac{x}{b}\right)^{\alpha} - \left(\frac{x}{b}\right)^{\beta}\right] \tag{4.11}$$

where

$$\alpha, \beta = -\frac{1}{2} \pm \frac{(1 + 4m^2b^2)^{1/2}}{2}$$

It is seen that $\beta < 0 < \alpha$ and that as $x \to 0, \theta(x \to 0) = \theta_a$. Thus, Eq. (4.11) may be employed to write

$$\theta(0) = \theta_a = \frac{\theta_b}{\alpha - \beta}\left[\alpha\left(\frac{x \to 0}{b}\right)^{\beta} - \beta\left(\frac{x \to 0}{b}\right)^{\alpha}\right] + \frac{q_b b}{k\delta_b L(\alpha - \beta)}\left[\left(\frac{x \to 0}{b}\right)^{\alpha} - \left(\frac{x \to 0}{b}\right)^{\beta}\right]$$

Because $\beta < 0, \theta_a$ will be unbounded unless the terms that involve $(x \to 0/b)^{\beta}$ cancel one another. Thus, the relationship

$$\frac{\alpha\theta_b}{\alpha - \beta}\left(\frac{x \to 0}{b}\right)^{\beta} = \frac{q_b b}{k\delta_b L(\alpha - \beta)}\left(\frac{x \to 0}{b}\right)^{\beta}$$

is implied and, after the cancellation of the common terms, the thermal transmission ratio for the longitudinal fin of concave parabolic profile is seen to be

$$\mu = \frac{q_b}{\theta_b} = \frac{\alpha k\delta_b L}{b} \tag{4.12}$$

Example 4.2: A longitudinal fin of concave parabolic profile is to have the same dimensions as the longitudinal fin of triangular profile in Ex. 4.1. It is fabricated from the same metal ($k = 217\,\text{W/m-K}$) and dissipates heat at the same base temperature excess ($\theta_b = 100°\text{C}$) to the surroundings, with an identical heat-transfer coefficient of $h = 100\,\text{W/m}^2\text{-K}$. Compare its heat dissipation to that of the triangular fin.

SOLUTION With

$$m = \left(\frac{2h}{k\delta_b}\right)^{1/2} = \left[\frac{2(100)}{217(0.009525)}\right]^{1/2} = 9.8368\,\text{m}^{-1}$$

and $b = 0.1016\,\text{m}$, the value of α becomes

$$\begin{aligned}\alpha &= -\frac{1}{2} + \frac{(1 + 4m^2b^2)^{1/2}}{2}\\ &= -\frac{1}{2} + \frac{[1 + 4(9.8368)^2(0.1016)^2]^{1/2}}{2}\\ &= -\frac{1}{2} + \frac{(1 + 3.9954)^{1/2}}{2}\\ &= 0.6175\end{aligned}$$

Then by Eq. (4.12)

$$\mu = \frac{q_b}{\theta_b} = \frac{\alpha k\delta_b L}{b}$$

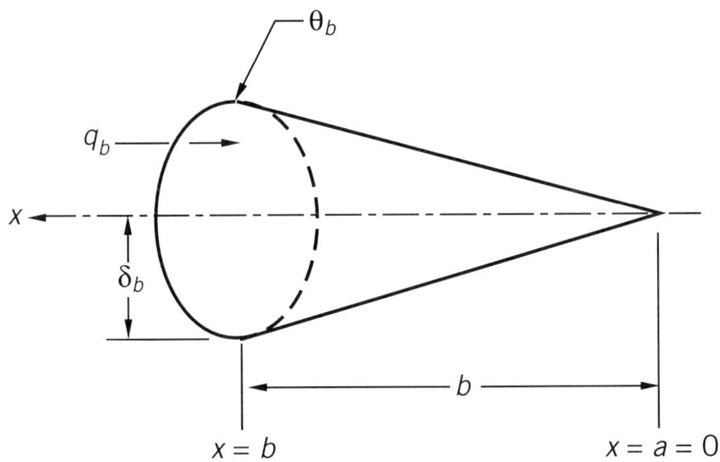

Figure 4.4 The conical spine.

$$= \frac{0.6175(217)(0.009525)(0.12)}{0.1016}$$
$$= 1.5074\,\text{W/K}$$

and, if $\theta_b = 100°\text{C}$,

$$q_b = 100(1.5074) = 150.74\,\text{W}$$

Here, the thermal resistance is

$$R_f = \frac{\theta_b}{q_b} = \frac{1}{\mu} = \frac{1}{1.5074} = 0.6634\,\text{K/W}$$

which shows that the longitudinal fin of concave parabolic profile has a higher thermal resistance than the longitudinal fin of triangular profile for the same height, length and base thickness, thermal properties, and operating conditions. This means that the fin of concave parabolic profile is, basically, not as good a heat dissipator as the longitudinal fin of triangular profile. However, the parabolic profile fin possesses only two-thirds of the profile area as the triangular fin and, hence, only two-thirds of the mass. This suggests that *a fin of least material* is more than just an item of interest. Chap. 9, in its entirety, is devoted to this very important consideration.

4.4 THE CONICAL SPINE

The conical spine shown in Fig. 4.4 is a special case of the truncated conical spine where the origin of the height coordinate is located at $x = a = 0$. Thus, the origin is at the tip of the spine, which tapers to a point, and the height coordinate is positive towards the fin base. Again, the height is designated as b.

The differential equation for temperature excess is that for the truncated conical spine:

$$x^2 \frac{d^2\theta}{dx^2} + 2x \frac{d\theta}{dx} - n^2 x\theta = 0 \tag{3.71}$$

where

$$n = \left[\frac{4h^2}{r_b^2 k^2} (b^2 + r_b^2) \right]^{1/4}$$

The particular solution in terms of the transformed variable

$$w = 2nx^{1/2}$$

is obtained using the initial value data

$$\theta(x = b) = \theta_b$$

and

$$q(x = b) = q_b$$

and is

$$\theta(x) = \frac{w_b^2}{w} [K_2(w_b) I_1(w) + I_2(w_b) K_1(w)] \theta_b$$

$$+ \frac{8n^2 b^2}{k\pi r_b^2 w_b w} [K_1(w_b) I_1(w) + I_1(w_b) K_1(w)] q_b \tag{4.13}$$

where

$$w_b = 2nb^{1/2}$$

and

$$w_a = 2na^{1/2}$$

As $x \to 0$, $w \to 0$, and $w_a \to 0$. In Eq. (4.13), $K_1(w)/w$ is indeterminant, but an application of the de l'Hospital rule shows that $K_1(w)/w$ is unbounded. In order for the tip temperature to remain finite, the terms involving $K_1(w)/w$ in Eq. (4.13) must cancel. Hence

$$\frac{w_b^2}{w} I_2(w_b) K_1(w) \theta_b = \frac{8n^2 b^2}{k\pi r_b^2 w_b w} I_1(w_b) K_1(w) q_b$$

or

$$\frac{q_b}{\theta_b} = \frac{k\pi r_b^2 w_b^3}{8n^2 b^2} \frac{I_2(w_b)}{I_1(w_b)}$$

But $w_b^3 = 8n^4 b^{3/2}$, so that

$$\mu = \frac{q_b}{\theta_b} = \frac{k\pi n r_b^2}{b^{1/2}} \frac{I_2(w_b)}{I_1(w_b)} \tag{4.14}$$

Example 4.3: A conical spine possesses a base diameter of 1 cm and is 4 cm high. It is fabricated from a metal with a thermal conductivity of 101.6 W/m-K, and it operates in an environment with a temperature of 40 °C and a heat-transfer coefficient of 56.71 W/m^2-K. If the base temperature of the spine is 120°C, what is its heat dissipation?

SOLUTION Here

$$\theta_b = 120 - 40 = 80°\text{C}$$

and

$$\begin{aligned} n &= \left\{ \frac{4h^2}{r_b^2 k^2} \left[b^2 + r_b^2 \right] \right\}^{1/4} \\ &= \left\{ \frac{4(56.71)^2}{(0.005)^2 (101.6)^2} \left[(0.04)^2 + (0.005)^2 \right] \right\}^{1/4} \\ &= (80.975)^{1/4} = 2.9998 \text{ or } 3.000 \end{aligned}$$

and

$$\begin{aligned} w_b &= 2nb^{1/2} \\ &= 2(3.000)(0.04)^{1/2} \\ &= 1.2000 \end{aligned}$$

Then with

$$I_1(w_b) = 0.71468$$

and

$$I_2(w_b) = 0.20260$$

Eq. (4.14) can be employed to give

$$\begin{aligned} \mu &= \frac{q_b}{\theta_b} \\ &= \left[\frac{k\pi n r_b^2}{b^{1/2}} \right] \left[\frac{I_2(w_b)}{I_1(w_b)} \right] \\ &= \left[\frac{101.6\pi(3.0000)(0.005)^2}{(0.04)^{1/2}} \right] \left[\frac{0.20260}{0.71468} \right] \\ &= 0.0339 \text{ W/K} \end{aligned}$$

and

$$q_b = \mu\theta_b = 0.0339(80) = 2.71 \text{ W}$$

The thermal resistance is

$$R_f = \frac{\theta_b}{q_b} = \frac{1}{\mu} = \frac{1}{0.0339} = 29.52 \text{ K/W}$$

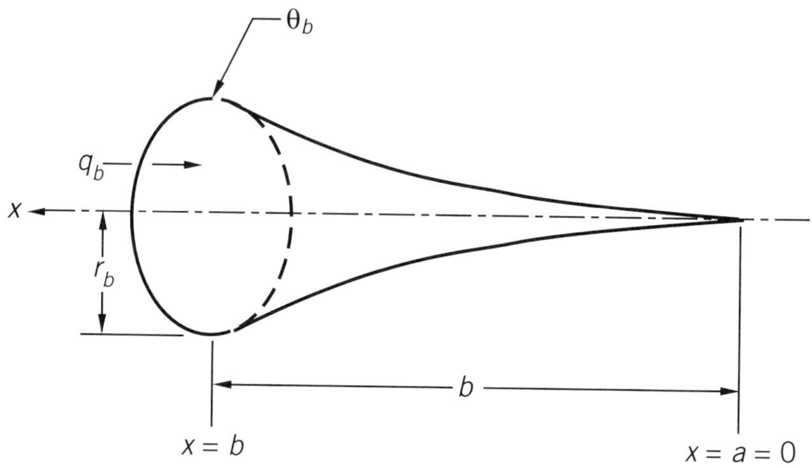

Figure 4.5 The spine of concave parabolic profile.

4.5 THE CONCAVE PARABOLIC SPINE

The concave parabolic spine shown in Fig. 4.5 is a special case of the truncated concave parabolic spine where the origin of the height coordinate is located at $x = a = 0$. Here too, the origin is at the tip of the spine, which tapers to a point, and the height coordinate is positive towards the fin base located at $x = b$.

The differential equation for temperature excess is that for the truncated concave parabolic spine:

$$x^2 \frac{d^2\theta}{dx^2} + 4x\frac{d\theta}{dx} - m^2 b^2 \theta = 0 \tag{3.84}$$

where

$$m = \left(\frac{2h}{kr_b}\right)^{1/2}$$

Here, the particular solution for $\theta(x)$ is also obtained using the initial value data

$$\theta(x = b) = \theta_b$$

and

$$q(x = b) = q_b$$

and is

$$\theta(x) = \frac{1}{\psi}\left[\alpha\left(\frac{x}{b}\right)^{\beta} - \beta\left(\frac{x}{b}\right)^{\alpha}\right]\theta_b + \frac{b}{k\pi r_b^2 \psi}\left[\left(\frac{x}{b}\right)^{\alpha} - \left(\frac{x}{b}\right)^{\beta}\right] q_b \tag{4.15}$$

where

$$\alpha, \beta = -\frac{3}{2} \pm \frac{(1 + 4m^2b^2)^{1/2}}{2}$$

and

$$\psi = \alpha\beta = -(mb)^2 = -\frac{2hb^2}{kr_b}$$

It is seen that $\beta < 0 < \alpha$ and that as $x \to 0, \theta_a = \theta(0)$, and Eq. (4.15) can be written in the form

$$\theta(0) = \theta_a = \frac{\theta_b}{\psi}\left[\alpha\left(\frac{x \to 0}{b}\right)^{\beta} - \beta\left(\frac{x \to 0}{b}\right)^{\alpha}\right]$$

$$+ \frac{q_b b}{k\pi r_b^2 \psi}\left[\left(\frac{x \to 0}{b}\right)^{\alpha} - \left(\frac{x \to 0}{b}\right)^{\beta}\right]$$

Because $\beta < 0, \theta_a$ will be unbounded unless the terms that involve $(x \to 0/b)^{\beta}$ cancel one another. Hence

$$\frac{\theta_b}{\psi}\alpha\left(\frac{x \to 0}{b}\right)^{\beta} = \frac{q_b b}{k\pi r_b^2 \psi}\left(\frac{x \to 0}{b}\right)^{\beta}$$

which leads to the thermal transmission ratio

$$\mu = \frac{q_b}{\theta_b} = \frac{k\pi r_b^2 \alpha}{b} \tag{4.16}$$

4.6 THE SINGLE SERIES RESISTANCE

In order to account properly for the effect of bond resistance using the algorithms for the combination of extended surface, which will be developed in Chap. 5, a linear transformation for a single series resistance must be formulated. Figure 4.6 shows such a representation, and it is proposed that the temperature drop across the thermal resistance is equal to the difference between the temperatures at the two ends. Notice that T_b is taken at the end of the resistance where the heat flow enters and that the thermal resistance, R, is in K/W.

Continuity demands that

$$q_b = q_a = q \tag{4.17a}$$

and the temperature drop will be equal to

$$T_b - T_a = qR$$

If the surrounding temperature is both added and subtracted to the left side of the latter, the result is

$$T_b - T_s - (T_a - T_s) = qR$$

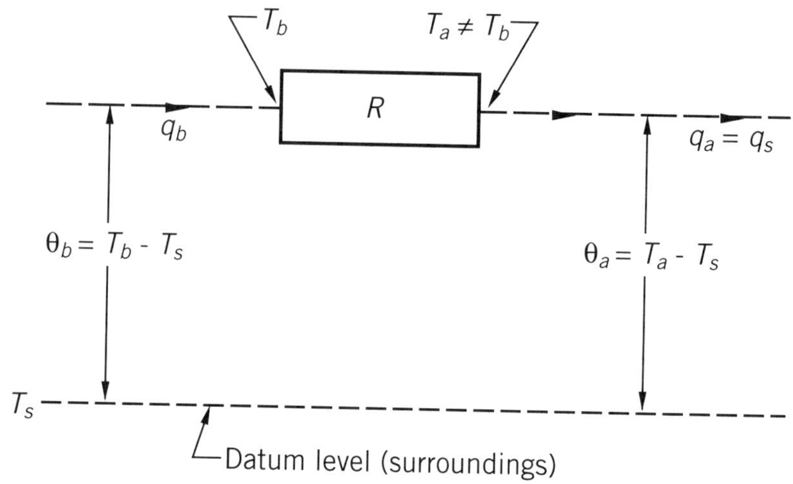

Figure 4.6 The single series resistance.

and this shows that

$$\theta_b - \theta_a = qR \tag{4.17b}$$

The **T**-matrix is defined by Eq. (2.17):

$$\begin{bmatrix} \theta_b \\ q_b \end{bmatrix} = \mathbf{T} \begin{bmatrix} \theta_a \\ q_a \end{bmatrix} = \begin{bmatrix} \tau_{11} & \tau_{12} \\ \tau_{21} & \tau_{22} \end{bmatrix} \begin{bmatrix} \theta_a \\ q_a \end{bmatrix} \tag{2.17}$$

A matrix multiplication yields

$$\begin{aligned} \theta_b &= \tau_{11}\theta_a + \tau_{12}q_a \\ q_b &= \tau_{21}\theta_a + \tau_{22}q_a \end{aligned}$$

and the τ's can be found by an application of superposition. If $q = q_a = 0$, Eq. (4.17b) shows that $\theta_a = \theta_b$ so that

$$\tau_{11} = \left.\frac{\theta_b}{\theta_a}\right|_{q_a=0} = 1 \tag{4.18a}$$

Next, with $\theta_a = 0$, Eq. (4.17b) shows that $\theta_b = Rq_b = Rq_a$. Hence

$$\tau_{12} = \left.\frac{\theta_b}{q_a}\right|_{\theta_a=0} = R \tag{4.18b}$$

Then with $q_a = 0$, Eq. (4.17a) indicates that both q's are zero and

$$\tau_{21} = \left.\frac{q_b}{\theta_a}\right|_{q_a=0} = 0 \tag{4.18c}$$

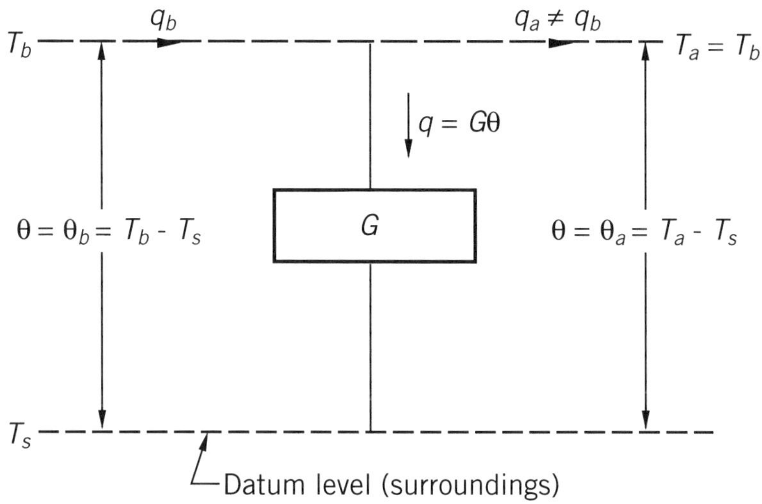

Figure 4.7 The single shunt conductance.

Finally, with $q_b = q_a = q$ by Eq. (4.17*a*)

$$\tau_{22} = \left.\frac{q_b}{q_a}\right|_{\theta_a=0} = 1 \tag{4.18d}$$

The **T**-matrix for the single series element is therefore

$$\mathbf{T} = \begin{bmatrix} 1 & R \\ 0 & 1 \end{bmatrix} \tag{4.19}$$

Table 2.2 shows that, because $\tau_{21} = 0$, the single series element does not possess a **Z**-matrix.

An example that demonstrates the evaluation of a single fin that possesses a bond resistance (via the single series element) is presented in Chap. 6.

4.7 THE SINGLE SHUNT CONDUCTANCE

Of equal importance is the single shunt conductance, which can be used to account for the fin tip heat loss. Because the single shunt conductance is somewhat more difficult to visualize, the details of the development of its **T**-matrix are somewhat more difficult to grasp. The single shunt element is the dual of the single series resistance and is displayed in Fig. 4.7. Notice that here the temperatures at the extremes are equal, that is,

$$T_b = T_a$$

and if T_s is subtracted from each side of the equation

$$T_b - T_s = T_a - T_s$$

or

$$\theta_b = \theta_a \tag{4.20a}$$

Of course, the shunt element is merely an artifice to account for the tip heat loss. For a tip surface area, $S(a)$, the shunt conductance will be $G = h_a S(a)$. Notice in Fig. 4.7, that this is *not* q_a.[2]

Continuity can be employed to show that with reference to Fig. 4.7,

$$q_b = G\theta_b + q_a = G\theta_a + q_a \tag{4.20b}$$

where G is the reciprocal of the thermal resistance

$$G = \frac{1}{R}$$

and is, therefore, a thermal conductance in W/K. The **T**-matrix, defined by Eq. (2.17), can be used here as well. Again, the matrix multiplication yields

$$\begin{aligned} \theta_b &= \tau_{11}\theta_a + \tau_{12}q_a \\ q_b &= \tau_{21}\theta_a + \tau_{22}q_a \end{aligned}$$

and the τ's can be found by another application of superposition. If $\theta_a = \theta_b$ as required by Eq. (4.20*a*), regardless of q_a

$$\tau_{11} = \left.\frac{\theta_b}{\theta_a}\right|_{q_a=0} = 1 \tag{4.21a}$$

Next, if $\theta_a = 0$, Eq. (4.20*a*) requires that $\theta_b = 0$. Hence

$$\tau_{12} = \left.\frac{\theta_b}{q_a}\right|_{\theta_a=0} = 0 \tag{4.21b}$$

Then with $q_a = 0$, Eq. (4.20*b*) indicates that $q_b = G\theta_b$ and

$$\tau_{21} = \left.\frac{q_b}{\theta_a}\right|_{q_a=0} = G \tag{4.21c}$$

Finally, with $\theta_b = \theta_a = 0$, Eq. (4.20*b*) shows that $q_a = q_b$ and

$$\tau_{22} = \left.\frac{q_b}{q_a}\right|_{\theta_a=0} = 1 \tag{4.21d}$$

The **T**-matrix for the single shunt element is therefore

$$\mathbf{T} = \begin{bmatrix} 1 & 0 \\ G & 1 \end{bmatrix} \tag{4.22}$$

Notice that because $\tau_{12} = 0$, Table 2.2 shows that the single series element does not possess a **Y**-matrix.

[2]The reason for the presence of the q_a will become more apparent in Chap. 5 where it is shown that the use of the single shunt element permits the tip to be treated as adiabatic, with $q_a = 0$.

4.8 SUMMARY AND CLOSURE

Table 4.1 provides a directory of the thermal transmission ratios for the singular fins and spines considered in this chapter.

The rather arduous task, which began at the beginning of Chap. 3, to develop a catalog of the elements of the linear transformation matrices and the thermal transmission ratios has now been completed. Six matrices were developed for each of the several regular fins and spines in Chap. 3. Four thermal transmission ratios for the singular fins and spines, as well as the **T**-matrix for the single series resistance and single shunt conductance, have been considered here in Chap. 4. These are the tools that permit rapid analysis of complicated finned arrays composed of many individual fins and spines in a variety of combinations.

Thus, attention now turns to the development of the algorithms for the combination of the fins and spines into finned arrays. This will be the task for Chapt. 5 and Chap. 6 will be devoted exclusively to applications of the use of the algorithms.

TABLE 4.1. Summary of Singular Fin and Spine thermal transmission ratios and thermal resistances

Longitudinal Fin of Triangular Profile

$$n = \left(\frac{h}{k \sin\phi}\right)^{1/2} \qquad \mu = \frac{2k\delta_b L n^2}{u_b}\frac{I_1(u_b)}{I_0(u_b)} \tag{4.10}$$

$$u_b = 2nb^{1/2} \qquad R_f = \frac{u_b}{2k\delta_b L n^2}\frac{I_0(u_b)}{I_1(u_b)}$$

Longitudinal Fin of Concave Parabolic Profile Profile

$$m = \left(\frac{2h}{k\delta_b}\right)^{1/2} \qquad \mu = \frac{\alpha\kappa\delta_b L}{b} \tag{4.12}$$

$$\alpha = -\frac{1}{2} + \frac{(1+4m^2b^2)^{1/2}}{2} \qquad R_f = \frac{b}{\alpha\kappa\delta_b L}$$

Conical Spine

$$n = \left[\frac{4h^2}{r_b^2 k^2}\left(b^2 + r_b^2\right)\right]^{1/4} \qquad \mu = \frac{k\pi n r_b^2}{b^{1/2}} \frac{I_2(w_b)}{I_1(w_b)} \tag{4.14}$$

$$w_b = 2nb^{1/2} \qquad R_f = \frac{b^{1/2}}{k\pi n r_b^2} \frac{I_1(w_b)}{I_2(w_b)}$$

Concave Parabolic Spine

$$m = \left(\frac{2h}{k_b^r}\right)^{1/2} \qquad \mu = \frac{k\pi r_b^2 \alpha}{b} \tag{4.16}$$

$$\alpha = -\frac{3}{2} + \frac{(1 + 4m^2 b^2)^{1/2}}{2} \qquad R_f = \frac{b}{k\pi r_b^2 \alpha}$$

Chapter

5

ALGORITHMS FOR FINNED ARRAY ASSEMBLY

5.1 INTRODUCTION

The growing availability of compact heat exchanger cores, as disclosed by Kays and London in 1984 and in their earlier editions in 1955 and 1964, stimulated interest in the fin efficiency of stacked heat exchanger cores. Kays (1960) treated as stack components two- and three-layer cores, termed the *double* and *triple sandwich*, and developed efficiency for equal heat input on both exterior plates. Kern and Kraus (1972) looked at the single, double, and triple stacks (*stacks*, rather than *sandwiches*, was the nomenclature used) with heat input on one side and heat input on two sides.

Kraus et al. (1978) showed that for longitudinal fins of rectangular, trapezoidal and triangular profiles, and in finned arrays (not arrays of fins) composed of these fins, conditions of heat flow and temperature excess at any point on a fin are induced by similar conditions at the fin base. In particular, it was shown[1] that there was a linear transformation that mapped conditions from the fin base to conditions at the fin tip:

$$\begin{bmatrix} \theta_a \\ q_a \end{bmatrix} = \mathbf{\Gamma} \begin{bmatrix} \theta_b \\ q_b \end{bmatrix} = \begin{bmatrix} \gamma_{11} & \gamma_{12} \\ \gamma_{21} & \gamma_{22} \end{bmatrix} \begin{bmatrix} \theta_b \\ q_b \end{bmatrix} \tag{2.18}$$

and vice-versa

$$\begin{bmatrix} \theta_b \\ q_b \end{bmatrix} = \mathbf{T} \begin{bmatrix} \theta_a \\ q_a \end{bmatrix} = \begin{bmatrix} \tau_{11} & \tau_{12} \\ \tau_{21} & \tau_{22} \end{bmatrix} \begin{bmatrix} \theta_a \\ q_a \end{bmatrix} \tag{2.17}$$

[1]See Chap. 2.

They claimed that the $\mathbf{\Gamma}$- and $\mathbf{T}$-matrices could be used to obtain a new parametization, called the *input admittance*, that would completely describe the performance of a fin. The conventional fin efficiency was abandoned, and it was proposed that single fins and finned arrays should be characterized by this single, yet important, parameter.

The input admittance was defined by Kraus and Snider (1980) as the ratio of the heat entering the base of a single fin or a finned array, to the temperature excess at the base of the fin or the finned array:

$$Y_{in} \equiv \frac{q_b}{\theta_b}$$

Algorithms were provided for combining the effects of individual fins in a finned array in what Kraus et al. (1978) called the *cascade, cluster* and *parallel connections*. It was also shown that a modification of the procedure developed led to an exact solution for the double stack, with unequal heat distribution on the two stack faces at opposite ends.

Section 2.8 provided the theory for the development of the linear transformations that focused on the solution of the differential equations for temperature excess as an initial value problem. The $\mathbf{\Gamma}$-matrix was developed for fins that were *regular,* which are those fins whose cross section is never zero. The longitudinal fins of triangular and concave parabolic profile and the conical and concave parabolic spines were termed *singular* because their cross section tapered to a point. These fins possess a $\mathbf{\Gamma}$-matrix that is singular and were characterized by yet another parameter called the *thermal transmission ratio.*

Kraus and Snider (1980) also elaborated on the status, validity, and virtue of these new parameters by discussing their apparent superiority over the classic notation of the fin efficiency with regard to adaptabiltity and versatility. Extracts from this paper have been provided here in Chap. 2. Kraus and Snider (1980) also provided the new parameters for several longitudinal and radial fins and spines. Many of these have been redeveloped in Chaps. 3 and 4.

The algorithms provided by Kraus et al. (1978) were insufficient for the analysis of finned arrays that contained loops in the graph theoretical sense. This limitation was removed by Snider and Kraus (1981) who identified the subclass of arrays that possessed these loops and who then extended their prior technique by deriving general equations for the efficient analysis of a perfectly arbitrary configuration of fins.

Kraus (1982) showed that the algorithms provided by Kraus et al. (1978) could be effectively employed in a different manner that relied on the fact that each individual fin in the array was analagous to an electrical two-port network. This was an extension of the transmission line analogy proposed by Kern and Kraus (1972). Two-port matrices were also developed to accommodate a single series element, to be used for bond resistance, and single shunt elements, to be used to account for tip heat loss. These have been considered here[2] at the end of Chap. 4. Kraus et al. (1982), working with these two ports, showed that all of the regular fins could be represented by a single *pi* network. They showed that the pi networks could be assembled in a manner that permitted a node analysis that would permit the determination of the temperature excess at all of the fin intersection points.

Snider and Kraus (1983) indicated a method for accommodating continuously distributed heat sources along the faces of a fin, and Kraus and Snider (1985) showed that when individual fins and spines are assembled into an array, the performance of the array may become limited by the performance of the fins or spines close to the base of the array. They provided a method for the determination of the width or diameter of the innermost fins or spines to avoid what they referred to as a "choking of the heat flow."

Attention will now be focussed on the algorithms for the combination of fins in a finned array. The task here is to develop rigorously the algorithms for fins in (1) cascade, (2) cluster, and (3) parallel. First, however, a discussion of the types of arrays is in order.

5.2 ARRAYS WITH AND WITHOUT LOOPS

Figure 5.1 displays two finned arrays. Each of them contains three longitudinal fin profiles, and in each, the fin of triangular profile is the most remote fin in the array. The array shown in Fig. 5.1*b* contains four loops in the graph theoretical sense, four loops that cannot be removed by a simple parallel combination. The array in Fig. 5.1*a* contains nine fins and no loops. Algorithms for the combination of fins in an array that does not contain loops in the graph theoretical sense will be developed in this chapter. Arrays with loops in the graph theoretical sense are discussed in Chaps. 7 and 8.

5.3 THE CASCADE ALGORITHM

Consider the simplest case of two fins in *cascade* shown in Fig. 5.2 Here, continuity dictates that the heat leaving fin 2 must enter fin 1 and that, at the point of intersection, the temperature excesses must match. Thus, one may set down a matrix equality embracing continuity and compatibility using a for tip conditions and b for base conditions:

$$\begin{bmatrix} \theta_{a2} \\ q_{a2} \end{bmatrix} = \begin{bmatrix} \theta_{b1} \\ q_{b1} \end{bmatrix} \tag{5.1}$$

In addition, each fin can be represented by a $\mathbf{T}$-matrix, which is a linear transformation from tip conditions to base conditions. For fin 1

$$\begin{bmatrix} \theta_{b1} \\ q_{b1} \end{bmatrix} = \mathbf{T}_1 \begin{bmatrix} \theta_{a1} \\ q_{a1} \end{bmatrix} \tag{5.2}$$

[2]But not in the two-port framework.

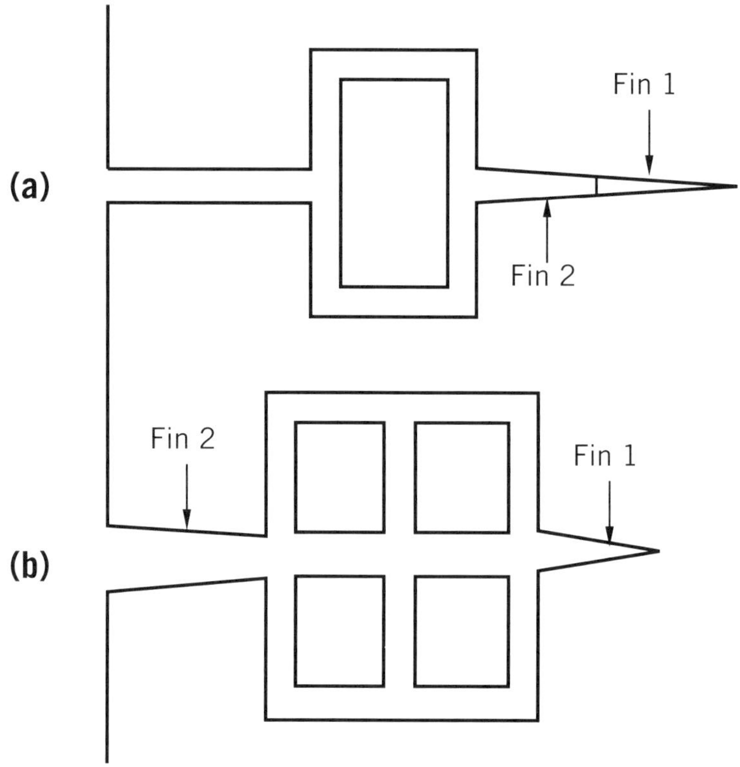

Figure 5.1 Two arrays of extended surface, each with three longitudinal fin profiles. The array in (*a*) contains no loops in a graph theoretical sense. The array in (*b*) contains four loops in a graph theoretical sense. In both arrays, the fin of triangular profile is fin 1: the fin of trapezoidal profile is fin 2, and all other fins are of rectangular profile.

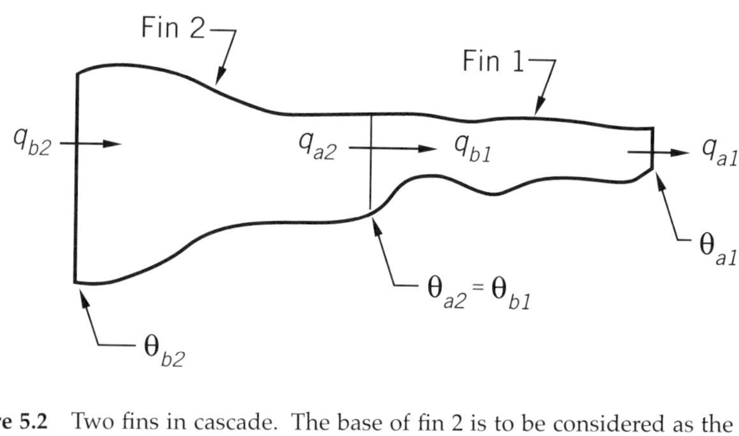

Figure 5.2 Two fins in cascade. The base of fin 2 is to be considered as the base of the combination, which can now be referred to as an *array*.

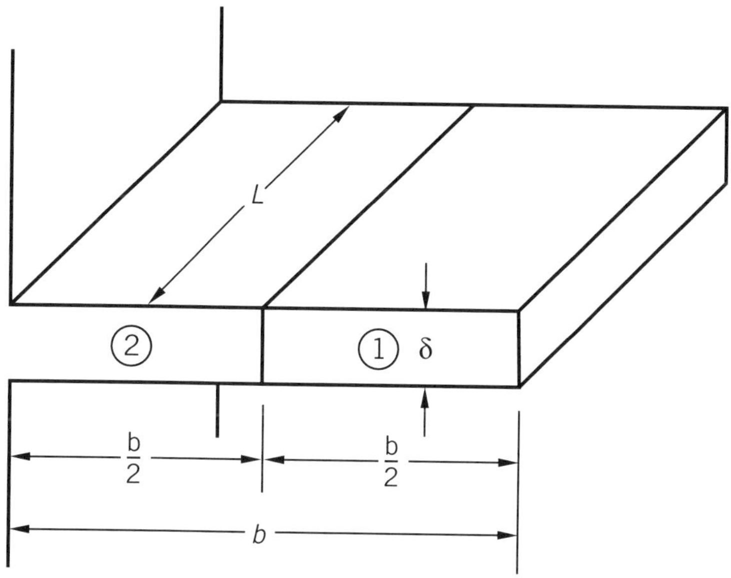

Figure 5.3 Two identical longitudinal fins of rectangular profile placed end-to-end.

and for fin 2

$$\begin{bmatrix} \theta_{b2} \\ q_{b2} \end{bmatrix} = \mathbf{T}_2 \begin{bmatrix} \theta_{a2} \\ q_{a2} \end{bmatrix} \tag{5.3}$$

A simple exercise in matrix algebra shows that there is an equivalent $\mathbf{T}$ representation that maps conditions from the tip of fin 1 to the base of fin 2, which is at the base of the array. First take Eq. (5.2) and modify it using Eq. (5.1):

$$\begin{bmatrix} \theta_{b1} \\ q_{b1} \end{bmatrix} = \begin{bmatrix} \theta_{a2} \\ q_{a2} \end{bmatrix} = \mathbf{T}_1 \begin{bmatrix} \theta_{a1} \\ q_{a1} \end{bmatrix}$$

and then use Eq. (5.3)

$$\begin{bmatrix} \theta_{b2} \\ q_{b2} \end{bmatrix} = \mathbf{T}_2 \begin{bmatrix} \theta_{a2} \\ q_{a2} \end{bmatrix} = \mathbf{T}_2\mathbf{T}_1 \begin{bmatrix} \theta_{a1} \\ q_{a1} \end{bmatrix} = \mathbf{T}_e \begin{bmatrix} \theta_{a1} \\ q_{a1} \end{bmatrix} \tag{5.4}$$

This shows the existence of an equivalent $\mathbf{T}$-matrix that is a simple matrix product of the two individual $\mathbf{T}$-matrices.[3]

This may be extended to n fins in cascade

$$\mathbf{T}_e = \mathbf{T}_n\mathbf{T}_{n-1}\mathbf{T}_{n-2}\cdots\mathbf{T}_2\mathbf{T}_1 \tag{5.5}$$

and Eq. (5.5) is the cascade algorithm.

[3]In the proper order because matrix multiplication, in general, is not commutative.

Example 5.1: Consider the two longitudinal fins of rectangular profile placed end-to-end (or in cascade) as shown in Fig. 5.3. The two fins are identical with heights, $b/2$, and no heat flows from the tip of fin 1. Show that the two fins in cascade are formally equivalent to a single fin of fin height, b.

SOLUTION This is the simplest possible cascade. Each of the two fins possesses a $\mathbf{T}$-matrix with elements given by Eqs. (3.9). With $u = b/2$,

$$\mathbf{T}_1 = \mathbf{T}_2 = \begin{bmatrix} \cosh mu & (1/Y_o)\sinh mu \\ Y_o \sinh mu & \cosh mu \end{bmatrix}$$

where $Y_o = (2hk\delta)^{1/2}L$ and $m = (2h/k\delta)^{1/2}$.

For the two fins in cascade, the cascade algorithm of Eq. (5.5) is employed to give an equivalent

$$\mathbf{T}_e = \begin{bmatrix} \cosh mu & (1/Y_o)\sinh mu \\ Y_o \sinh mu & \cosh mu \end{bmatrix}^2$$

or

$$\mathbf{T}_e = \begin{bmatrix} \cosh^2 mu + \sinh^2 mu & (2/Y_o)\cosh mu \sinh mu \\ 2Y_o \cosh mu \sinh mu & \cosh^2 mu + \sinh^2 mu \end{bmatrix}$$

There are two hyperbolic function identities that can be employed here:

$$\cosh 2mu = \cosh^2 mu + \sinh^2 mu$$

and

$$\sinh 2mu = 2\cosh mu \sinh mu$$

Thus

$$\mathbf{T}_e = \begin{bmatrix} \cosh 2mu & (1/Y_o)\sinh 2mu \\ Y_o \sinh 2mu & \cosh 2mu \end{bmatrix}$$

and because $mb = 2mu$,

$$\mathbf{T}_e = \begin{bmatrix} \cosh mb & (1/Y_o)\sinh mb \\ Y_o \sinh mb & \cosh mb \end{bmatrix}$$

5.4 THE ARRAY INPUT ADMITTANCE

The *input admittance* applies to *both* individual fins or spines or to arrays of fins and spines. As indicated in Sec. 2.10, it is defined as the ratio of the heat entering the base to the temperature excess of the individual fin or spine or the array. It can be obtained from the expanded form of the $\mathbf{T}$-matrix:

$$\begin{aligned} \theta_b &= \tau_{11}\theta_a + \tau_{12}q_a \\ q_b &= \tau_{21}\theta_a + \tau_{22}q_a \end{aligned}$$

If the second of these is divided by the first, the result is the input admittance:

$$Y_{in} = \frac{\tau_{21}\theta_a + \tau_{22}q_a}{\tau_{11}\theta_a + \tau_{12}q_a}$$

and if a θ_a is factored in both numerator and denominator, one obtains

$$Y_{in} = \frac{\tau_{21} + (q_a/\theta_a)\tau_{22}}{\tau_{11} + (q_a/\theta_a)\tau_{12}} \tag{5.6}$$

Equation (5.6) is a *bilinear transformation*[4] and is called the *reflection relationship*. It is computationally efficient and can be used to provide a word description of how an array performs.

For example, consider an array of n fins in cascade. Call the base of fin n the base of the array, and the tip of fin 1, the tip of the array. At the tip of fin 1, $q_{a,1}$ has a value. Either $q_{a,1} = 0$, or $q_{a,1} = h_a S_a \theta_{a,1}$ where h_a is the tip heat transfer coefficient and S_a is the tip surface area. If $q_{a,1} = 0$, the input admittance at the base of fin 1 is simply

$$Y_{in,1} = \frac{\tau_{21}}{\tau_{11}}$$

If $q_{a,1} = h_a S_a \theta_{a,1}$, then the ratio

$$\frac{q_{a,1}}{\theta_{a,1}} = h_a S_a$$

is known and the input admittance of fin 1 is obtained from Eq. (5.6) as

$$Y_{in,1} = \frac{\tau_{21,1} + (q_{a,1}/\theta_{a,1})\tau_{22,1}}{\tau_{11,1} + (q_{a,1}/\theta_{a,1})\tau_{12,1}}$$

Proceed now from fin 1 to fin 2, which is in a direction from array tip to array base. By continuity, the tip heat flow for fin 2 is equal to the base heat flow of fin 1. Moreover, the temperature excesses must match, $\theta_{a,2} = \theta_{b,1}$. Then, the input admittance of fin 2 may also be obtained from Eq. (5.6):

$$Y_{in,2} = \frac{\tau_{21,2} + (q_{a,2}/\theta_{a,2})\tau_{22,2}}{\tau_{11,2} + (q_{a,2}/\theta_{a,2})\tau_{12,2}}$$

because $q_{a,2}/\theta_{a,2} = Y_{in,1} = q_{b,1}/\theta_{b,1}$.

This procedure can be extended until fin n is reached where the input admittance can be obtained from Eq. (5.6) as

$$Y_{in,n} = \frac{\tau_{21,n} + (q_{a,n}/\theta_{a,n})\tau_{22,n}}{\tau_{11,n} + (q_{a,n}/\theta_{a,n})\tau_{12,n}}$$

This step-by-step procedure points to two significant facts:

[4]A designation that need not concern the reader and which will not be mentioned again.

1. The heat dissipation at the tip of any fin in a cascade of n fins is *always* known. In the most remote fin, it is either zero or given by the tip dissipation. In all other fins, it is the heat injection into the next fin.

2. The input admittance at any particular fin can be determined by repeated applications of the *reflection relationship* [Eq. (5.6)].

This illustrates a computationally efficient procedure as an alternative to the matrix multiplication of the n **T**-elements that represent the n fins in the cascade:

$$\mathbf{T_e} = \mathbf{T_n T_{n-1} \cdots T_3 T_2 T_1}$$

In this case, Eq. (5.6) gives

$$Y_{in,n} = \frac{\tau_{21,e} + (q_{a,1}/\theta_{a,1})\tau_{22,e}}{\tau_{11,e} + (q_{a,1}/\theta_{a,1})\tau_{12,e}}$$

The foregoing seems to indicate an apparent *toss-up* between Eq. (5.6) and the use of $\mathbf{T_e}$. Indeed, the usefulness of the cascade algorithm may be concealed by the simplicity of Eq. (5.6).

Consider, however, that the determination of an array input admittance for any convective heat-transfer coefficient variation over the faces of any or all of the fins in the array is easily accomplished by an application of the cascade algorithm. One merely makes a piecewise continuous approximation (to the accuracy that is required) of the heat-transfer coefficient. This dictates the number of subfins that must be employed to represent the original fin. The elements of **T** (via $\mathbf{\Gamma}$) can then be obtained for each subfin using the average value of h in the individual segments of the h-curve. A repeated matrix multiplication then gives the elements of $\mathbf{T_e}$, and then Eq. (5.6), with the appropriate value of the tip q_a, will yield the required Y_{in}.

To be sure, efficiencies have been developed for various heat-transfer coefficient variations on longitudinal fins of rectangular profile. Gardner (1951) proposed a power series variation, Han and Lefkowitz (1960) used a parabolic variation, and Chen and Zyskowski (1963) employed an exponential variation. However, the cascade algorithm will efficiently handle convective coefficients of any distribution on the faces of any fin or spine. The Han and Lefkowitz study was verified, in part, by Kraus (1990) using the cascade algorithm. This will be reviewed in Chap. 6.

A summary of the input admittances for fins with and without tip heat loss and for each of the six matrix transformations is presented in Table 5.1.

Example 5.2: It was demonstrated in Ex. 3.2 that for a single longitudinal fin of rectangular profile dissipating heat from its tip under a heat-transfer coefficient, h_a, the heat entering the fin base is

$$q_b = \frac{k\delta L m\theta_b \left[\dfrac{h_a}{mk} + \tanh mb\right]}{1 + \dfrac{h_a}{mk}\tanh mb}$$

TABLE 5.1 Summary of input admittance relationships.

Matrix Representation	With tip Heat Loss	Without tip Heat Loss
Z	$\dfrac{1 - z_{22}Y_a}{z_{11} - \Delta_Z Y_a}$	$\dfrac{1}{z_{11}}$
Y	$\dfrac{\Delta_Y + y_{11}Y_a}{-y_{22} + Y_a}$	$\dfrac{\Delta_Y}{y_{22}}$
T	$\dfrac{\tau_{21} + \tau_{22}Y_a}{\tau_{11} + \tau_{12}Y_a}$	$\dfrac{\tau_{21}}{\tau_{11}}$
Γ	$\dfrac{-\gamma_{21} + \gamma_{11}Y_a}{\gamma_{22} - \gamma_{12}Y_a}$	$-\dfrac{\gamma_{21}}{\gamma_{22}}$
H	$\dfrac{-h_{22} + Y_a}{-\Delta_H + h_{11}Y_a}$	$\dfrac{h_{22}}{\Delta_H}$
G	$\dfrac{g_{11} - \Delta_G Y_a}{1 - g_{22}Y_a}$	g_{11}

Verify this result using the cascade algorithm.

SOLUTION The **T**-matrix for the longitudinal fin of rectangular profile is defined by Eq. (3.9):

$$\mathbf{T} = \begin{bmatrix} \cosh mb & Z_o \sinh mb \\ Y_o \sinh mb & \cosh mb \end{bmatrix} \tag{3.9}$$

where $Y_o = 1/Z_o = (2hk\delta)^{1/2}L$, and $m = (2h/k\delta)^{1/2}$. The tip heat dissipation is $q_a = Y_a\theta_a$, where $Y_a = h_a\delta L$

The tip-to-base transformation is in accordance with the cascade algorithm, with the tip heat dissipation handled by the single shunt conductance **T**-matrix introduced in Sec. 4.7:

$$\mathbf{T}_e = \begin{bmatrix} \tau_{11,e} & \tau_{12,e} \\ \tau_{21,e} & \tau_{22,e} \end{bmatrix} = \begin{bmatrix} \cosh mb & Z_o \sinh mb \\ Y_o \sinh mb & \cosh mb \end{bmatrix} \begin{bmatrix} 1 & 0 \\ h_a\delta L & 1 \end{bmatrix}$$

or, after the matrix multiplication,

$$\mathbf{T}_e = \begin{bmatrix} \tau_{11,e} & \tau_{12,e} \\ \tau_{21,e} & \tau_{22,e} \end{bmatrix} = \begin{bmatrix} \cosh mb + (h_a\delta L)Z_o \sinh mb & Z_o \sinh mb \\ Y_o \sinh mb + (h_a\delta L)\cosh mb & \cosh mb \end{bmatrix}$$

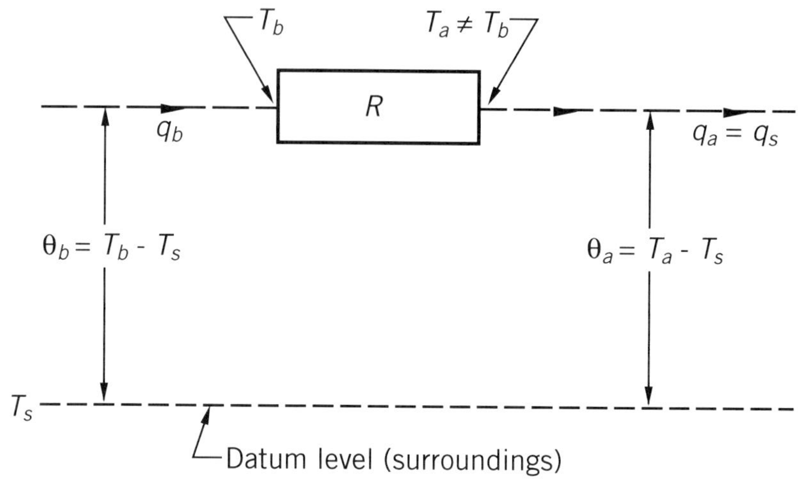

Figure 5.4 The single series resistance.

which makes

$$\tau_{11,e} = \cosh mb + (h_a \delta L) Z_o \sinh mb$$

and

$$\tau_{21,e} = Y_o \sinh mb + (h_a \delta L) \cosh mb$$

The discussion surrounding the development of the **T**-matrix for the single shunt conductance was based on the configuration shown in in Fig. 4.7, repeated here as Fig. 5.4. Notice that the actual conductance, G, accounts for the tip heat loss. However, the derivation in Sec. 4.7 was based on the fact that $q_a = 0$, so that, if the single shunt element is employed at the tip of a remote regular fin in the array, its use implies that the total equivalent entity of the fin and its tip heat loss is equivalent to a fin with no tip heat loss. Thus, as indicated in Table 5.1, the input admittance for the fin of this example can be written as

$$Y_{in} = \frac{\tau_{21,e}}{\tau_{22,e}} = \frac{Y_o \sinh mb + (h_a \delta L) \cosh mb}{\cosh mb + (h_a \delta L) Z_o \sinh mb}$$

Then with $h_a \delta L Z_o = h_a \delta L / Y_o = h_a / mk$ and with a factoring of $\cosh mb$ in both numerator and denominator, one obtains

$$Y_{in} = \frac{Y_o \tanh mb + h_a \delta L}{1 + (h_a/mk) \tanh mb}$$

and with $Y_o = (2hk\delta)^{1/2} L = k\delta L m$

$$Y_{in} = \frac{q_b}{\theta_b} = \frac{k\delta L m \left[\dfrac{h_a}{mk} + \tanh mb \right]}{1 + \dfrac{h_a}{mk} \tanh mb}$$

or

$$q_b = \frac{k\delta L m\theta_b\left[\dfrac{h_a}{mk} + \tanh mb\right]}{1 + \dfrac{h_a}{mk}\tanh mb}$$

5.4.1 The Fin Efficiency

If the array fin efficiency is desired, it can be obtained from the input admittance. The definition can be extended to an array, and it may be recalled that the fin efficiency is defined as the ratio of the actual heat dissipated to the ideal heat dissipated if the entire array were to operate at the base temperature excess.[5] The actual heat dissipation may be written in terms of the input admittance, $q_b = Y_{in}\theta_b$, and with the actual heat dissipation given by $q_b = hS\theta_b$, the relationship giving the connection between the two quantitites is

$$\eta = \frac{Y_{in}\theta_b}{hS\theta_b} = \frac{Y_{in}}{hS} \tag{5.7}$$

where S is the dissipating surface area of the array.

5.4.2 Bond Resistance

The bond resistance of several types of interference fit, high fin tubes (tubes with radial fins of pronounced rectangular profile), was studied by Gardner and Carnavos (1960). Kraus and Specht (1974) developed equations that permitted an adjustment of the fin efficiency when the thermal bond resistance between a longitudinal fin of rectangular profile and its base or prime surface is known. This study was based on the analogy between the fin and the electrical transmission line treated by Kern and Kraus (1972).

Manzoor et al. (1984) investigated the validity of the perfect contact assumption in the context of fin assembly heat transfer. Attention was restricted to the case of longitudinal fins of curved or tapered profile. Serebryanyi (1985) gave recommendations on the evaluation of the contact resistance for a bundle of rectangular fins attached to a module containing dissipating semiconductors.

Sheffield et al. (1985) examined surface characteristics of finned tube heat exchangers in order to help predict the contact geometry between the metal surfaces. Primary characteristics evaluated include the surface roughness, the surface waviness, and the material hardness. A scanning electron microscope was used to observe the surface typography of the fins as well as the actual contact between the aluminum fins and a copper tube. Tests revealed that only a portion of the fin collar was in actual contact with the tube. The results

[5]Or if the array were to be fabricated from some material that possessed an infinite thermal conductivity.

were subsequently used to predict a mathematical model for predicting the heat transfer between tube and fin.

If the bond resistance is known, the use of the single series element matrix permits an efficient evaluation of the effect. This is demonstrated in the following example.

***Example* 5.3:** Consider a longitudinal fin of rectangular profile with a height of 8 cm, a width of 0.3125 cm, and a length of 20 cm. The fin base is at 100°C, and the surroundings are at 20°C. The fin has a thermal conductivity of 100 W/m-K, and the heat-transfer coefficient between fin and surroundings is 40 W/m^2-K. If the tip of the fin is not dissipating, determine the heat flow through its base and the efficiency. Repeat these calculations using a single series element to account for the bond resistance between the fin and the prime surface if the fin has been bonded to the prime surface in a manner such that a the bond resistance is 0.325 K/W.

SOLUTION First, obtain m and Y_o:

$$m = \left(\frac{2h}{k\delta}\right)^{1/2} = \left[\frac{2(40)}{100(.003125)}\right]^{1/2} = 16.00\,\mathrm{m}^{-1}$$

and

$$Y_o = (2hk\delta)^{1/2}L = [2(40)(100)(.003125)]^{1/2}(0.20) = 1.00\,\mathrm{W/K}$$

Then with

$$mb = 16.00(0.08) = 1.28$$

the $\mathbf{T}$-matrix for the fin will be[6]

$$\mathbf{T} = \begin{bmatrix} \cosh mb & Z_o \sinh mb \\ Y_o \sinh mb & \cosh mb \end{bmatrix} = \begin{bmatrix} 1.9373 & 1.6593 \\ 1.6593 & 1.9373 \end{bmatrix}$$

With no bond resistance and with an adiabatic tip, use Table 5.1:

$$Y_{in} = \frac{q_b}{\theta_b} = \frac{\tau_{21}}{\tau_{11}} = \frac{1.6593}{1.9373} = 0.8565\,\mathrm{W/^\circ K}$$

and with $\theta_b = 100 - 20 = 80\,^\circ\mathrm{C}$,

$$q_b = Y_{in}\theta_b = 0.8565(80) = 68.52\,\mathrm{W}$$

The surface area of the fin is

$$S = 2bL = 2(0.08)(0.20) = 0.0320\,\mathrm{m}^2$$

[6]Because $Y_o = 1.00$ W/K, $\tau_{12} = \tau_{21}$.

so that

$$\eta = \frac{Y_{in}}{hS} = \frac{0.8565}{40(0.0320)} = 0.669$$

To determine the effect of the bond resistance, use the cascade algorithm with the single series element **T**-matrix that represents the bond resistance. With $R_b = 0.3250\,\text{K/W}$,

$$\mathbf{T}_e = \begin{bmatrix} 1.0000 & 0.3250 \\ 0 & 1.0000 \end{bmatrix} \begin{bmatrix} \cosh mb & Z_o \sinh mb \\ Y_o \sinh mb & \cosh mb \end{bmatrix}$$

$$= \begin{bmatrix} 1.0000 & 0.3250 \\ 0 & 1.0000 \end{bmatrix} \begin{bmatrix} 1.9373 & 1.6593 \\ 1.6593 & 1.9373 \end{bmatrix}$$

or

$$\mathbf{T}_e = \begin{bmatrix} 2.4766 & 2.2889 \\ 1.6593 & 1.9373 \end{bmatrix}$$

Then

$$Y_{in} = \frac{q_b}{\theta_b} = \frac{\tau_{21,e}}{\tau_{11,e}} = \frac{1.6593}{2.4776} = 0.6697\,\text{W/K}$$

and the thermal resistance is

$$R = \frac{1}{Y_{in}} = \frac{1}{0.6697} = 1.493\,\text{K/W}$$

The heat flow through the bond into the fin will be

$$q_b = Y_{in}\theta_b = 0.6697(80) = 53.58\,\text{W}$$

and

$$\eta = \frac{Y_{in}}{hS} = \frac{0.6697}{40(0.0320)} = 0.523$$

Clearly, the bond resistance is destroying the performance of the fin.

An alternative approach may be taken. The thermal resistance of the fin may be determined from the **T**-matrix for the fin:

$$\begin{bmatrix} 1.9373 & 1.6593 \\ 1.6593 & 1.9373 \end{bmatrix}$$

With an adiabatic fin tip, Eq. (5.6) gives

$$Y_{in} = \frac{\tau_{21}}{\tau_{22}} = \frac{1.6593}{1.9373} = 0.8565\,\text{K/W}$$

and this converts to a thermal resistance of

$$R_f = \frac{1}{Y_{in}} = \frac{1}{0.8565} = 1.168 < 1.493\,\text{K/W}$$

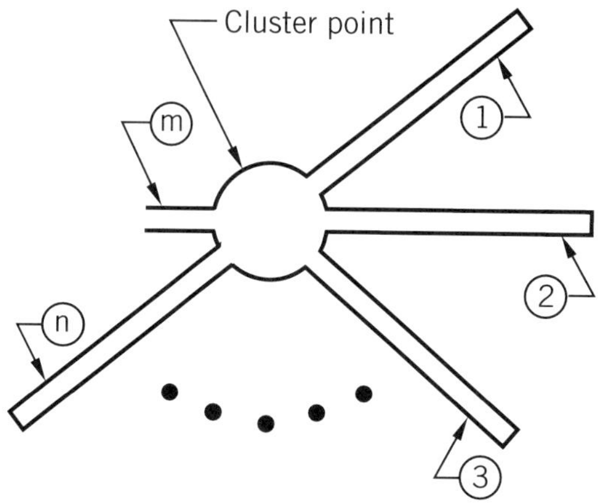

Figure 5.5 Fins in cluster. There are n fins attached to the tip of fin m, and the tip of fin m is designated as the *cluster point*.

The bond resistance is $R_b = 0.3250\,\text{K/W}$, and the the total thermal resistance for the bond-fin entity is the sum of the two resistances in series:

$$R = R_b + R_f = 0.325 + 1.168 = 1.493\,\text{K/W}$$

This confirms the previous result obtained by the cascade algorithm using the **T**-matrix for the single series element. It shows that there are two methods for the evaluation of the bond resistance. Both depend on the evaluation of the **T**-matrix for the fin.

5.4.3 The Thermal Transmission Ratio

The input admittance and the thermal transmission ratio are defined in exactly the same manner: the ratio of the heat entering the base of the fin to the temperature excess existing at the base of the fin. The difference between them lies in their usage. The input admittance may be used to describe the performance of an individual fin or a finned array. The thermal transmission ratio is meant to be used to describe the performance of a singular fin or spine that, because of a taper to a zero cross section at the tip, can only be used as the most remote entity in an array or portion of an array.

5.5 FINS IN CLUSTER

Figure 5.5 shows n fins that are appended to fin m. These fins are said to be in a *cluster arrangement*, and the tip of fin m is called a *cluster point*. Here, continuity dictates that the heat that leaves fin m must divide among fins 1 through n. Moreover, compatibility dictates that the temperature excesses of all fins involved in the cluster must match. Thus, with the use of subscripts a and b, which has now become customary, it is observed that

$$q_{am} = q_{b1} + q_{b2} + q_{b3} + \cdots + q_{bn} \tag{5.8a}$$

and

$$\theta_{am} = \theta_{b1} = \theta_{b2} = \theta_{b3} = \cdots = \theta_{bn} \tag{5.8b}$$

Consider the ratio of heat flow to the temperature excess at the tip of fin m and call this the cluster admittance, Y_c:

$$Y_c = Y_{am} = \frac{q_{am}}{\theta_{am}}$$

and employ Eqs. (5.8). This yields

$$Y_c = Y_{am} = \frac{q_{b1} + q_{b2} + q_{b3} + \ldots + q_{bn}}{\theta_{am}}$$

or

$$Y_c = Y_{am} = \frac{q_{b1}}{\theta_{b1}} + \frac{q_{b2}}{\theta_{b2}} + \frac{q_{b3}}{\theta_{b3}} + \ldots + \frac{q_{bn}}{\theta_{bn}} \tag{5.9}$$

Each term on the right side of Eq. (5.9) represents an individual fin input admittance or a thermal transmission ratio. Thus, a cluster admittance may be defined as

$$Y_c = \sum_{i=1}^{n} Y_i = \sum_{i=1}^{n} \mu_i \tag{5.10}$$

This addition of input admittances to form the cluster input admittance is the cluster algorithm.

Example 5.4: An analysis of the single compact heat exchanger stack with heat input on one side shown in Fig. 5.6 was given via a laborious procedure by Kern and Kraus (1972). An extension of this material gives the expression for the heat entering the base of the stack for *unit length* as

$$q_{b3} = k\delta_3 m\theta_{b3} \left[\frac{\sinh mb_3 + (2\delta_2/\delta_3)^{1/2} \tanh nb_1 \cosh mb_3}{\cosh mb_3 + (2\delta_2/\delta_3)^{1/2} \tanh nb_1 \sinh mb_3} \right]$$

where $m = (2h/k\delta_3)^{1/2}$, $n = (h/k\delta_1)^{1/2}$, and $b_1 = b_2$. If $\delta_1 = \delta_2 = \delta_3 = \delta$ and $b_1 = b_2 = b_3 = b$, use the cluster and cascade algorithms to show that

$$q_{b3} = k\delta m\theta_{b3} \left[\frac{\sinh mb + \sqrt{2} \tanh nb \cosh mb}{\cosh mb + \sqrt{2} \tanh nb \sinh mb} \right]$$

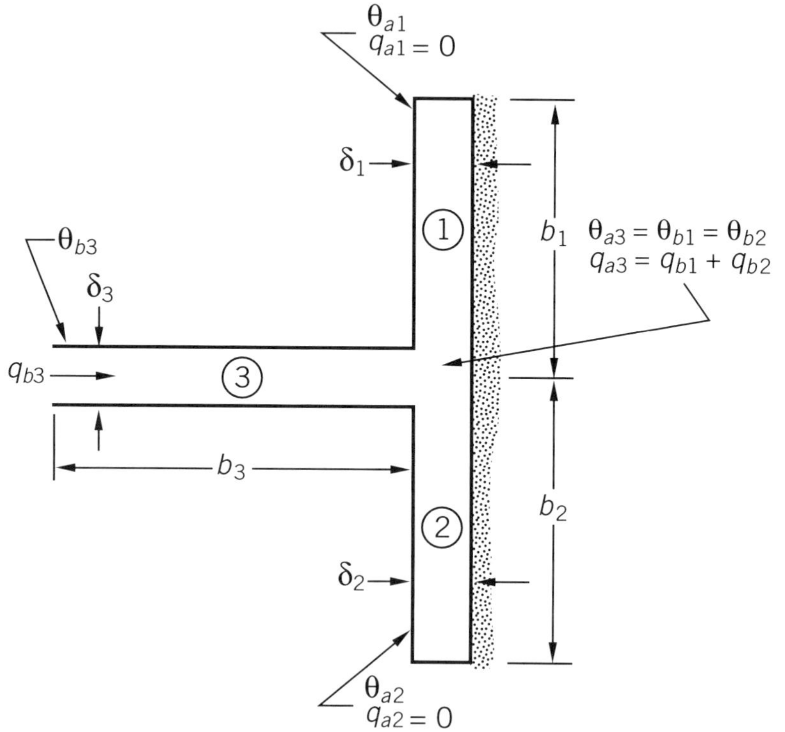

Figure 5.6 A single compact heat exchanger stack with heat input at the base. The shaded area is used to show that fins 1 and 2 dissipate heat from only one face.

SOLUTION For fin 3, the **T**-matrix for the longitudinal fin of rectangular profile is given by Eq. (3.9):

$$\mathbf{T}_3 = \begin{bmatrix} \tau_{11,3} & \tau_{12,3} \\ \tau_{21,3} & \tau_{22,3} \end{bmatrix} = \begin{bmatrix} \cosh mb & Z_{o3} \sinh mb \\ Y_{o3} \sinh mb & \cosh mb \end{bmatrix}$$

where, for unit length, $Y_{o3} = 1/Z_{o3} = (2hk\delta)^{1/2}$ and $m = (2h/k\delta)^{1/2}$.

For fins 1 and 2, Eq. (3.9) also applies but, because these fins dissipate from only one lateral surface, $n = (h/k\delta)^{1/2}$ must be substituted for m. Thus, for unit length,

$$\mathbf{T}_1 = \mathbf{T}_2 = \begin{bmatrix} \cosh nb & Z_{o2} \sinh nb \\ Y_{o2} \sinh nb & \cosh nb \end{bmatrix}$$

but here, again for unit length, $Y_{o2} = 1/Z_{o2} = (hk\delta)^{1/2}$.

Fins 1 and 2 are in cluster at the tip of fin 3 and they have non-dissipating tips. Table 5.1 gives their input admittance as

$$Y_{in,1} = Y_{in,2} = \frac{\tau_{21,2}}{\tau_{11,2}} = (hk\delta)^{1/2} \tanh nb$$

By the cluster algorithm of Eq. (5.10),

$$Y_{a3} = Y_{in,1} + Y_{in,2} = 2(hk\delta)^{1/2} \tanh nb$$

and because

$$2(hk\delta)^{1/2} = \sqrt{2}(2hk\delta)^{1/2} = \sqrt{2}Y_{o3}$$

$$Y_{a3} = \frac{q_{a3}}{\theta_{a3}} = Y_{in,1} + Y_{in,2} = \sqrt{2}Y_{o3} \tanh nb$$

The rest of the procedure is straightforward. Either an equivalent **T**-matrix can be developed after putting the foregoing cluster admittance into a single shunt matrix or the reflection relationship of Eq. (5.6) can be used. Adopting the latter idea,

$$Y_{in,3} = \frac{\tau_{21,3} + Y_{a3}\tau_{22,3}}{\tau_{11,3} + Y_{a3}\tau_{12,3}}$$

and with the τ's and Y_{a3} substituted,

$$Y_{in,3} = \frac{q_{b3}}{\theta_{b3}} = \frac{Y_{o3} \sinh mb + \sqrt{2}Y_{o3} \tanh nb \cosh mb}{\cosh mb + \sqrt{2}Z_{o3}Y_{o3} \tanh nb \sinh mb}$$

When $Y_{o3} = (2hk\delta)^{1/2} = k\delta m$ is factored in the numerator, the desired result is obtained

$$q_{b3} = k\delta m\theta_{b3} \left[\frac{\sinh mb + \sqrt{2} \tanh nb \cosh mb}{\cosh mb + \sqrt{2} \tanh nb \sinh mb} \right]$$

Of course, an identical result would come from the use of Table 5.1.

5.6 FINS IN PARALLEL

A parallel arrangement of n fins is shown in Fig. 5.7. Notice that a single fin, designated as fin m is delivering heat through its tip to the parallel combination, and that a single fin, designated as fin p, is picking up, at its base, all of the heat leaving the parallel combination. In this case, continuity demands that

$$q_{am} = q_{b1} + q_{b2} + q_{b3} + \ldots + q_{bn} \tag{5.11a}$$

and

$$q_{bp} = q_{a1} + q_{a2} + q_{a3} + \ldots + q_{an} \tag{5.11b}$$

In addition, compatibility requires that at point m

$$\theta_{am} = \theta_{b1} = \theta_{b2} = \theta_{b3} = \ldots = \theta_{bn} \tag{5.12a}$$

and at point p

$$\theta_{bp} = \theta_{a1} = \theta_{a2} = \theta_{a3} = \ldots = \theta_{an} \tag{5.12b}$$

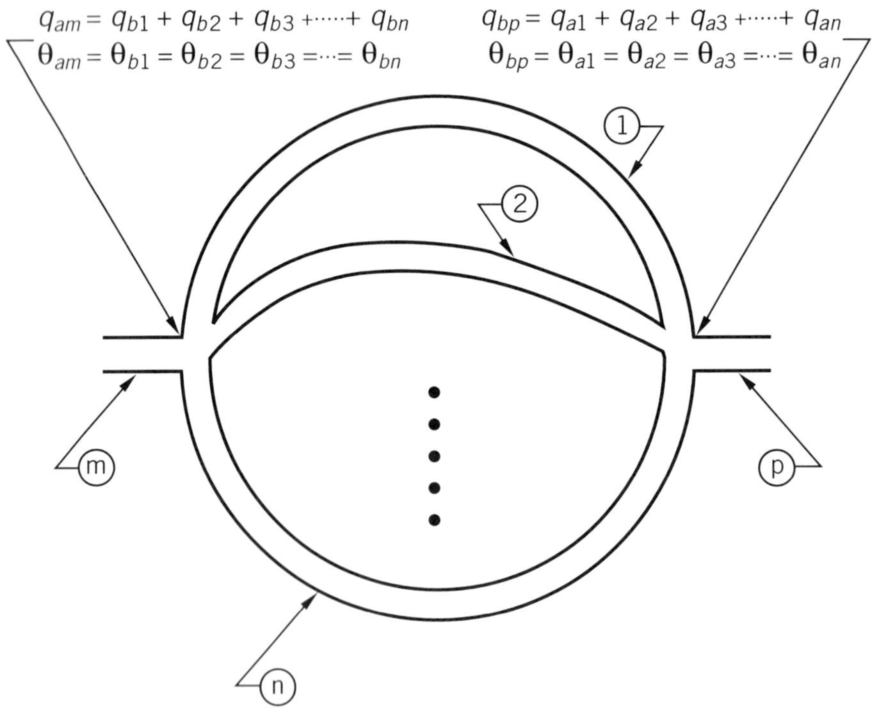

Figure 5.7 Fins in parallel.

Use of the **Y**-matrix, which will be referred to as *the thermal admittance matrix*, defined by Eq. (2.13) [reproduced here as Eq. (5.13)]

$$\begin{bmatrix} q_b \\ q_a \end{bmatrix} = \mathbf{Y} \begin{bmatrix} \theta_b \\ \theta_a \end{bmatrix} = \begin{bmatrix} y_{11} & y_{12} \\ y_{21} & y_{22} \end{bmatrix} \begin{bmatrix} \theta_b \\ \theta_a \end{bmatrix} \tag{5.13}$$

permits a linear transformation from temperature excesses to heat flows for any individual fin. Thus, from Eqs. (5.11), using fins 1, 2, $\cdots n$

$$\begin{bmatrix} q_{am} \\ q_{bp} \end{bmatrix} = \begin{bmatrix} q_{b1} \\ q_{a1} \end{bmatrix} + \begin{bmatrix} q_{b2} \\ q_{a2} \end{bmatrix} + \ldots + \begin{bmatrix} q_{bn} \\ q_{an} \end{bmatrix}$$

and the use of Eq. (5.13) provides

$$\begin{bmatrix} q_{am} \\ q_{bp} \end{bmatrix} = \mathbf{Y}_1 \begin{bmatrix} \theta_{b1} \\ \theta_{a1} \end{bmatrix} + \mathbf{Y}_2 \begin{bmatrix} \theta_{b2} \\ \theta_{a2} \end{bmatrix} + \ldots + \mathbf{Y}_n \begin{bmatrix} \theta_{bn} \\ \theta_{an} \end{bmatrix} \tag{5.14}$$

But, by Eqs. (5.12)

$$\begin{bmatrix} \theta_{am} \\ \theta_{bp} \end{bmatrix} = \begin{bmatrix} \theta_{b1} \\ \theta_{a1} \end{bmatrix} = \begin{bmatrix} \theta_{b2} \\ \theta_{a2} \end{bmatrix} = \ldots = \begin{bmatrix} \theta_{bn} \\ \theta_{an} \end{bmatrix}$$

and when this is incorporated into Eq. (5.14)

$$\begin{bmatrix} q_{am} \\ q_{w\alpha} \end{bmatrix} = [\mathbf{Y}_1 + \mathbf{Y}_2 + \ldots + \mathbf{Y}_n] \begin{bmatrix} \theta_{am} \\ \theta_{bp} \end{bmatrix} \tag{5.15}$$

This shows that there is an equivalent thermal admittance matrix, $\mathbf{Y}_e$, that relates the total heat flows to the temperature excesses at the base and tip of a parallel combination of fins:

$$\begin{bmatrix} q_b \\ q_a \end{bmatrix} = \mathbf{Y}_e \begin{bmatrix} \theta_b \\ \theta_a \end{bmatrix}$$

where

$$\mathbf{Y}_e = \mathbf{Y}_1 + \mathbf{Y}_2 + \mathbf{Y}_3 + \ldots + \mathbf{Y}_n \tag{5.16}$$

Equation (5.16) is the parallel algorithm.

Example 5.5: Show that two identical fins with width δ, with adjacent faces not dissipating, is formally equivalent to a single fin of width 2δ, with both faces dissipating.

SOLUTION First consider the *double width* fin whose $\mathbf{Y}$-matrix is given by Eq. (3.11):

$$\mathbf{Y} = \begin{bmatrix} Y_o \coth mb & -Y_o \operatorname{csch} mb \\ Y_o \operatorname{csch} mb & -Y_o \coth mb \end{bmatrix}$$

where, because the double-width fin is dissipating from both faces,

$$m = \left(\frac{2h}{k(2\delta)} \right)^{1/2} = \left(\frac{2h}{\delta} \right)^{1/2}$$

and

$$Y_o = (4hk\delta)^{1/2} L = 2(hk\delta)^{1/2} L$$

Each of the two fins of width δ and one face insulated is governed by the same $\mathbf{Y}$-matrix with $m = (h/k\delta)^{1/2}$, which is the same m as for the *double fin*. Moreover, for the case of the fin with width δ, $Y_o = (hk\delta)^{1/2} L$, and this is observed to be just one-half of the value for the double fin. Thus, if Y_o is assigned as the value of the double-thickness fin, the value for the single-thickness fin will be $Y_o/2$, and the $\mathbf{Y}$-matrix for the single-thickness fin will be

$$\mathbf{Y} = \frac{1}{2} \begin{bmatrix} Y_o \coth mb & -Y_o \operatorname{csch} mb \\ Y_o \operatorname{csch} mb & -Y_o \coth mb \end{bmatrix}$$

and by the parallel algorithm, the $\mathbf{Y}$-matrix for the double-thickness fin is just twice that for the single-thickness fin.

5.7 THE CHOKING PHENOMENON AND ARRAY OPTIMIZATION

The fin or spine that accommodates a cluster at its tip may exhibit a heat flow choke, because the fin or spine dimensions may not permit, in a cooling application, the transfer of a sufficient amount of heat to the cluster at its tip.

In the case of the longitudinal fin of rectangular profile, Eq. (5.6) can be written with $Y_a = q_a/\theta_a$ as

$$Y_{in} = \frac{Y_o \sinh mb + Y_a \cosh mb}{\cosh mb + (Y_a/Y_o) \sinh mb} \tag{5.17}$$

If numerator and denominator are divided by $\cosh mb$, the result is

$$Y_{in} = \frac{Y_o \tanh mb + Y_a}{1 + (Y_a/Y_o) \tanh mb}$$

and it may be observed that

$$\lim_{b \to 0} Y_{in} = Y_a$$

and

$$\lim_{b \to \infty} Y_{in} = Y_o$$

These provide the extremes for Y_{in} for the range of values of the fin height b.

In order to find the fin height that makes Y_{in} a maximum, the customary method where $\partial Y_{in}/\partial b$ vanishes is employed. In Eq. (5.17), once again observe that neither Y_o nor Y_a is a function of b, and let the numerator and denominator be designated by u and v:

$$\begin{aligned} u &= Y_o \sinh mb + Y_a \cosh mb \\ v &= \cosh mb + Y \sinh mb \end{aligned}$$

where $Y = Y_a/Y_o$. Then

$$\frac{\partial Y_{in}}{\partial b} = \frac{vm\,[Y_o \cosh mb + Y_a \sinh mb] - um\,[\sinh mb + Y \cosh mb]}{v^2} = 0$$

The m's in the numerator and the v^2 in the denominator may be canceled and thus

$$v\,[Y_o \cosh mb + Y_a \sinh mb] = u\,[\sinh mb + Y \cosh mb]$$

and with v and u substituted, the result, after some algebraic tedium is

$$Y_o = YY_a = \frac{Y_a^2}{Y_o}$$

or

$$Y_o = Y_a \tag{5.18}$$

Because Eq. (5.18) does not involve b, it implies that the slope of $Y_{in} = f(b)$ is never zero unless $Y_o = Y_a$, in which case it is always zero. Thus, there is no question of optimizing with respect to fin height and three cases are evident.

Case I $Y_o < Y_a$: When the characteristic admittance is less than the tip admittance, the fin is choking the heat flow regardless of its height. In this case, any fix cannot involve the fin height b. But, to save metal, b should be as low as possible ($b \to 0$).

Case II $Y_o > Y_a$: In this case, no choking occurs, and b should be made as high as possible because a benefit exists for high b ($b \to \infty$).

Case III $Y_o = Y_a$: In this case, the value of b has no effect on the heat-transfer process.

This conclusion may appear somewhat surprising, but it is certainly useful as a design guideline. For the longitudinal fin of rectangular profile, $Y_o = (2hk\delta)^{1/2}L$, so that to achieve the optimum

$$\delta = \frac{Y_a^2}{2hkL^2} \tag{5.19}$$

and for the cylindrical spine where $Y_o = (\pi^2 hkd^3/4)^{1/2}$, the optimum diameter will be given by

$$d = \left[\frac{4Y_a^2}{\pi^2 hk}\right]^{1/3} \tag{5.20}$$

5.8 CLOSURE

The algorithms developed in this chapter can be used to quickly analyze a complicated finned array via a computationally efficient procedure. Chap. 6, which now follows, will be devoted to some examples of their use. These examples may indicate an example of the choking phenomenon somewhere in the array and will consider the remedies that can be taken to eliminate the *heat flow choke*.

It must be remembered that the algorithms presented in this chapter for the combination of fins in an array apply only to arrays that do not contain loops in the graph theoretical sense. For such arrays, the method of node analysis, to be discussed in Chap. 7 or the general array algorithm, considered in Chap. 8, must be employed.

Chapter

6

EXAMPLES OF FINNED ARRAY ANALYSIS

6.1 INTRODUCTION

As heat rejection systems become more and more complicated, computationally efficient methods must be developed for their analysis and evaluation. The first five chapters of this book form the basis of the analysis of complex finned arrays that do not contain loops in the graph theoretical sense and do not have multiple heat inputs (inputs at several points in the array). This chapter provides some examples of the use of the algorithms for array combination in Chap. 5. Several comprehensive examples will be presented, and these examples include a discussion of the choking phenomena. The chapter concludes with a discussion of individual fin analysis with a variable heat-transfer coefficient.

6.2 AN EXAMPLE OF HEAT SINK ANALYSIS

Typical of a wide selection of commercial heat sinks available for the temperature control of electronic equipment is the array displayed in Fig. 6.1. The electronic components are mounted on a shelf to which two sinks of the form of Fig. 6.1 are attached, and air is forced through the finned array into the plane of the paper.

Example 6.1: Consider the finned array containing 10 fins shown in Fig. 6.1. The tips of fins 1, 4, 5, 7, and 8 may be considered as adiabatic. The array is fabricated of aluminum ($k = 202\,\mathrm{W/m\text{-}K}$), and it is used in an application where $h = 101\,\mathrm{W/m^2\text{-}K}$. All fins are longitudinal fins of rectangular profile with a length (into the plane of the paper) of 7.62 cm. All other dimensions

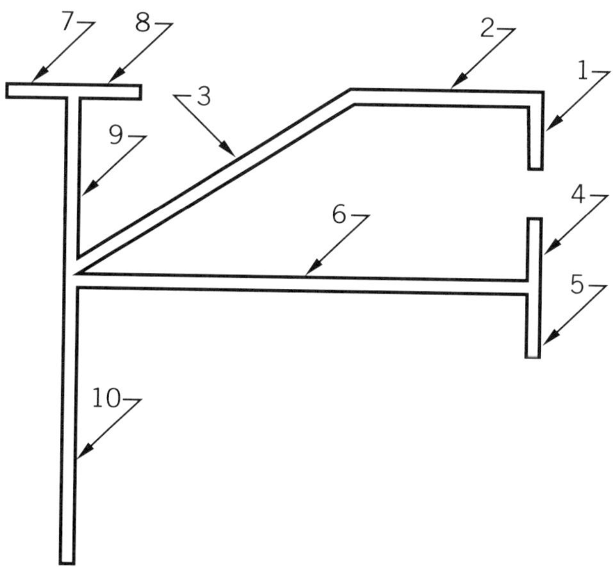

Figure 6.1 Electronic Equipment Heat Sink.

are shown in Table 6.1. A design decision is required as to whether this array can dissipate 91.2 W at a temperature excess of $\theta_b = 38\,^\circ\text{C}$. The temperature excess is measured between the base of fin 10 and the environment.

SOLUTION The first step is to determine the **T**-matrix for all fins. This can be done by using the representation of Eqs. (3.9):

$$\begin{aligned}
\tau_{11} &= \cosh mb \\
\tau_{12} &= (1/Y_o)\sinh mb \\
\tau_{21} &= Y_o \sinh mb \\
\tau_{22} &= \cosh mb
\end{aligned}$$

and writing a short program for a hand held calculator or a computer or by using some commercially available software. The parameters for all of the fins in Fig. 6.1 are shown in Table 6.2. They require a computation of the characteristic admittance

$$Y_o = (2hk\delta)^{1/2}L$$

and the parameter

$$mb = \left(\frac{2h}{k\delta}\right)^{1/2} b$$

These are also shown for each fin in Table 6.2.

The detailed procedure involves a repeated use of the reflection relationship of Eq. (5.6), with appropriate cognizance of the cluster points:

TABLE 6.1 Heights and Thicknesses of Fins in Fig. 6.1. (All dimensions are in centimeters)

Fin No.	b, cm	δ, cm
1	1.250	0.3175
2	3.000	0.3175
3	6.000	0.3175
4	1.250	0.3175
5	1.250	0.3175
6	8.196	0.3175
7	1.250	0.3175
8	1.250	0.3175
9	3.000	0.3175
10	6.000	0.4572

TABLE 6.2 Characteristic Admittance and τ-parameters for all Fins in Fig. 6.1.

Fin	mb	Y_o, $^oC/W$	$\tau_{11} = \tau_{22}$	τ_{12}, $^oC/W$	τ_{21}, $W/^oC$
1	0.2218	0.8673	1.0247	0.2579	0.1940
2	0.5324	0.8673	1.1451	0.6433	0.4839
3	1.0648	0.8673	1.6226	1.4732	1.1082
4	0.2218	0.8673	1.0247	0.2579	0.1940
5	0.2218	0.8673	1.0247	0.2579	0.1940
6	1.4546	0.8673	2.2580	2.3342	1.7559
7	0.2218	0.8673	1.0247	0.2579	0.1940
8	0.2218	0.8673	1.0247	0.2579	0.1940
9	0.5324	0.8673	1.1451	0.6433	0.4839
10	0.8874	1.0408	1.4202	0.9690	1.0496

1. Fins 1, 2, and 3: Fin 1 is remote and has an adiabatic tip. While the use of the reflection relationship is called for, the input admittance at the base of fin 3 can be obtained from the cascade algorithm of Eq. (5.5):

$$\mathbf{T} = \mathbf{T}_3\mathbf{T}_2\mathbf{T}_1$$

$$= \begin{bmatrix} 1.6226 & 1.4732 \\ 1.1082 & 1.6226 \end{bmatrix} \begin{bmatrix} 1.1451 & 0.6433 \\ 0.4839 & 1.1451 \end{bmatrix} \begin{bmatrix} 1.0247 & 0.2579 \\ 0.1940 & 1.0247 \end{bmatrix}$$

$$= \begin{bmatrix} 2.5709 & 2.7308 \\ 2.0542 & 2.5709 \end{bmatrix} \begin{bmatrix} 1.0247 & 0.2579 \\ 0.1940 & 1.0247 \end{bmatrix}$$

$$= \begin{bmatrix} 3.1642 & 3.4612 \\ 2.6037 & 3.1642 \end{bmatrix}$$

Then

$$Y_{in,3} = \frac{\tau_{21,e}}{\tau_{11,e}} = \frac{2.6037}{3.1642} = 0.8229\,\text{W/K}$$

This is easily confirmed by the reflection relationship of of Eq. (5.6). With $q_{a,1} = 0$,

$$Y_{in,1} = \frac{\tau_{21,1}}{\tau_{11,1}} = \frac{0.1940}{1.0247} = 0.1893\,\text{W/K}$$

so that

$$\begin{aligned} Y_{in,2} &= \frac{\tau_{21,2} + Y_{a,2}\tau_{22,2}}{\tau_{11,2} + Y_{a,2}\tau_{12,2}} \\ &= \frac{0.4839 + 0.1893(1.1451)}{1.1451 + 0.1893(0.6433)} \\ &= \frac{0.7007}{1.2669} = 0.5531\,\text{W/K} \end{aligned}$$

and

$$\begin{aligned} Y_{in,3} &= \frac{\tau_{21,3} + Y_{a,3}\tau_{22,3}}{\tau_{11,3} + Y_{a,3}\tau_{12,3}} \\ &= \frac{1.1082 + 0.5531(1.6226)}{1.6226 + 0.5531(1.4732)} \\ &= \frac{2.0056}{2.4374} = 0.8229\,\text{W/K} \end{aligned}$$

2. Fins 4, 5, and 6: Fins 4 and 5, with presumed adiabatic tips, are in cluster at the tip of fin 6. Each has an input admittance of

$$Y_{in,4} = Y_{in,5} = \frac{\tau_{21,4}}{\tau_{11,4}} = \frac{0.1940}{1.0247} = 0.1893\,\text{W/K}$$

Thus, at the tip of fin 6, by the cluster algorithm of Eq. (5.10),

$$\begin{aligned} Y_{a,6} &= \frac{\theta_{a,6}}{q_{a,6}} \\ &= Y_{in,4} + Y_{in,5} \\ &= 2(0.1893) = 0.3786\,\text{W/K} \end{aligned}$$

and then by the reflection relationship

$$\begin{aligned} Y_{in,6} &= \frac{\tau_{21,6} + Y_{a,6}\tau_{22,6}}{\tau_{11,6} + Y_{a,6}\tau_{12,6}} \\ &= \frac{1.7559 + 0.3786(2.2580)}{2.2580 + 0.3786(2.3342)} \\ &= \frac{2.6109}{3.1418} = 0.8310\,\text{W/K} \end{aligned}$$

3. Fins 7, 8, and 9: Fins 7 and 8, with presumed adiabatic tips, are in cluster at the tip of fin 6. Each has an input admittance of

$$Y_{in,7} = Y_{in,8} = \frac{\tau_{21,7}}{\tau_{11,7}} = \frac{0.1940}{1.0247} = 0.1893\,\text{W/K}$$

Thus, at the tip of fin 9, by the cluster algorithm of Eq. (5.10),

$$\begin{aligned} Y_{a,9} &= \frac{\theta_{a,9}}{q_{a,9}} \\ &= Y_{i7} + Y_{i8} \\ &= 2(0.1893) = 0.3786\,\text{W/K} \end{aligned}$$

and again by the reflection relationship

$$\begin{aligned} Y_{in,9} &= \frac{\tau_{21,9} + Y_{a,9}\tau_{22,9}}{\tau_{11,9} + Y_{a,9}\tau_{12,9}} \\ &= \frac{0.4839 + 0.3786(1.1451)}{1.1451 + 0.3786(0.6433)} \\ &= \frac{0.9175}{1.3887} = 0.6607\,\text{W/K} \end{aligned}$$

4. Fin 10: Fins 3, 6, and 9 are clustered at the tip of fin 10. Here the cluster algorithm of Eq. (5.10) is employed to give the cluster admittance:

$$\begin{aligned} Y_{a,10} &= \frac{\theta_{a,10}}{q_{a,10}} \\ &= Y_{in,3} + Y_{in,6} + Y_{in,9} \\ &= 0.8229 + 0.8310 + 0.6607 = 2.3145\,\text{W/K} \end{aligned}$$

Note from Table 6.2 that

$$Y_{o,10} = 1.0408\,\text{W/K}$$

so that there will be a heat flow choke. This can be verified by an application of the reflection relationship of Eq. (5.6) to find the input admittance for the entire array.

$$\begin{aligned} Y_{in,10} &= \frac{\tau_{21,10} + Y_{a,10}\tau_{22,10}}{\tau_{11,10} + Y_{a,10}\tau_{12,10}} \\ &= \frac{1.0496 + 2.3145(1.4202)}{1.4202 + 2.3145(0.9690)} \\ &= \frac{4.3368}{3.6629} = 1.1840\,\text{W/°C} \end{aligned}$$

5. Assessment: The specification requires that the array dissipate 91.2 W at a base temperature excess of 38°C. This requires an input admittance of

$$Y_{in} = \frac{q_b}{\theta_b} = \frac{91.2}{38} = 2.4000\,\text{W/K}$$

This can almost be attained *without the presence of fin 10.*

Some Observations The array, constituted as shown in Fig. 6.1, but without fin 10, almost satisfies the specification. The requirement for the entire array is $q_b/\theta_b = 2.4000\,\text{W/K}$. The array input admittance at the cluster of fins 3, 6 and 9 is $Y_{in,9} = 2.3145\,\text{W/K}$. Thus, there is no thought of removing fin 10 entirely, because the array composed of fins 1 through 9 will not quite meet the specification.

The problem, therefore, is to make an adjustment at the cluster point to force the array to meet the specification. The calculations certainly indicate that the presence of fin 10, in its present form, only makes matters worse because of the heat flow choke.

There are two choices for fin 10. Either make fin 10 thicker or add fins in parallel with fin 10. Both of these alternatives are examined in the next section.

6.3 REMOVAL OF THE CHOKE

The first step in an attempt to remove the heat flow choke in the array of Ex. 6.1 is to try to increase the thickness of fin 10. The relationship to be used is Eq. (5.19):

$$\delta = \frac{Y_a^2}{2hkL^2} \tag{5.19}$$

which provides the fin thickness that removes the choke. However, this value of δ is nothing more that the "break-even" value that permits $Y_{in} = Y_a = Y_o$.

With the parameters of Example 6.1

$$\delta = \frac{(2.3145)^2}{2(101)(202)(0.0762)^2} = 0.02261\,\text{m}$$

It is obvious that an attempt to increase the fin thickness to 2.261 cm will be treated with disdain. Among other practical aspects, a fin this thick will violate the Murray Gardner assumptions. Of course, considerable benefit would accrue from an increase of the length of all fins to a value above 7.62 cm.

When an increase of the thickness of the fin that is the culprit in a choking situation is impossible, a change of the fin configuration may have merit. Consider Fig. 6.2, which shows a modification of the array of Fig. 6.1. Notice that the single fin 10 has been augmented by four fins in parallel and these are designated as fins 11 through 14. These are all identical and possess a height of 4.2426 cm and a thickness of 0.3175 cm. Each will have a value for mb:

$$\begin{aligned} mb &= \left(\frac{2h}{k\delta}\right)^{1/2} b \\ &= \left[\frac{2(101)}{202(0.003175)}\right]^{1/2} (0.04243) \\ &= 0.7529 \end{aligned}$$

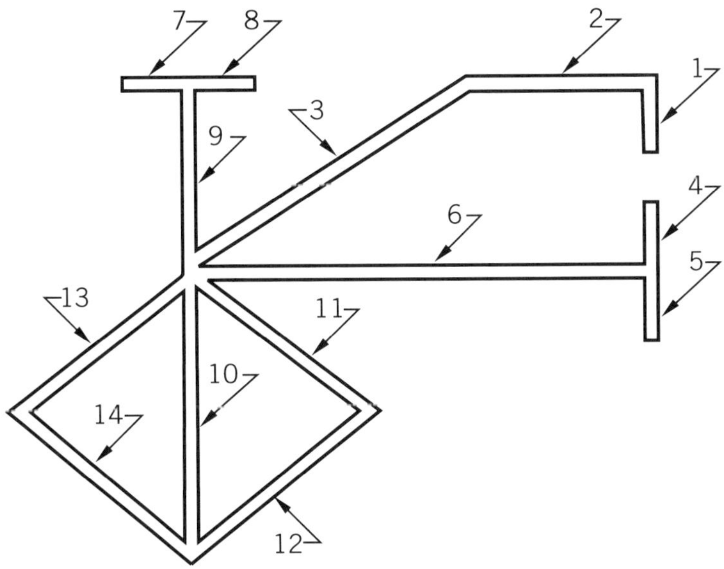

Figure 6.2 The electronic equipment heat sink of Fig. 5.1 with a modification. In this configuration, the parallel combination at the base of the array includes fin 10 in Fig. 5.1. The additional fins are fins 11 through 14. It should be noted that fin 10 now has a width of 0.3175cm.

and a characteristic admittance of

$$\begin{aligned} Y_o &= (2hk\delta)^{1/2}L \\ &= [2(101)(202)(0.003175)]^{1/2}(0.0762) = 0.8673\,\mathrm{W/K} \end{aligned}$$

With these in hand, Eqs. (3.9) provide the **T**-matrices for each of the fins in the parallel configuration. All of these will be equal:

$$\mathbf{T} = \begin{bmatrix} 1.2971 & 0.9525 \\ 0.7165 & 1.2971 \end{bmatrix}$$

The strategy is straightforward:

1. Use two **T**-matrices in cascade to evaluate an equivalent **T**-matrix for the two parallel legs containing two fins.

2. Use Table 2.2 to convert the equivalent **T**-matrix to an equivalent **Y**-matrix.

3. There are two identical parallel legs. Therefore, an equivalent **Y**-matrix for both legs can be obtained by multiplying the **Y**-matrix found in step 2 by 2.

4. The **Y**-matrix for fin 10 can be obtained directly from Eqs. (3.11).

5. Add the **Y**-matrix for fin 10 to the equivalent **Y**-matrix for fins 11 to 14. This will give an equivalent **Y**-matrix for the combination of fins 10 to 14.

6. Once again, use Table 2.2. This time go from the equivalent **Y**-matrix to an equivalent **T**-matrix for the entire parallel combination of all five fins.

7. Repeat the last step in Ex. 6.1 and note the improvement. One has no guarantee that this procedure will completely eliminate the choke.

The foregoing strategy is implemented quite easily:

1. The equivalent **T**-matrix for each leg is obtained from a simple matrix multiplication. Call the equivalent $\mathbf{T_X}$,

$$\mathbf{T_X} = \mathbf{T}^2 = \begin{bmatrix} 1.2971 & 0.9525 \\ 0.7165 & 1.2971 \end{bmatrix}^2 = \begin{bmatrix} 2.3650 & 2.4710 \\ 1.8588 & 2.3650 \end{bmatrix}$$

2. Use Table 2.2 to obtain $\mathbf{Y_X}$. First find det $\mathbf{T_X}$:

$$\det \mathbf{T_X} = \Delta_T = \begin{vmatrix} 2.3650 & 2.4710 \\ 1.8588 & 2.3650 \end{vmatrix} = 1.0000$$

Then from Table 2.2,

$$y_{11,x} = \frac{\tau_{22}}{\tau_{12}} = \frac{2.3650}{2.4710} = 0.9571\,\mathrm{W/K}$$

$$y_{12,x} = -\frac{\Delta_T}{\tau_{12}} = -\frac{1}{2.4710} = -0.4047\,\mathrm{W/K}$$

$$y_{21,x} = \frac{1}{\tau_{12}} = \frac{1}{2.4710} = 0.4047\,\mathrm{W/K}$$

and

$$y_{22,x} = -\frac{\tau_{11}}{\tau_{12}} = -\frac{2.3650}{2.4710} = -0.9571\,\mathrm{W/K}$$

This makes

$$\mathbf{Y_X} = \begin{bmatrix} 0.9571 & -0.4047 \\ 0.4047 & -0.9571 \end{bmatrix} \mathrm{W/K}$$

3. The equivalent **Y**-matrix for the entire parallel combination is

$$\mathbf{Y_e} = 2\mathbf{Y_X} = 2\begin{bmatrix} 0.9571 & -0.4047 \\ 0.4047 & -0.9571 \end{bmatrix}$$

$$= \begin{bmatrix} 1.9142 & -0.8094 \\ 0.8094 & -1.9142 \end{bmatrix} \mathrm{W/K} \qquad (a)$$

4. Use Eqs. (3.11) to obtain the **Y**-matrix for fin 10. Here

$$mb = \left(\frac{2h}{k\delta}\right)^{1/2} b = \left[\frac{2(101)}{202(0.003175)}\right]^{1/2} (0.06) = 1.0648$$

and, because the fin thickness has been reduced to $\delta = 0.003175\,\text{m}$, the characteristic admittance will be the same as for the other fins in the array:

$$Y_o = 0.8673\,\text{W/K}$$

Then from Eqs. (3.11),

$$y_{11,10} = Y_o \coth mb = 0.8673(1.2698) = 1.1014\,\text{W/K}$$

$$y_{12,10} = -Y_o \,\text{csch}\, mb = -0.8673(0.7826) = -0.6788\,\text{W/K}$$

$$y_{21,10} = Y_o \,\text{csch}\, mb\,\text{W/°C} = 0.8673(0.7826) = 0.6788\,\text{W/K}$$

and

$$y_{22,10} = -Y_o \coth mb = -0.8673(1.2698) = -1.1014\,\text{W/K}$$

Thus

$$\mathbf{Y}_{10} = \begin{bmatrix} 1.1014 & -0.6788 \\ 0.6788 & -1.1014 \end{bmatrix} \text{W/K} \qquad (b)$$

5. A new equivalent **Y**-matrix comes from an addition of Eqs. (a) and (b)

$$\mathbf{Ye} = \begin{bmatrix} 3.0155 & -1.4881 \\ 1.4881 & -3.0155 \end{bmatrix} \text{W/K}$$

6. Obtain the equivalent **T**-matrix using Table 2.2. First, the determinant of **Ye** will be

$$\begin{aligned} \Delta_Y &= \begin{vmatrix} 3.0155 & -1.4881 \\ 1.4881 & -3.0155 \end{vmatrix} \\ &= -(3.0155)^2 + (1.4881)^2 \\ &= -6.8788 \end{aligned}$$

Then from Table 2.2,

$$\tau_{11,e} = -\frac{y_{22,e}}{y_{21,e}} = -\frac{-3.0155}{1.4881} = 2.0264$$

$$\tau_{12,e} = \frac{1}{y_{21,e}} = -\frac{1}{1.4881} = 0.6720\,\text{K/W}$$

$$\tau_{21,e} = -\frac{\Delta_Y}{y_{21,e}} = -\frac{-6.8788}{1.4881} = 4.6224\,\mathrm{W/K}$$

and

$$\tau_{22,e} = \frac{y_{11,e}}{y_{21,e}} = \frac{3.0155}{1.4881} = 2.0264$$

These are the elements of the equivalent **T**-matrix that represents the entire entity of fins 10 through 14.

7. The performance with the equivalent for fin 10: With $Y_{a,10} = 2.3145\,\mathrm{W/K}$ and $\mathbf{T_e}$ written as

$$\mathbf{T_e} = \begin{bmatrix} 2.0264 & 0.6720 \\ 4.6224 & 2.0264 \end{bmatrix}$$

the reflection relationship of Eq. (5.6) is used to obtain the new value of Y_{in}:

$$\begin{aligned} Y_{in} &= \frac{\tau_{21,e} + Y_{a,10}\tau_{22,e}}{\tau_{11,e} + Y_{a,10}\tau_{12,e}} \\ &= \frac{4.6224 + 2.3145(2.0264)}{2.0264 + 2.3145(0.6720)} \\ &= \frac{9.3125}{3.5817} = 2.6000\,\mathrm{W/K} \end{aligned}$$

8. Assessment: The specification requires that the array have an input admittance of $Y_{in} = 2.4000\,\mathrm{W/^\circ C}$. The design with 14 fins shown in Fig. 6.2 is acceptable because

$$Y_{in} = 2.6000 > 2.4000\,\mathrm{W/K}$$

Observation: There is still the gnawing feeling that perhaps the choke could have been removed using only fins 11 to 14 and *without* fin 10. This can be checked by returning to Eq. (*a*), which represents the equivalent **Y**-matrix for fins 11 through 14:

$$\mathbf{Y_e} = \begin{bmatrix} 1.9142 & -0.8094 \\ 0.8094 & -1.9142 \end{bmatrix} \mathrm{W/K} \qquad (a)$$

This matrix possesses a determinant:

$$\det \mathbf{Y_e} = \Delta_Y = \begin{vmatrix} 1.9142 & -0.8094 \\ 0.8094 & -1.9142 \end{vmatrix} = -3.0090\,\mathrm{W^2K^2}$$

Then from Table 2.2,

$$\tau_{11,e} = -\frac{y_{22,e}}{y_{21,e}} = -\frac{-1.9142}{0.8094} = 2.3650$$

$$\tau_{12,e} = \frac{1}{y_{21,e}} = -\frac{1}{0.8094} = 1.2355\,\mathrm{K/W}$$

$$\tau_{21,e} = -\frac{\Delta_Y}{y_{21,e}} = -\frac{-3.0090}{0.8094} = 3.7176\,\mathrm{W/K}$$

and

$$\tau_{22,e} = \frac{y_{11,e}}{y_{21,e}} = \frac{1.9142}{0.8094} = 2.3650$$

These are the elements of the equivalent **T**-matrix that represents the entity of fins 11 through 14, without the inclusion of fin 10.

Now the reflection relationship of Eq. (5.6) can be used to find Y_{in} using $Y_{in,9} = 2.3145\,\mathrm{W/K}$ for Y_a:

$$\begin{aligned} Y_{in} &= \frac{\tau_{21,e} + Y_a\tau_{12,e}}{\tau_{11,e} + Y_a\tau_{12,e}} \\ &= \frac{3.7176 + 2.3145(2.3650)}{2.3650 + 2.3145(1.2355)} \\ &= \frac{9.1915}{5.2247} = 1.7593\,\mathrm{W/K} \end{aligned}$$

and, because $Y_{in} = 1.7593 < 2.4000\,\mathrm{W/K}$, it can be concluded that fin 10 in Fig. 6.2 is needed.

TABLE 6.3 Physical data for heat exchanger array in Fig 6.3. (All dimensions are in centemeters)

Fin No.	b, cm	δ, cm
1	0.254	0.0254
2	0.254	0.0254
3	1.270	0.0152
4	0.254	0.0254
5	0.254	0.0254
6	1.270	0.0152

6.4 AN EXAMPLE OF A FINNED HEAT EXCHANGER

Example 6.2 Consider the finned array consisting of six fins, as shown in Fig. 6.3. Suppose that fins 1 and 2 are dissipating from one face only and that the physical dimensions are as listed in Table 6.3. It is desired to obtain the

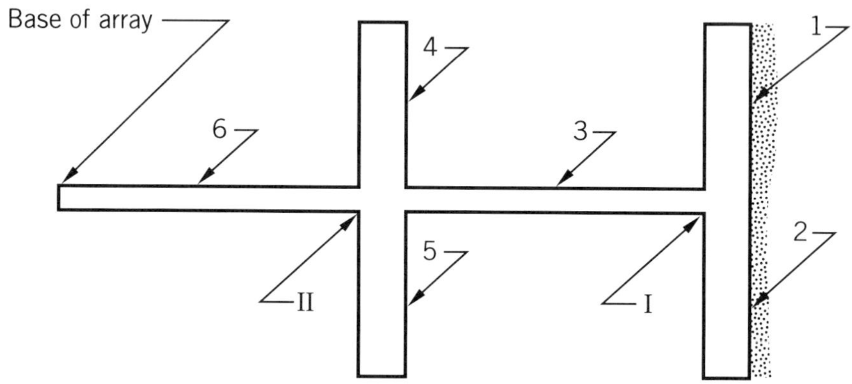

Figure 6.3 A repeating section of an array of six fins. Fins 1 and 2 dissipate on one face only.

input admittance at the base of fin 6 and to determine if a heat flow choke occurs anywhere in the array if $k = 160\,\text{W/m-K}$ and $h = 80\,\text{W/m}^2\text{-K}$. All fins are longitudinal fins of rectangular profile with a length (into the plane of the paper) of 1 m.

SOLUTION Here is the overall strategy:

1. Determine the input admittance for fins 1 and 2, which possess adiabatic tips. Note well that fins 1 and 2 dissipate from *one face only*.

2. Fins 3 and 6 will be identical, and all four elements of the **T**-matrix for fins 3 and 6 must be determined.

3. Use fins 1 and 2 in cluster at the tip of fin 3, and use the reflection relationship of Eq. (5.6) to determine the input admittance at the base of fin 3.

4. Determine the input admittance for fins 4 and 5, which possess adiabatic tips. While the dimensions of fins 4 and 5 are identical to those of fins 1 and 2, the input admittances will be different because fins 4 and 5 dissipate from *two faces*.

5. Fins 3, 4, and 5 are in cluster at the tip of fin 6. Determine the cluster admittance, and use the reflection relationship of Eq. (5.6) to determine the input admittance at the base of fin 6.

Now follow the foregoing procedure:

1. For fins 1 and 2, in accordance with Table 3.1,

$$m_1 = m_2 = \left(\frac{h}{k\delta}\right)^{1/2} = \left[\frac{80}{160(.000254)}\right]^{1/2} = 44.3678\,\text{m}^{-1}$$

$$m_1 b_1 = m_2 b_2 = 44.3678(0.00254) = 0.1127$$

and

$$\begin{aligned} Y_{o,1} = Y_{o,2} = Y_o &= (hk\delta)^{1/2}L \\ &= [80(160)(0.000254)]^{1/2}(1.0000) \\ &= 1.8031\,\mathrm{W/K} \end{aligned}$$

The input admittance is determined from the reflection relationship of Eq. (5.6), with $Y_a = q_a/\theta_a = 0$ because of the adiabatic tips. Only $\tau_{11,1}$ and $\tau_{21,1}$ need be determined:

$$\begin{aligned} \tau_{11,1} &= \tau_{11,2} = \cosh mb \\ &= 1.0064 \end{aligned}$$

and

$$\begin{aligned} \tau_{21,1} &= \tau_{21,2} = Y_o \sinh mb \\ &= 1.8031(0.1129) \\ &= 0.2036\,\mathrm{W/K} \end{aligned}$$

Then by Eq. (5.6),

$$\begin{aligned} Y_{in,1} = Y_{in,2} &= \frac{\tau_{21,1}}{\tau_{11,1}} \\ &= \frac{0.2036}{1.0064} \\ &= 0.2023\,\mathrm{W/K} \end{aligned}$$

2. For fins 3 and 6, which have identical dimensions,

$$m_3 = m_6 = \left(\frac{2h}{k\delta}\right)^{1/2} = \left[\frac{2(80)}{160(.000152)}\right]^{1/2} = 81.1107\,\mathrm{m}^{-1}$$

$$m_3 b_3 = m_6 b_6 = 81.1107(0.0127) = 1.0301$$

and

$$\begin{aligned} Y_{o,3} = Y_{o,6} = Y_o &= (2hk\delta)^{1/2}L \\ &= [2(80)(160)(0.000152)]^{1/2}(1.0000) \\ &= 1.9726\,\mathrm{W/K} \end{aligned}$$

By Eqs. (3.9), which define the τ-parameters,

$$\tau_{11,3} = \tau_{11,6} = \cosh mb = 1.5792$$

$$\begin{aligned}\tau_{12,3} &= \tau_{12,6} = \frac{\sinh mb}{Y_o} \\ &= \frac{1.2222}{1.9726} = 0.6196\,\mathrm{K/W}\end{aligned}$$

$$\begin{aligned}\tau_{21,3} &= \tau_{21,6} = Y_o \sinh mb \\ &= 1.9726(1.2222) = 2.4109\,\mathrm{W/K}\end{aligned}$$

and

$$\tau_{22,3} = \tau_{22,6} = \cosh mb = 1.5792$$

Fins 1 and 2 are clustered at the tip of fin 3. Thus

$$Y_{a,3} = Y_{in,1} + Y_{in,2} = 0.2023 + 0.2023 = 0.4046\,\mathrm{W/K}$$

and

$$\begin{aligned}Y_{in,3} &= \frac{\tau_{21,3} + Y_{a,3}\tau_{12,3}}{\tau_{11,3} + Y_{a,3}\tau_{12,3}} \\ &= \frac{2.4109 + 0.4046(1.5792)}{1.5792 + 0.4046(0.6196)} \\ &= \frac{3.0500}{1.8299} = 1.6667\,\mathrm{W/K}\end{aligned}$$

3. Fins 4 and 5 dissipate heat from both faces and have adiabatic tips:

$$m_4 = m_5 = \left(\frac{2h}{k\delta}\right)^{1/2} = \left[\frac{2(80)}{160(0.000254)}\right]^{1/2} = 62.7456\,\mathrm{m}^{-1}$$

$$m_4 b_4 = m_5 b_5 = 62.7456(0.00254) = 0.1594$$

and

$$\begin{aligned}Y_o = Y_{o,4} = Y_{o,5} &= (2hk\delta)^{1/2} L \\ &= [2(80)(160)(0.000254)]^{1/2}(1.0000) \\ &= 2.5500\,\mathrm{W/K}\end{aligned}$$

The input admittance is determined from Eq. (5.6), with $Y_a = q_a/\theta_a = 0$ because of the adiabatic tips. Only $\tau_{11,4} = \tau_{11,5}$ and $\tau_{21,4} = \tau_{21,5}$ need be determined:

$$\begin{aligned}\tau_{11,4} = \tau_{11,5} &= \cosh mb \\ &= 1.0127\end{aligned}$$

and

$$\begin{aligned}\tau_{21,4} = \tau_{22,4} &= Y_o \sinh mb \\ &= 2.5500(0.1600) \\ &= 0.4081\,\mathrm{W/K}\end{aligned}$$

Then by Eq. (5.6),

$$\begin{aligned}Y_{in,4} = Y_{in,5} &= \frac{\tau_{21,4}}{\tau_{11,5}} \\ &= \frac{0.4081}{1.0127} = 0.4030\,\mathrm{W/K}\end{aligned}$$

4. There is a three-fin cluster at the tip of fin 6 consisting of fins 3, 4, and 5. The cluster admittance will be

$$\begin{aligned}Y_{a,6} &= Y_{in,3} + Y_{in,4} + Y_{in,5} \\ &= 1.6667 + 0.4030 + 0.4030 \\ &= 2.4727\,\mathrm{W/K}\end{aligned}$$

It is observed immediately that fin 6 will choke the heat flow because its characteristic admittance ($Y_{o,6} = 1.9726\,\mathrm{W/K}$) is less than the cluster admittance at its tip ($Y_{a,6} = 2.4727\,\mathrm{W/K}$). This may be confirmed by the reflection relationship of Eq. (5.6)

$$\begin{aligned}Y_{in,6} &= \frac{\tau_{21,6} + Y_{a,6}\tau_{12,6}}{\tau_{11,6} + Y_{a,6}\tau_{12,6}} \\ &= \frac{2.4109 + 2.4727(1.5792)}{1.5792 + 2.4727(0.6196)} \\ &= \frac{6.3158}{3.1112} = 2.0300\,\mathrm{W/K}\end{aligned}$$

The choke can be avoided by adjusting the thickness of fin 6 in accordance with Eq. (5.19):

$$\delta > \frac{Y_a^2}{2hk\mathrm{L}^2} \tag{4.19}$$

and with the parameters for fin 6:

$$\delta > \frac{(2.4727)^2}{2(80)(160)(1.00)^2} = 0.0002388\,\mathrm{m}$$

or

$$\delta > 0.02388\,\mathrm{cm}$$

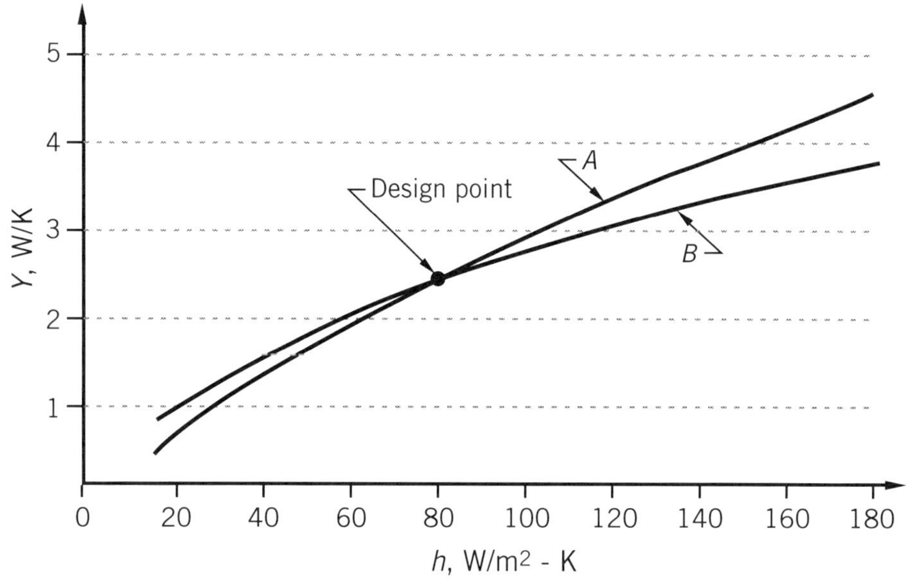

Figure 6.4 Input admittances as a function of heat-transfer coefficient, h, for the array of Fig. 6.3. Curve A represents the input admittance at the base of fin 5 and curve B represents the input admittance at the base of fin 6.

6.5 A CAUTION WITH REGARD TO CHOKING

The heat flow choke at fin 6 in Ex. 6.2 (see Fig. 6.3) was removed by adjusting the thickness of fin 6 from 0.01520 cm to 0.02388 cm. This adjustment permits fin 6 to *"break even"*; that is, with the adjustment, fin 6 does not enhance or degrade the performance of the balance of the array. Whether or not this is desireable depends on the point of view. The objective is to prevent a degradation, and there is no reason why the magnitude of the adjusted fin thickness cannot be greater than that dictated by Eq. (5.19) after it is modified to show the minimum width required to prevent a choke:

$$\delta = \frac{Y_a^2}{2hk\mathrm{L}^2}$$

Observe, however, that the magnitude of the heat-transfer coefficient can have an impact on the heat flow choke. If the fin thickness is set to eliminate the choke *at a particular value of h*, the choke may reoccur at other values of h. In Ex. 6.2, the array was analyzed for a value of $h = 80\,\mathrm{W/m^2\text{-}K}$. If the array is analyzed with the adjusted value of thickness for fin 6 ($\delta = 0.02388$ cm) for other values of h, one may plot $Y_{in,5}$ and $Y_{in,6}$ as a function of h, as indicated in Fig. 6.4. The intersection of the $Y_{in,5}$ and $Y_{in,6}$ curves indicates the *design point* and shows that the choke has been eliminated at the *design value* of

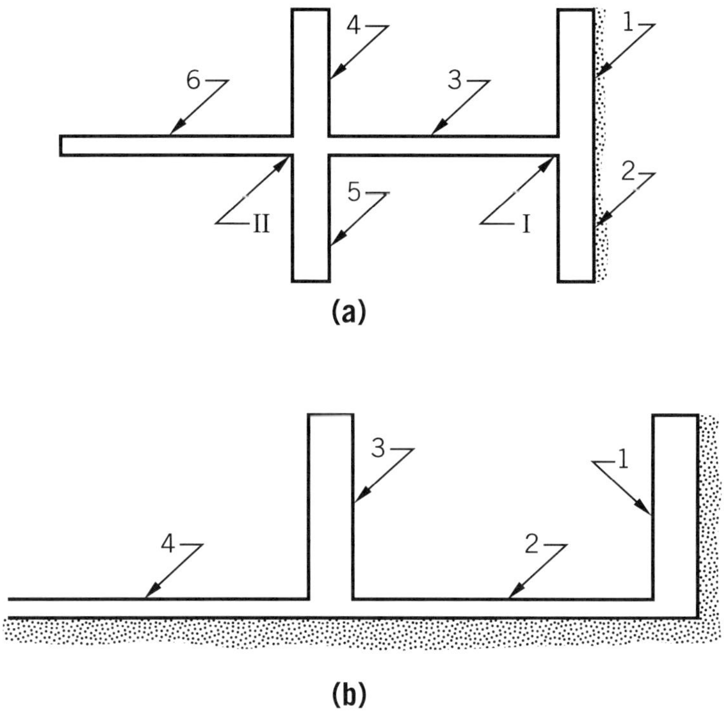

Figure 6.5 (*a*) Repeating section of an array of six fins and (*b*) a modification for analysis that exploits symmetry of the array in (*a*). In (*a*), fins 1 and 2 dissipate on one face only, and (*b*) is derived from the fact that fins 1, 2, 4, and 5, and fins 3 and 6 have the same dimensions.

$h = 80\,\text{W/m}^2$-K. Below this value of h, fin 6 does not choke the heat flow because $Y_{in,5} < Y_{in,6}$. However, above this value it does choke the heat flow because $Y_{in,5} > Y_{in,6}$.

One is led to the conclusion that care must be exercised in the selection of the design point and that the designer must anticipate the range of h values and proceed accordingly.

6.6 THE EXPLOITATION OF SYMMETRY

The array in Fig. 6.3, which is repeated in Fig. 6.5*a* is symmetrical because fins 1, 2, 4, and 5 and fins 3 and 6 have the same dimensions. The labor involved in the determination of the input admittance at the base of fin 6 can be reduced[1] by exploiting the fact that the array is symmetrical. The array shown in Fig. 6.5*b* contains four fins. A consideration of the parallel algorithm

[1]Somewhat in this case and substantially in the case of more-complicated arrays.

of Eq. (5.16) shows that as long as fins 2 and 4 are treated as longitudinal fins of rectangular profile with heat dissipation from one face with a thickness of $\delta = 0.0076\,\text{cm}$, the value of $Y_{in,6}$ in Fig. 6.5*a* will be just twice the value of $Y_{in,4}$ in Fig. 6.5*b*.

Once again, the strategy is straightforward:

1. Fins 1 and 2: The input admittance for fin 1 is the same as that in Ex. 6.2. The **T**-matrix for fin 2 must be determined using

$$m = \left(\frac{h}{k\delta}\right)^{1/2}$$

   and with $L = 1\,\text{m}$

$$Y_o = (hk\delta)^{1/2}$$

   The input admittance at the base of fin 2 may be evaluated using the reflection relationship of Eq. (5.6) with $Y_a = Y_{in,1}$.

2. Fin 3: The input admittance for fin 3 is the same as that for fin 4 in Ex. 6.2.

3. Fin 4: The cluster admittance at the tip of fin 4 is equal to the sum of the input admittances at fin 2 and fin 3. The reflection relationship of Eq. (5.6) can be used to yield $Y_{in,4}$.

4. The array input admittance: In accordance with the parallel algorithm of Eq. (5.16), the input admittance for the entire array (fin 6 in Fig. 6.5*a*) will be equal to just twice the value of $Y_{in,4}$.

The configuration is that of Fig. 6.5*b* and with $h = 80\,\text{W/m}^2\text{-K}$ and $k = 160\,\text{W/m-K}$. The calculations now follow:

1. For fin 1 from Ex. 6.2:

$$Y_{in,1} = 0.2023\,\text{W/K}$$

   For fin 2 with $\delta = 0.0076\,\text{cm}$, $b = 1.27\,\text{cm}$, and $L = 1\,\text{m}$, for a longitudinal fin of rectangular profile with one face insulated,

$$m_2 = \left(\frac{h}{k\delta}\right)^{1/2} = \left[\frac{80}{160(.000076)}\right]^{1/2} = 81.1107\,\text{m}^{-1}$$

$$m_2 b_2 = 81.1107(0.0127) = 1.0301$$

   and

$$\begin{aligned} Y_o &= (hk\delta)^{1/2} L \\ &= [80(160)(0.000076)]^{1/2}(1.0000) \\ &= 0.9863\,\text{W/K} \end{aligned}$$

Notice that the value of mb here is identical to the value obtained for fin 3 in Ex. 6.2 because

$$\left[\frac{2h}{k\delta}\right]^{1/2} = \left[\frac{h}{k(\delta/2)}\right]^{1/2}$$

Moreover, observe that the value of Y_o here is half the value for fin 3 in Ex. 6.2 because

$$\left[hk\frac{\delta}{2}\right]^{1/2} L = \frac{1}{2}(2hk\delta)^{1/2} L$$

By Eqs. (3.9), which define the **T**-matrix,

$$\begin{aligned}
\tau_{11,2} &= \cosh mb = 1.5792 \\
\tau_{12,2} &= \frac{\sinh mb}{Y_o} \\
&= \frac{1.2222}{0.9863} = 1.2392\,\mathrm{K/W} \\
\tau_{21,2} &= Y_o \sinh mb \\
&= 0.9863(1.2222) = 1.2055\,\mathrm{W/K}
\end{aligned}$$

and

$$\tau_{22,4} = \cosh mb = 1.5792$$

Then by the reflection relationship of Eq. (5.6) with $Y_{a,2} = Y_{in,1} = 0.2023\,\mathrm{W/K}$,

$$\begin{aligned}
Y_{in,2} &= \frac{\tau_{21,2} + Y_{a,2}\tau_{12,2}}{\tau_{11,2} + Y_{a,2}\tau_{12,2}} \\
&= \frac{1.2055 + 0.2023(1.5792)}{1.5792 + 0.2023(1.2392)} \\
&= \frac{1.5250}{1.8299} = 0.8334\,\mathrm{W/K}
\end{aligned}$$

2. For fin 3 from fin 4 of Ex. 6.2,

$$Y_{in,3} = 0.4030\,\mathrm{W/K}$$

3. There is a cluster at the tip of fin 4 consisting of fins 2 and 3. Fin 4 is identical to fin 2, and the cluster admittance will be

$$\begin{aligned}
Y_{a,4} &= Y_{in,2} + Y_{in,3} \\
&= 0.8334 + 0.4030 \\
&= 1.2364\,\mathrm{W/K}
\end{aligned}$$

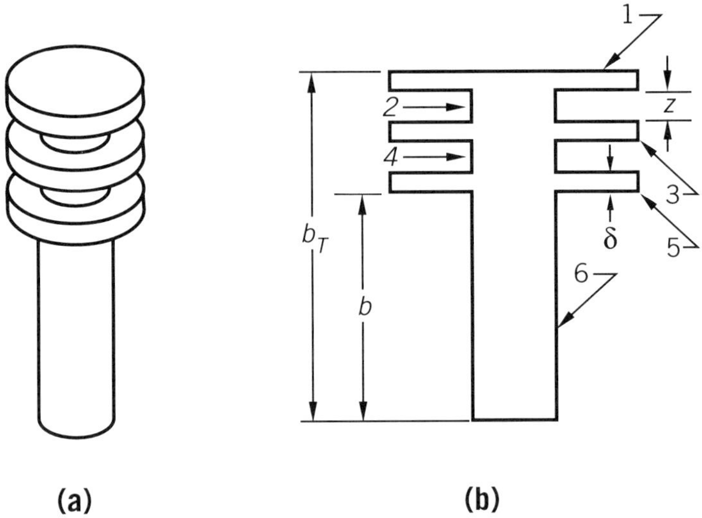

Figure 6.6 (*a*) Artist's conception of a small finned array used to cool flatpacks and (*b*) detail showing terminology. This array has achieved a certain notoriety and is often called the *fin bush problem*.

and a choke can be expected. Then employment of the reflection relationship of Eq. (5.6) with the τ-parameters from fin 2 gives

$$\begin{aligned} Y_{in,4} &= \frac{\tau_{21,4} + Y_{a,4}\tau_{12,4}}{\tau_{11,4} + Y_{a,4}\tau_{12,4}} \\ &= \frac{1.2055 + 1.2364(1.5792)}{1.5792 + 1.2364(1.2392)} \\ &= \frac{3.1579}{3.1112} = 1.0150\,\text{W/K} \end{aligned}$$

4. An application of the parallel algorithm of Eq. (5.16) then gives the input admittance of the array in Fig. 6.5*a* as

$$Y_{in} = 2Y_{in,4} = 2(1.0150) = 2.0300\,\text{W/K}$$

This is the result of Ex. 6.2.

6.7 AN ARRAY CONTAINING RADIAL FINS AND SPINES

It is useful to consider an array containing spines and radial fins. An artist's conception of such a configuration is shown in Fig. 6.6*a*, and the terminology and dimensions are indicated in Fig. 6.6*b*. Fins 1, 3, and 5 are radial fins of rectangular profile with thickness $\delta = 0.3175$ cm, inner diameter $d_b = 1.27$ cm, and outer diameter $d_a = 2.54$ cm. The spacing between the fins on a cylinder of $d = 1.27$ cm is designated as z, and $z = 0.3175$ cm. The cylinder between the fins is considered as a spine of height, $b = z$, with fin 2 between radial fins 1 and 3, and fin 4 between radial fins 3 and 5. Fin 6 contains the base of the array, which is attached to a flatpack, multichip module, or other microelectronic device.

Example 6.3: Consider the finned array in Fig. 6.6 with the specification that it must dissipate 2.75 W when at a base temperature excess of 10°C. The heat transfer coefficient for fins 1 through 5 is $h = 106.3\,\mathrm{W/m^2\text{-}K}$, and that for fin 6 is $h = 137.9\,\mathrm{W/m^2\text{-}K}$. The thermal conductivity for all fins in the configuration is $k = 120\ \mathrm{W/m\text{-}K}$. The height of fin 6 is to be determined and must be such that the specification is satisfied.

SOLUTION The strategy is similar to that used in the other examples:

1. Determine the input admittance for fins 1, 3, and 5. Because the tips of these fins are presumed adiabatic, only the parameters τ_{11} and τ_{21} need be calculated. Call this[2] $Y_{in,r}$.

2. Obtain the elements of the **T**-matrix for fins 2 and 4.

3. Use the reflection relationship of Eq. (5.6) to obtain $Y_{in,2}$, with $Y_{a,2} = Y_{in,r}$.

4. Use the cluster algorithm of Eq. (5.10) to determine the cluster admittance at the tip of fin 4. Then use the reflection relationship of Eq. (5.6) to obtain $Y_{in,4}$ with $Y_{a,4} = Y_{in,2} + Y_{in,r}$.

5. Use the cluster algorithm of Eq. (5.10) to determine the cluster admittance at the tip of fin 6. If no heat flow choke is apparent, then use the reflection relationship of Eq. (5.6) to obtain $Y_{in,6} = q_{b6}/\theta_{b6} = 2.75/10 = 0.2750\,\mathrm{W/K}$ by trial-and-error using $Y_{a,6} = Y_{in,4} + Y_{in,r}$. The trial and error process will involve the assumption of the spine height, b, and calculation until $Y_{in,6} = 0.2750\,\mathrm{W/K}$

The application of the foregoing procedure now follows:

1. For radial fins 1, 3, and 5 with adiabatic tips, Eqs. (3.56*a*) and (3.56*c*) are needed to evaluate $Y_{in,r}$. First

$$m = \left(\frac{2h}{k\delta}\right)^{1/2} = \left[\frac{2(106.3)}{120(0.003175)}\right]^{1/2} = 23.6221\,\mathrm{m}^{-1}$$

so that

$$\alpha = mr_b = 23.6221(0.00635) = 0.1500$$

[2] r for *radial fin*.

and

$$\beta = mr_a = 23.6221(0.0127) = 0.3000$$

The values of six Bessel functions are needed. From tables or software

$$\begin{aligned}
I_0(\alpha) &= 1.00563 \\
I_1(\alpha) &= 0.07521 \\
K_0(\alpha) &= 2.03003 \\
K_1(\alpha) &= 6.47750 \\
I_1(\beta) &= 0.15169 \\
K_1(\beta) &= 3.05599
\end{aligned}$$

Now using Eqs. (3.56*a*) and (3.56*c*) with the subscript r designating a *radial fin*,

$$\begin{aligned}
\tau_{11,r} &= \beta[I_0(\alpha)K_1(\beta) + I_1(\beta)K_0(\alpha)] \\
&= 0.3000[1.00563(3.05599) + 0.15169(2.03003)] \\
&= 0.3000(3.38113) = 1.0143
\end{aligned}$$

and with

$$\begin{aligned}
2k\pi\delta\alpha\beta &= 2(120)(\pi)(0.003175)(0.1500)(0.3000) \\
&= 0.10773\,\mathrm{W/K} \\
\tau_{21,r} &= 2k\pi\delta\alpha\beta[I_1(\beta)K_1(\alpha) - I_1(\alpha)K_1(\beta)] \\
&= 0.10773[0.15169(6.47750) - 0.07521(3.05599)] \\
&= 0.10773(0.75273) = 0.0811\,\mathrm{W/K}
\end{aligned}$$

The input admittance for the radial fins will be

$$Y_{in,r} = \frac{\tau_{21,r}}{\tau_{11,r}} = \frac{0.0811}{1.0143} = 0.0799\,\mathrm{W/K}$$

2. For fins 2 and 4, the **T**-matrix is given by Eqs. (3.64), with

$$m = \left(\frac{4h}{kd}\right)^{1/2} = \left[\frac{4(106.3)}{120(0.0127)}\right]^{1/2} = 16.7034\,\mathrm{m}^{-1}$$

and with b equal to the clear space between the radial fins, z,

$$mb = 16.7034(0.003175) = 0.0530$$

Moreover,

$$\begin{aligned}
Y_{o,2} = Y_{o,4} = Y_o &= \frac{1}{2}\left[\pi^2 hkd^3\right]^{1/2} \\
&= \frac{1}{2}\left[\pi^2(106.3)(120)(0.0127)^3\right]^{1/2} \\
&= 0.2539\,\mathrm{W/K}
\end{aligned}$$

Then

$$\begin{aligned}
\tau_{11,2} = \tau_{11,4} &= \cosh mb \\
&= 1.0014 \\
\tau_{12,2} = \tau_{12,4} &= (1/Y_o)\sinh mb \\
&= \frac{0.0531}{0.2539} = 0.2090\,\text{K/W} \\
\tau_{21,2} = \tau_{21,4} &= Y_o \sinh mb \\
&= 0.2539(0.0531) = 0.0135\,\text{W/K}
\end{aligned}$$

and

$$\begin{aligned}
\tau_{22,2} = \tau_{22,4} &= \cosh mb \\
&= 1.0014
\end{aligned}$$

3. Use of the reflection relationship of Eq. (5.6), with $Y_{a,2} = Y_{in,r}$, gives $Y_{in,2}$:

$$\begin{aligned}
Y_{in,2} &= \frac{\tau_{21,2} + Y_{a,2}\tau_{12,2}}{\tau_{11,2} + Y_{a,2}\tau_{12,2}} \\
&= \frac{0.0135 + 0.0799(1.0014)}{1.0014 + 0.0799(0.2090)} \\
&= \frac{0.0935}{1.0181} = 0.0919\,\text{W/K}
\end{aligned}$$

4. By Eq. (5.10), the cluster admittance at the tip of fin 4 is

$$\begin{aligned}
Y_{a,4} &= Y_{in,2} + Y_{in,r} \\
&= 0.0919 + 0.0799 \\
&= 0.1718\,\text{W/K}
\end{aligned}$$

With $Y_{a,4} = 0.1718\,\text{W/K}$, use of the reflection relationship of Eq. (5.6) gives $Y_{in,4}$

$$\begin{aligned}
Y_{in,4} &= \frac{\tau_{21,4} + Y_{a,2}\tau_{12,4}}{\tau_{11,4} + Y_{a,2}\tau_{12,4}} \\
&= \frac{0.0135 + 0.1718(1.0014)}{1.0014 + 0.1718(0.2090)} \\
&= \frac{0.1855}{1.0373} = 0.1788\,\text{W/K}
\end{aligned}$$

and it is gratifying to observe that there is no heat flow choke.

5. The cluster algorithm of Eq. (5.10) is now employed at the tip of fin 6.

$$\begin{aligned} Y_{a,6} &= Y_{in,4} + Y_{in,r} \\ &= 0.1788 + 0.0799 \\ &= 0.2588\,\text{W/K} \end{aligned}$$

To determine whether fin 6 chokes the heat flow, $Y_{o,6}$ must be evaluated.

$$\begin{aligned} Y_{o,6} &= \frac{1}{2}(\pi^2 hkd^3)^{1/2} \\ &= \frac{1}{2}[\pi^2(137.9)(120)(0.0127)^3]^{1/2} \\ &= 0.2892\,\text{W/K} \end{aligned}$$

and because $Y_{a,6} < Y_{o,6}$, there is no choke, and the height of fin 6 can be adjusted to meet the specification that $Y_{in,6} = 0.275\,\text{W/K}$.

But the heat-transfer coefficient for fin 6 is not equal to that for fins 2 and 4. Thus with $h = 137.9\,\text{W/m}^2\text{-K}$,

$$m_6 = \left(\frac{4h}{kd}\right)^{1/2} = \left[\frac{4(137.9)}{120(0.0127)}\right]^{1/2} = 19.0248\,\text{m}^{-1}$$

so that $m_6 b_6 = 19.0248 b_6$ and

$$Y_{in,6} = \frac{\tau_{21,6} + 0.2588\tau_{22,6}}{\tau_{11,6} + 0.2588\tau_{12,6}}$$

where the τ's are functions of $m_6 b_6$.

The computations can be organized to be readily programmed for a hand-held calculator. Table 6.4 presents a summary of the computations for four values of spine height. It may be noted that four decimal figures were carried and there is room for argument that three would have been more than sufficient.

6.8 FINS WITH VARIABLE HEAT-TRANSFER COEFFICIENT

Ever since the publication of the limiting assumptions for ideal fin analysis by Murray (1938) and Gardner (1945), investigators have been considering the effect of relaxing the requirement that the heat-transfer coefficient between the fin faces and the environment be uniform and constant. Ghai (1951) was apparently the first to take exception to this assumption. This was followed in short order by Gardner (1951) in a discussion of Ghai's work, Melese (1957), Fortescue (1957), and Hughes and Slack (1958). Han and Lefkowitz (1960) assumed a power law variation of h over the height of the fin; Chen and

TABLE 6.4 Computations for the height of fin 6 in Ex. 6.3. The computations are based on the reflection relationship of Eq. (5.6).

b cm	3.000	2.000	2.050	2.075
Y_o W/°C	0.2892	0.2892	0.2892	0.2892
Y_a W/°C	0.2588	0.2588	0.2588	0.2588
mb	0.5707	0.3805	0.3901	0.3948
τ_{11}	1.1673	1.0733	1.0770	1.0789
τ_{12}	2.0824	1.3477	1.3830	1.4029
τ_{21}	0.1742	0.1127	0.1157	0.1172
τ_{22}	1.1673	1.0733	1.0770	1.0789
$Y_{in,6}$ W/°C	0.2791	0.2746	0.2749	0.2500

Zyskowski (1963) considered an exponential variation; and Cumo et al. (1965) published a two-dimensional study that allowed for the variation of the heat-transfer coefficient, as well as an internal heat generation. Snider and Kraus (1982) demonstrated how perturbation theory could be used to estimate the effect of a variation in the heat-transfer coefficient.

The **T**-matrix and the cascade algorithm of Eq. (5.5) [or the reflection relationship of Eq. (5.6)] can be used to evaluate the performance of a combination of individual fins and spines of almost any shape that are attached end-to-end to form an array. The ultimate goal is, of course, the computation of a numerical value for the input admittance, Y_{in}. A fin with a variation of the heat-transfer coefficient on its faces may be decomposed into n subfins that are attached end-to-end. The **T**-matrix can be developed for each subfin, each with its own local heat-transfer coefficient, and a computationally efficient procedure involving the cascade algorithm can then be employed to determine the overall equivalent **T**-matrix for the single fin composed of the n subfins. Then, from this equivalent **T**-matrix, the input admittance or the fin efficiency can then be computed.

The theory behind this approach has been presented in Sec. 5.3, which involves the cascade algorithm. It is important, however to consider the accuracy of the results obtained using this approach. Four-figure accuracy will now be demonstrated in a comparison to the classical solution of Han and Lefkowitz (1960) for the linear variation of the heat-transfer coefficient on the faces of the longitudinal fin of rectangular profile.

6.8.1 The Han and Lefkowitz Solution

Han and Lefkowitz (1960) considered a heat-transfer coefficient variation of the form

$$h(x) = h_o(\gamma + 1)\left(\frac{x}{b}\right)^{\gamma} \tag{6.1}$$

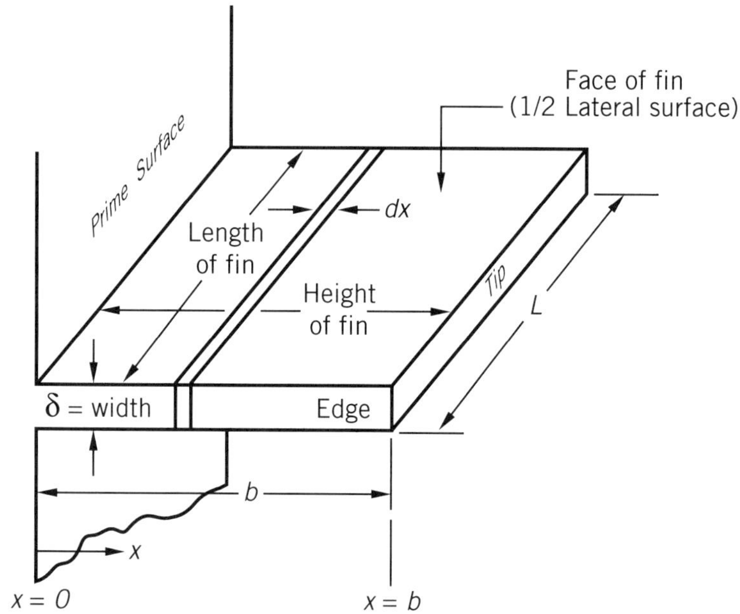

Figure 6.7 Terminology and coordinate system for the longitudinal fin of rectangular profile. Observe that the origin of the height coordinate, x, is at the fin base.

on the faces of the longitudinal fin of rectangular profile shown with its terminology and coordinate system in Fig. 6.7. It should be noted in Fig. 6.7 that the height coordinate has its origin at the fin base and that its positive orientation is from fin base to fin tip.

The differential equation for the temperature excess is

$$\frac{d^2\theta}{dx^2} - (\gamma + 1)m^2 \left(\frac{x}{b}\right)^{\gamma} \theta = 0 \tag{6.2}$$

where

$$m = \left(\frac{2h_o}{k\delta}\right)^{1/2}$$

After a rather involved transformation of variables, Han and Lefkowitz showed that the solution to the differential equation with the boundary conditions of fin base temperature excess,

$$\theta(x = 0) = \theta_b$$

and the condition for the insulated tip,

$$\left.\frac{d\theta}{dx}\right|_{x=a} = 0$$

is

$$\theta = \frac{[(\gamma+1)/(\gamma+2)]}{2^{1/(\gamma+2)}}\theta_b u^{[1/(\gamma+2)]} \tag{6.3}$$

where

$$u = \frac{2\sqrt{\gamma+1}}{\gamma+2} m b^{-\gamma/2} x^{[(\gamma+2)/2]} \tag{6.4}$$

$$u_b = \frac{2\sqrt{\gamma+1}}{\gamma+2} mb \tag{6.5}$$

and where

$$= I_{[1/(\gamma+2)]}(u) + \frac{I_{[(\gamma+1)/(\gamma+2)]}(u_b)}{I_{-[(\gamma+1)/(\gamma+2)]}(u_b)} I_{[1/(\gamma+2)]}(u)$$

With the fin efficiency defined by

$$\eta = \frac{-k\delta L \left.\dfrac{d\theta}{dx}\right|_{x=0}}{2h_o L b \theta_b} \tag{6.6}$$

a term-wise differentiation of Eq. (6.3) with subsequent evaluation at $x = 0$ gives

$$\eta = \left[\frac{(\gamma+2)^{\gamma}(\gamma+1)}{(mb)^{2(\gamma+1)}}\right]^{1/(\gamma+1)} \left[\frac{I_{[(\gamma+1)/(\gamma+2)]}(u_b)}{I_{-[(\gamma+2)/(\gamma+1)]}(u_b)}\right] \left[\frac{\Gamma[(\gamma+1)/(\gamma+2)]}{\Gamma[1/(\gamma+2)]}\right] \tag{6.7}$$

For a linear variation of the heat-transfer coefficient on the faces of the fin, $\gamma = 1$ and Eq. (6.1) becomes

$$h(x) = 2h_o \frac{x}{b} \tag{6.8}$$

where it is assumed that $L >> \delta$, so that

$$m = \left(\frac{2h_o}{k\delta}\right)^{1/2}$$

and with

$$u_b = \frac{2\sqrt{2}}{3} mb$$

the efficiency is

$$\eta = \left[\frac{6}{(mb)^4}\right]^{1/2} \frac{I_{2/3}(u_b)}{I_{-2/3}(u_b)} \frac{\Gamma(2/3)}{\Gamma(1/3)} \tag{6.9}$$

6.8.2 The Use of the Cascade Algorithm

The connection of fins in cascade pertains to the configuration where at least two fins are connected tip-to-base. For n fins in cascade (connected tip-to-base), the cascade algorithm of Eq. (5.5) shows that there will be an equivalent $\mathbf{T}$-matrix that relates conditions at the tip of fin 1 to the base of fin n. Clearly, care must be exercised in the selection of the fin numbering system and in the order of multiplication, because matrix multiplication is, in general, not commutative. Every effort should be made to number the fins, in sequence, from the tip up through the base, where the base is defined as the point where the heat to be dissipated by the fin or by the entity of fins is injected.

The thrust here is to consider that any single fin with any arbitrary, piecewise continuous and single valued heat-transfer coefficient distribution on its faces, can be modeled as n fins in cascade. Each of the n fins will possess its own heat-transfer coefficient, which may be assumed constant and uniform over its faces. The cascade algorithm guarantees that the equivalent $\mathbf{T}$-matrix will represent the entire configuration. Only the number of subfins needed to guarantee the required accuracy is in question.

6.8.3 The Number of Subfins Required

An example has been contrived to provide an idea as to the number of subfins required to match the classical solution of Han and Lefkowitz (1960). For a longitudinal fin of rectangular profile operating with the thermal and physical data:

$$\begin{aligned} b &= 10\,\text{cm} \\ L &= 10\,\text{cm} \\ \delta &= 0.1975\,\text{cm} \\ h_o &= 50.5\,\text{W/m}^2\text{-K} \\ k &= 202\,\text{W/m-K} \end{aligned}$$

and

$$\gamma = 1$$

These numbers, which give an aspect ratio for the fin profile of approximately 50:1, are for a fin that is unusually high for its thickness and will yield a value for the fin efficiency that is quite low. However, they may be used to provide a point of comparison between the two methods.

With the foregoing numbers and with $\gamma = 1$ indicating a linear variation of the heat-transfer coefficient over the fin faces,

$$h(x) = 2h_o(x/b) = 2(50.5)x/(0.100) = 1010x\ \text{W/m}^2\text{-}^\circ\text{K}$$

The efficiency is obtained from Eq. (6.9):

$$\eta = 0.4474$$

and with the fin face area equal to

$$S = 2bL = 2(0.100)(0.100) = 0.0200\,\text{m}^2$$

the input admittance is obtained from Eq. (2.45):

$$Y_{in} = h_o S\eta = 50.5(0.0200)(0.4474) = 0.4519\,\text{W/K}$$

Now, in what follows, let it be stressed that how a discretized analysis converges to a continuous analysis depends on the specific nature of the problem and not on the nature of the variation in h. Thus, the following discussion represents an example and not a generalization.

The number of subfins required to yield an input admittance of $Y_{in} = 0.4519\,\text{W/K}$ has been determined in a computationally efficient procedure using a hand-held calculator with a huge memory. To be sure, the use of the hand-held calculator is not relevant here, and perhaps few people would use one for this study. Nevertheless, all numbers computed were stored in the memory, and four-figure accuracy, although not necessary, is ensured. One program took the thermal and physical data as input and, using Eq. (6.6), quickly determined Y_{in} as a function of n, the total number of subfins. The results are shown in Table 6.5, where it is seen that 50 subfins will provide the sought-after accuracy.

TABLE 6.5 Number of subfins required to yield $Y_{in} = 0.4519\,\text{W/K}$ when $b = L = 10\,\text{cm}$, $\delta = 0.1975\,\text{cm}$, $h_o = 50.5\,\text{W/m}^2\text{-K}$ and $k = 202\,\text{W/m-K}$.

NO OF SUBFINS	Y_{in}, W/K
10	0.4534
20	0.4522
30	0.4520
40	0.4520
50	0.4519
75	0.4519

Table 6.5 shows that, if three-decimal accuracy is all that is required because the value of h is not known to an accuracy that demands four-decimal precision, as few as 20 subfins may be employed. But, if 50 subfins are to be used, Table 6.6 provides an insight as to just what happens along the fin surface as the base of the fin is approached. While Table 6.6 shows data for every even-numbered subfin, data for subfin 31 are also included. It is at subfin 31 that the heat flow through the fin begins to choke, and this indicates that close to 40 percent of the fin is actually acting as a partial insulator.

TABLE 6.6 Cascade calculations for 50 subfins. Thermal and physical data are the same as for Table 6.2. Except for fin 31, only data for even-numbered fins are presented. Fin 31 is included because it is the location of a heat flow choke.

i	x	h	Y_o	$\tau_{11} = \tau_{22}$	τ_{12}	τ_{21}	Y_{in}
2	0.097	97.97	0.8842	1.0010	0.0501	0.0392	0.0790
4	0.093	93.93	0.8658	1.0009	0.0501	0.0376	0.1535
6	0.089	89.89	0.8470	1.0009	0.0501	0.0360	0.2226
8	0.085	85.85	0.8277	1.0009	0.0501	0.0343	0.2856
10	0.081	81.81	0.8080	1.0008	0.0501	0.0327	0.3419
12	0.077	77.77	0.7877	1.0008	0.0501	0.0311	0.3914
14	0.073	73.73	0.7670	1.0007	0.0501	0.0295	0.4341
16	0.069	69.69	0.7457	1.0007	0.0501	0.0279	0.4701
18	0.065	65.65	0.7238	1.0007	0.0501	0.0263	0.4998
20	0.061	61.61	0.7011	1.0006	0.0501	0.0246	0.5236
22	0.057	57.57	0.6778	1.0006	0.0501	0.0230	0.5419
24	0.053	53.53	0.6535	1.0005	0.0501	0.0214	0.5554
26	0.049	49.49	0.6284	1.0005	0.0501	0.0198	0.5643
28	0.045	45.45	0.6022	1.0005	0.0501	0.0182	0.5693
30	0.041	41.41	0.5748	1.0004	0.0501	0.0166	0.5706
31	0.039	39.39	0.5606	1.0004	0.0501	0.0158	0.5700
32	0.037	37.37	0.5461	1.0004	0.0501	0.0149	0.5687
34	0.033	33.33	0.5157	1.0003	0.0501	0.0133	0.5640
36	0.029	29.29	0.4834	1.0003	0.0501	0.0117	0.5568
38	0.025	25.25	0.4489	1.0003	0.0501	0.0101	0.5472
40	0.021	21.21	0.4114	1.0002	0.0501	0.0085	0.5356
42	0.017	17.17	0.3701	1.0002	0.0501	0.0069	0.5221
44	0.013	13.13	0.3237	1.0001	0.0501	0.0053	0.5068
46	0.009	9.09	0.2693	1.0001	0.0501	0.0036	0.4900
48	0.005	5.05	0.2007	1.0001	0.0501	0.0020	0.4717
50	0.001	1.01	0.0898	1.0000	0.0501	0.0004	0.4519

Chapter

7

RECIPROCITY AND NODE ANALYSIS

7.1 INTRODUCTION

Finned arrays containing loops in the graph theoretical sense such as the array shown in Fig. 7.1, may be analyzed by the general array algorithm proposed by Snider and Kraus (1981), to be considered in detail in Chap. 8 or by a method of node analysis proposed by Kraus et al. (1982). The node analysis method is described here in a matrix-oriented procedure believed to be more general than the more specific procedure given in the 1982 work.

It is the intent of this chapter to develop the concept of *reciprocity*, which allows the representation of the distributed thermal configuration of a single fin in terms of a *pi-network* of three lumped thermal admittances. When the fins are assembled into an array, the pi-networks are arranged to yield a lumped admittance representation that may then be analyzed by node analysis, which in itself can be organized into a computionally efficient procedure.

7.2 A STEP FORWARD

The procedure of Chap. 5 is also applicable to configurations where the q/θ ratios at the tips of the most remote fins are not known a priori. Consider, for instance, the repeating section of the unevenly loaded *double stack* shown in Fig. 7.2. Here, two surfaces are dissipating heat to the coolant fluid. The operating parameters are the left and right prime surface temperature excesses, θ_ℓ and θ_r, and the associated heat flows, q_ℓ and q_r. If one prescribes, say, unequal temperature excesses, θ_ℓ and θ_r, then the asymmetry forbids an a priori assessment of the q/θ ratios at the two remote fin tips.

One could, in fact, follow the procedures of Chap. 5 by treating the ratio,

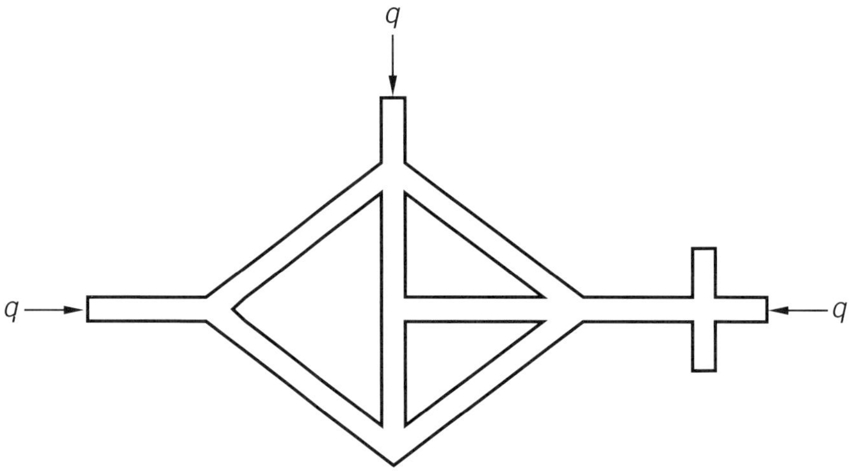

Figure 7.1 A finned array containing loops in the graph theoretical sense.

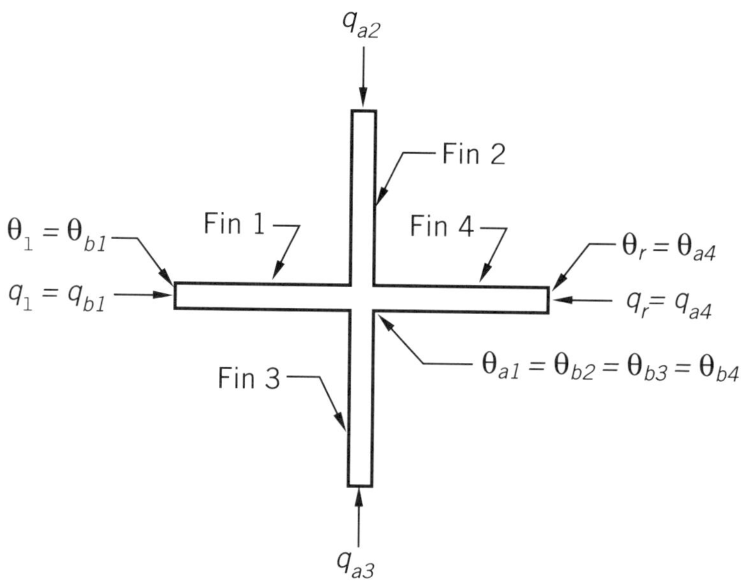

Figure 7.2 A repeating section of a compact heat exchanger double stack.

q_r/θ_r as an unknown parameter and formally carrying it through the cascade process from right to left. The result would be a relationship[1] between the two q/θ ratios. However, the following matrix procedure is somewhat more illuminating.

Using the $\mathbf{\Gamma}$-matrices, one may write, keeping in mind that a positive q indicates a heat flow from right to left,

$$\begin{bmatrix} \theta_{a,1} \\ q_{a,1} \end{bmatrix} = \mathbf{\Gamma_1} \begin{bmatrix} \theta_\ell \\ q_\ell \end{bmatrix} \tag{7.1a}$$

and

$$\begin{bmatrix} \theta_r \\ q_r \end{bmatrix} = \mathbf{\Gamma_4} \begin{bmatrix} \theta_{b,4} \\ q_{b,4} \end{bmatrix} \tag{7.1b}$$

where opposite signs are to be expected for q_ℓ and q_r. Furthermore, it is known that the temperature excess must match at the point of intersection of all four fins:

$$\theta_{a,1} = \theta_{b,2} = \theta_{b,3} = \theta_{b,4} \tag{7.2a}$$

and that continuity requires that

$$q_{b,4} = q_{a,1} - q_{b,2} - q_{b,3} \tag{7.2b}$$

Moreover, because the tips of fins 2 and 3 are adiabatic,

$$\frac{q_{b,2}}{\theta_{b,2}} = Y_{in,2} \tag{7.3a}$$

and

$$\frac{q_{b,3}}{\theta_{b,3}} = Y_{in,3} \tag{7.3b}$$

A relationship that links the base conditions of fin 4 to the tip conditions of fin 1 can be established by a mathematical artifice:

$$\begin{aligned} \begin{bmatrix} \theta_{b,4} \\ q_{b,4} \end{bmatrix} &= \begin{bmatrix} \theta_{a,1} \\ q_{a,1} \end{bmatrix} - \begin{bmatrix} 0 \\ q_{b,2} \end{bmatrix} - \begin{bmatrix} 0 \\ q_{b,3} \end{bmatrix} \\ &= \begin{bmatrix} \theta_{a,1} \\ q_{a,1} \end{bmatrix} - \begin{bmatrix} 0 \\ Y_{in,2}\theta_{b,2} \end{bmatrix} - \begin{bmatrix} 0 \\ Y_{in,3}\theta_{b,3} \end{bmatrix} \end{aligned}$$

Equation (7.2*a*) permits this to be written as

$$\begin{bmatrix} \theta_{b,4} \\ q_{b,4} \end{bmatrix} = \begin{bmatrix} \theta_{a,1} \\ q_{a,1} \end{bmatrix} - \begin{bmatrix} 0 \\ Y_{in,2}\theta_{a,1} \end{bmatrix} - \begin{bmatrix} 0 \\ Y_{in,3}\theta_{a,1} \end{bmatrix}$$

or as

$$\begin{bmatrix} \theta_{b,4} \\ q_{b,4} \end{bmatrix} = \begin{bmatrix} 1 & 0 \\ -(Y_{in,2} + Y_{in,3}) & 1 \end{bmatrix} \begin{bmatrix} \theta_{a,1} \\ q_{a,1} \end{bmatrix} \equiv \mathbf{\Upsilon} \begin{bmatrix} \theta_{a,1} \\ q_{a,1} \end{bmatrix} \tag{7.4}$$

[1] A bilinear relationship.

where

$$\Upsilon = \begin{bmatrix} 1 & 0 \\ -(Y_{in,2} + Y_{in,3}) & 1 \end{bmatrix}$$

Hence, by Eqs. (7.1)

$$\begin{bmatrix} \theta_r \\ q_r \end{bmatrix} = \mathbf{\Gamma_4 \Upsilon \Gamma_1} \begin{bmatrix} \theta_\ell \\ q_\ell \end{bmatrix} \tag{7.5}$$

Thus, any two of the parameters can be related to the other two. If it is kept in mind that q_r is negative, a closed-form solution can be obtained. This will be demonstrated in Ex. 7.1.

Example 7.1: Suppose that the configuration shown in Fig. 7.2 represents a repeating section of an 11.1 plain plate fin heat exchanger with *splitter plates* (fins 2 and 3)[2] having a thickness of 0.0254 cm and a height of 0.1144 cm. fins 1 and 4 have a thickness of 0.01524 cm and a height of 0.635 cm. The exchanger is fabricated of aluminum ($k = 200\,\text{W/m-K}$), and it is used in an application where $h = 50\,\text{W/m}^2\text{-K}$. All fins may be treated as longitudinal fins of rectangular profile with a length (into the plane of the paper) of 40 cm. Determine $\theta_{b,1}$ and $\theta_{a,4}$ if $|q_{b,1}| = 20\,\text{W}$ and $|q_{a,4}| = 16\,\text{W}$.

SOLUTION 1. For fins 2 and 3,

$$m_2 = m_3 = \left(\frac{2h}{k\delta}\right)^{1/2} = \left[\frac{2(50)}{200(.000254)}\right]^{1/2} = 44.3678\,\text{m}^{-1}$$

$$m_2 b_2 = m_3 b_3 = 44.3678(0.001144) = 0.0508$$

and

$$\begin{aligned} Y_{o,2} &= Y_{o,3} = (2hk\delta)^{1/2} L \\ &= [2(50)(200)(0.000254)]^{1/2}(0.4000) = 0.9016\,\text{W/K} \end{aligned}$$

The input admittance is determined from the reflection relationship of Eq. (5.6) with $Y_a = q_a/\theta_a = 0$ because of the adiabatic tips. Only τ_{11} and τ_{21} need be determined from Eqs. (3.9):

$$\begin{aligned} \tau_{11,2} &= \tau_{11,3} = \cosh mb \\ &= 1.0013 \end{aligned}$$

and

$$\begin{aligned} \tau_{21,2} &= \tau_{21,3} = Y_o \sinh mb \\ &= 0.9016(0.0508) \\ &= 0.0458\,\text{W/K} \end{aligned}$$

[2]The data for the 11.1 plain plate fin heat exchanger have been extracted from W. M. Kays and A. L. London, *Compact Heat Exchangers*, McGraw-Hill Book Company, New York 1984.

Then, by the reflection relationship of Eq. (5.6),

$$\begin{aligned} Y_{in,2} = Y_{in,3} &= \frac{\tau_{21,2}}{\tau_{11,2}} \\ &= \frac{0.0458}{1.0013} = 0.0457\,\mathrm{W/K} \end{aligned}$$

2. For fins 1 and 4, which have identical dimensions,

$$m_1 = m_4 = \left(\frac{2h}{k\delta}\right)^{1/2} = \left[\frac{2(50)}{200(.0001524)}\right]^{1/2} = 57.2786\,\mathrm{m}^{-1}$$

$$m_1 b_1 = m_4 b_4 = 57.2786(0.00635) = 0.3637$$

and

$$\begin{aligned} Y_{o,1} &= Y_{o,4} = (2hk\delta)^{1/2} L \\ &= [2(50)(200)(0.0001524)]^{1/2}(0.4000) = 0.6983\,\mathrm{W/K} \end{aligned}$$

By Eqs. (3.8),

$$\gamma_{11,1} = \gamma_{11,4} = \cosh mb = 1.0669$$

$$\begin{aligned} \gamma_{12,1} &= \gamma_{12,4} = -\frac{\sinh mb}{Y_o} \\ &= -\frac{0.3718}{0.6983} = -0.5324\,\mathrm{K/W} \end{aligned}$$

$$\begin{aligned} \gamma_{21,1} &= \gamma_{21,4} = -Y_o \sinh mb \\ &= -0.6983(0.3718) = -0.2596\,\mathrm{W/K} \end{aligned}$$

and

$$\gamma_{22,1} = \gamma_{22,4} = \cosh mb = 1.0669$$

This makes

$$\mathbf{\Gamma} = \begin{bmatrix} 1.0669 & -0.5324 \\ -0.2596 & 1.0669 \end{bmatrix}$$

3. With

$$\mathbf{\Upsilon} = \begin{bmatrix} 1 & 0 \\ -(Y_{in,2} + Y_{in,3}) & 1 \end{bmatrix} = \begin{bmatrix} 1 & 0 \\ -0.0915 & 1 \end{bmatrix}$$

one obtains via Eq. (7.5)

$$\begin{bmatrix} \theta_r \\ q_r \end{bmatrix} = \mathbf{\Gamma_4 \Upsilon \Gamma_1} \begin{bmatrix} \theta_\ell \\ q_\ell \end{bmatrix} \tag{7.5}$$

$$\begin{bmatrix} \theta_r \\ q_r \end{bmatrix} = \begin{bmatrix} 1.0669 & -0.5324 \\ -0.2596 & 1.0669 \end{bmatrix} \begin{bmatrix} 1 & 0 \\ -0.0914 & 1 \end{bmatrix} \bullet$$

$$\begin{bmatrix} 1.0669 & -0.5324 \\ -0.2596 & 1.0669 \end{bmatrix} \begin{bmatrix} \theta_\ell \\ q_\ell \end{bmatrix}$$

or

$$\begin{bmatrix} \theta_r \\ q_r \end{bmatrix} \begin{bmatrix} 1.3284 & -1.1619 \\ -0.6581 & 1.3284 \end{bmatrix} \begin{bmatrix} \theta_\ell \\ q_\ell \end{bmatrix}$$

This can be expanded. With $q_\ell = 20\,\text{W}$ and $q_r = -16\,\text{W}$,

$$\theta_r = 1.3284\theta_\ell - 1.1619(20.0000)$$

or

$$\theta_r = 1.3284\theta_\ell - 23.2384$$

and

$$-16.0000 = -0.6581\theta_\ell + 1.3284(20.0000)$$

or

$$0.6581\theta_\ell = 42.5682$$

From the second of these,

$$\theta_\ell = \frac{42.5682}{0.6581} = 64.7°\text{C}$$

and with this result in the first,

$$\theta_r = 1.3284(64.7) - 23.2 = 62.7°\text{C}$$

The procedure developed here holds some interest, but it is readily observed that its employment may become more difficult as the array becomes more complex and as the number of heat inputs increases. A more general procedure is desireable, and attention now turns to the node analysis of finned arrays. There are, however, some very important preliminaries.

7.3 RECIPROCITY OF REGULAR FINS

7.3.1 Introduction

The temperature excess, θ, in any longitudinal or radial fin or spine is governed, in the steady state, by a differential equation of the form

$$\frac{d}{dx}\left[k(x)A(x)\frac{d\theta(x)}{dx}\right] - h(x)\frac{dS(x)}{dx}\theta(x) = 0 \tag{7.6}$$

In most cases, x is a height coordinate with origin set arbitrarily at some point remote from the fin or spine tip, which is located at $x = a$. For the specific cases where the origin of the height coordinate is set at the fin tip, $a = 0$, and for radial fins, x is replaced by r and is measured from fin base to fin tip. The fin surface area, $S(x)$, the fin cross-sectional area, $A(x)$, and the fin thickness, δ, are all functions of x. The Murray-Gardner assumptions still apply, and the temperature excess is defined as

$$\theta(x) = T(x) - T_s$$

where T_s is the temperature of the surroundings.

Recall that in Chap. 2, it was shown that the theory of differential equations states that Eq. (7.6) is *regular* at all points where

$$k(x)A(x) \neq 0 \tag{7.7}$$

and *singular* when $k(x)A(x)$ vanishes. The latter situation can only occur physically when the fin has zero thickness. This happens at the tips of certain fins or spines such as the longitudinal fins of triangular and concave parabolic profiles and the conical and concave parabolic spines.

The theory also guarantees that Eq. (7.6) possesses two solutions, $\lambda_1(x)$ and $\lambda_2(x)$, satisfying the *initial conditions* at the *base* of the fin:

$$\lambda_1(b) = 1; \qquad \frac{d\lambda_1(b)}{dx} = 0 \tag{7.8a}$$

and

$$\lambda_2(b) = 0; \qquad \frac{d\lambda_2(b)}{dx} = \frac{1}{k(b)A(b)} \tag{7.8b}$$

The heat flow in the fin, $q(x)$, is always taken as positive from base to tip. For nonradial fins, $q(x)$ is given by

$$q(x) = k(x)A(x)\frac{d\theta(x)}{dx} \tag{7.9}$$

Thus, one can use the solutions $\lambda_1(x)$ and $\lambda_2(x)$ to assemble the expressions for the temperature excess, $\theta(x)$, and the heat flow, $q(x)$, at any point in the fin in terms of their values, θ_b and q_b, at the fin base:

$$\theta(x) = \theta_b\lambda_1(x) + q_b\lambda_2(x) \tag{7.10}$$

and

$$q(x) = k(x)A(x)[\theta_b\lambda_1'(x) + q_b\lambda_2'(x)] \tag{7.11}$$

In matrix form, this becomes

$$\begin{bmatrix} \theta(x) \\ q(x) \end{bmatrix} = \begin{bmatrix} 1 & 0 \\ 0 & k(x)A(x) \end{bmatrix} \begin{bmatrix} \lambda_1(x) & \lambda_2(x) \\ \lambda_1'(x) & \lambda_2'(x) \end{bmatrix} \begin{bmatrix} \theta_b \\ q_b \end{bmatrix} \tag{7.12}$$

Setting $x = a$ produces a linear transformation that maps the temperature excess and heat flow at the fin tip in terms of conditions at the fin base:

$$\begin{bmatrix} \theta_a \\ q_a \end{bmatrix} = \mathbf{\Gamma} \begin{bmatrix} \theta_b \\ q_b \end{bmatrix} = \begin{bmatrix} \gamma_{11} & \gamma_{12} \\ \gamma_{21} & \gamma_{22} \end{bmatrix} \begin{bmatrix} \theta_b \\ q_b \end{bmatrix} \tag{7.13}$$

where

$$\gamma_{11} = \lambda_1(a) \tag{7.14a}$$

$$\gamma_{12} = \lambda_2(a) \tag{7.14b}$$

$$\gamma_{21} = k(a)A(a)\lambda_1'(a) \tag{7.14c}$$

and

$$\gamma_{22} = k(a)A(a)\lambda_2'(a) \tag{7.14d}$$

For radial fins, a minus sign must be introduced into Eq. (7.9) because, x, now the radial coordinate, r, is measured from base to tip.

The $\mathbf{\Gamma}$-matrix depends only on the fin or spine geometry and the heat-transfer properties of the fin and system and, as shown in Chap. 3, has been developed for many of the useful regular fin shapes. For convenience, it is useful to recast Eq. (7.13) into a form employing a different linear transformation, the $\mathbf{Y}$-matrix,

$$\begin{bmatrix} q_b \\ q_a \end{bmatrix} = \mathbf{Y} \begin{bmatrix} \theta_b \\ \theta_a \end{bmatrix} = \begin{bmatrix} y_{11} & y_{12} \\ y_{21} & y_{22} \end{bmatrix} \begin{bmatrix} \theta_b \\ \theta_a \end{bmatrix} \tag{7.15}$$

where the y's can be obtained from Table 2.2. These conversions are repeated here for the reader's convenience.

$$\begin{aligned} y_{11} &= -\frac{\gamma_{11}}{\gamma_{12}} \\ y_{12} &= \frac{1}{\gamma_{12}} \\ y_{21} &= -\frac{\det \mathbf{\Gamma}}{\gamma_{12}} \end{aligned}$$

and

$$y_{22} = \frac{\gamma_{22}}{\gamma_{12}}$$

7.3.2 Reciprocity

The determinant of $\mathbf{\Gamma}$ is observed to be the product of two determinants:

$$\det \mathbf{\Gamma} = \det \begin{bmatrix} 1 & 0 \\ 0 & k(a)A(a) \end{bmatrix} \begin{bmatrix} \lambda_1(a) & \lambda_2(a) \\ \lambda_1'(a) & \lambda_2'(a) \end{bmatrix} \tag{7.16}$$

One may identify the second determinant in this expression as the *Wronskian* and then form

$$\det \mathbf{\Gamma} = k(a)A(a)W(\lambda_1, \lambda_2; a) \tag{7.17}$$

Abel's formula[3] can be applied to the differential equation of Eq. (7.6) to evaluate the Wronskian. In terms of the parameters in Eq. (7.6), Abel's formula states that

$$W(\lambda_1, \lambda_2; x) = \frac{C}{k(x)A(x)} \tag{7.18}$$

where C is some constant. If Eq. (7.18) is true at any x, it must be true at $x = b$. Then

$$W(\lambda_1, \lambda_2; b) = \det \begin{bmatrix} \lambda_1(b) & \lambda_2(b) \\ \lambda_1'(b) & \lambda_2'(b) \end{bmatrix}$$

and by invoking the conditions of Eqs. (7.8)

$$W(\lambda_1, \lambda_2; b) = \det \begin{bmatrix} 1 & 0 \\ 0 & 1/k(b)A(b) \end{bmatrix} = \frac{1}{k(b)A(b)} \tag{7.19}$$

A comparison of Eq. (7.19) with Eq. (7.18) with $x = b$ shows that

$$W(\lambda_1, \lambda_2; b) = \frac{C}{k(b)A(b)} = \frac{1}{k(b)A(b)}$$

which indicates that $C = 1$ and that, from Eq. (7.17)

$$\det \mathbf{\Gamma} = k(a)A(a)\frac{1}{k(a)A(a)} = 1 \tag{7.20}$$

The relation $\det \mathbf{\Gamma} = 1$ is also valid for radial fins because the change of sign of both off-diagonal entries in $\mathbf{\Gamma}$ does not change its determinant.

With $\det \mathbf{\Gamma} = 1$, the off-diagonal elements of $\mathbf{Y}$ are observed (Table 2.2) to be the negative of one another.

$$y_{12} = \frac{1}{\gamma_{12}}$$

and

$$y_{21} = -\frac{1}{\gamma_{12}}$$

Consider Fig. 7.3, which shows a hypothetical fin of arbitrary profile with a heat source, q, provided first at the the base (the left end) and then at the tip (the right end). Observe that in each case, a temperature excess (θ_a and θ_b) is also indicated. Moreover, a height coordinate, x, having origin at the fin tip and positive sense toward the fin base appears in both cases.

A fin will be termed *reciprocal* if an identical heat injection, q, introduced at either end with the other end insulated produces an equal temperature excess

[3]See Hildebrand, F. B., *Advanced Calculus for Applications*, Prentice-Hall, Englewood Cliffs, N. J. (1976).

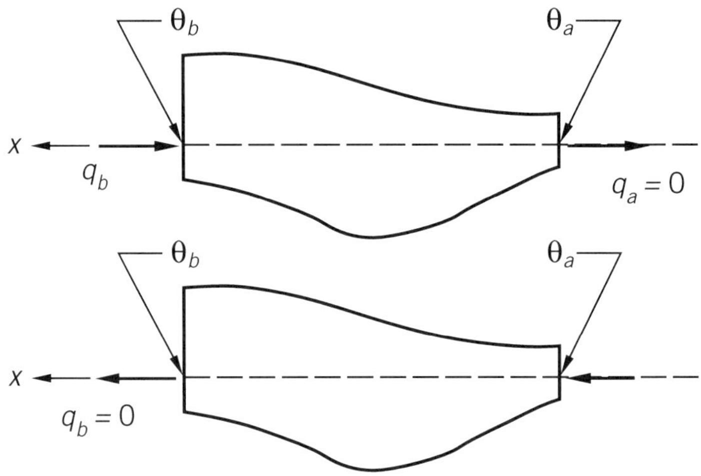

Figure 7.3 Hypothetical fin of arbitrary profile.

at the opposite end. This mouthful can be stated mathematically: If $q_b = q$ produces θ_a, and if $q_a = -q$ (the heat is inserted in the opposite direction) produces θ_b, then the fin is said to be reciprocal if $\theta_a = \theta_b$.

The heat flow and temperature exesses are related by Eq. (7.15), and one may obtain θ_b and θ_a by merely taking the inverse:

$$\begin{bmatrix} \theta_b \\ \theta_a \end{bmatrix} = \frac{1}{\det \mathbf{Y}} \begin{bmatrix} y_{22} & -y_{12} \\ -y_{21} & y_{11} \end{bmatrix} \begin{bmatrix} q_b \\ q_a \end{bmatrix} \tag{7.21}$$

In the first case, with $q_b = q$ and the fin tip insulated so that $q_a = 0$,

$$\theta_a = \frac{-y_{21}}{\det \mathbf{Y}} q \tag{7.22a}$$

In the second case, with $q_a = -q$ and the fin base insulated so that $q_b = 0$,

$$\theta_b = \frac{-y_{12}}{\det \mathbf{Y}}(-q) = \frac{y_{12}}{\det \mathbf{Y}} q \tag{7.22b}$$

Reciprocity requires that $\theta_a = \theta_b$. Thus, it is seen that for a fin to be reciprocal,

$$y_{12} = -y_{21}$$

It is this reciprocity principle that permits a single regular fin to be represented as a *pi-network* containing three thermal admittances as displayed in Fig. 7.4.

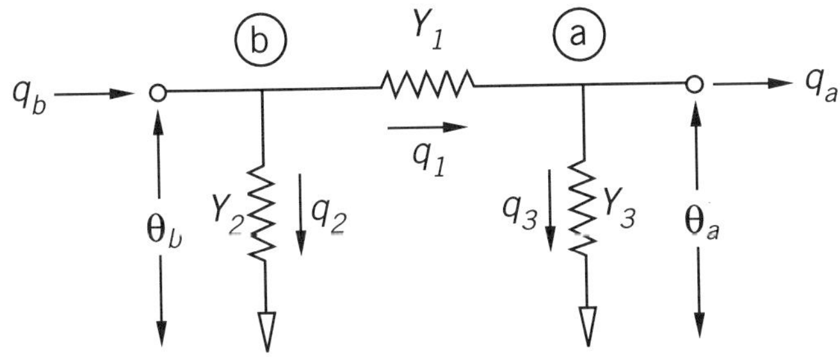

Figure 7.4 Equivalent pi-network.

7.3.3 Conditions for Reciprocity

With $y_{12} = -y_{21}$, the conditions for reciprocity for the other five parameters can be found from Table 2.2.

$$\begin{aligned}
\det \mathbf{\Gamma} &= 1 \\
\det \mathbf{T} &= 1 \\
z_{12} &= -z_{21} \\
y_{12} &= -y_{21} \\
h_{12} &= h_{21} \\
g_{12} &= g_{21}
\end{aligned}$$

7.4 THE EQUIVALENT PI-NETWORK

In Fig. 7.4, application of continuity at points a and b, called *node points* or merely *nodes*, gives

$$\begin{aligned}
q_b &= q_1 + q_2 \\
q_a &= q_1 - q_3
\end{aligned}$$

However, each heat flow, q_1, q_2, and q_3 in Fig. 7.4, can be represented as a product of an admittance and a temperature excess:

$$\begin{aligned}
q_1 &= Y_1(\theta_b - \theta_a) \\
q_2 &= Y_2\theta_b \\
q_3 &= Y_3\theta_a
\end{aligned}$$

so that

$$\begin{aligned}
q_b &= Y_1(\theta_b - \theta_a) + Y_2\theta_b \\
q_a &= Y_1(\theta_b - \theta_a) - Y_3\theta_a
\end{aligned}$$

or

$$q_b = (Y_1 + Y_2)\theta_b - Y_1\theta_a \tag{7.23a}$$

and

$$q_a = Y_1\theta_b - (Y_1 + Y_3)\theta_a \tag{7.23b}$$

Equations (7.23) may be put into matrix form:

$$\begin{bmatrix} q_b \\ q_a \end{bmatrix} = \begin{bmatrix} (Y_1 + Y_2) & -Y_1 \\ Y_1 & -(Y_1 + Y_3) \end{bmatrix} \begin{bmatrix} \theta_b \\ \theta_a \end{bmatrix}$$

and then compared to Eq. (7.15) which defines the $\mathbf{Y}$-matrix:

$$\begin{bmatrix} q_b \\ q_a \end{bmatrix} = \begin{bmatrix} y_{11} & y_{12} \\ y_{21} & y_{22} \end{bmatrix} \begin{bmatrix} \theta_b \\ \theta_a \end{bmatrix} \tag{7.15}$$

The comparison shows that

$$\begin{aligned} Y_1 + Y_2 &= y_{11} \\ Y_1 &= -y_{12} = y_{21} \end{aligned}$$

and

$$Y_1 + Y_3 = -y_{22}$$

Then, it is only a matter of algebra to show that

$$Y_1 = -y_{12} = y_{21} \tag{7.24a}$$

$$Y_2 = y_{11} + y_{12} = y_{11} - y_{21} \tag{7.24b}$$

and

$$Y_3 = y_{12} - y_{22} = -y_{21} - y_{22} \tag{7.24c}$$

Example 7.2: In Ex. 6.1 during the procedure for the elimination of the heat flow choke, an equivalent $\mathbf{Y}$-matrix,

$$\mathbf{Y} = \begin{bmatrix} 3.0155 & -1.4881 \\ 1.4881 & -3.0155 \end{bmatrix}$$

was obtained. Develop an equivalent pi-network for this $\mathbf{Y}$-matrix.

SOLUTION First observe that $y_{12} = -y_{21}$ and that the matrix represents a reciprocal fin. Then, by Eqs. (7.24),

$$Y_1 = -y_{12} = y_{21} = 1.4881\,\mathrm{W/K}$$

$$Y_2 = y_{11} + y_{12} = 3.0155 + (-1.4881) = 1.5274\,\mathrm{W/K}$$

and

$$Y_3 = y_{12} - y_{22} = -1.4881 - (-3.0155) = 1.5274\,\mathrm{W/K}$$

It is noted that as long as $y_{12} = -y_{21}$, any fin may be represented as an equivalent pi. All regular fins are reciprocal and possess this important property. Hence, arrays composed of regular fins may be treated as a combination of equivalent pi-networks representing the individual fins in the array and connected as the array is connected. An example of a four-fin array with unequal heat inputs provided at the opposite ends is displayed, along with its network representation, in Fig. 7.5. This is the double-stack heat exchanger of Ex. 7.1.

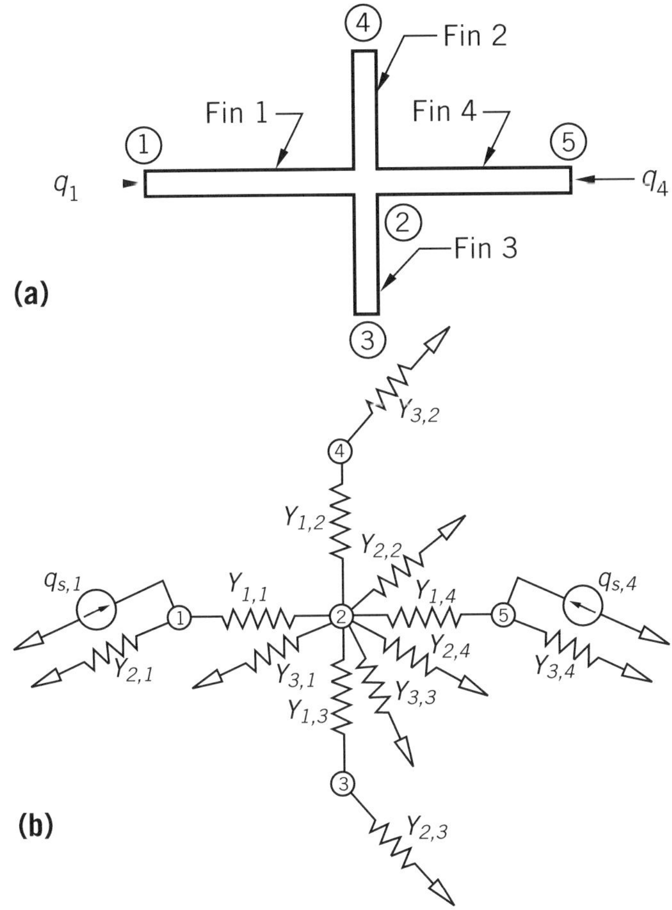

Figure 7.5 (*a*) An array of four fins with heat input at opposite ends, and (*b*) an equivalent network composed of four pi-networks to represent the array.

7.4.1 The Array Graph

The several lumped admittances, arranged as in Fig. 7.5*b*, can be considered as a network of admittances, and reference may be made to Fig. 7.5*b* as a network. *Network topology* derives from a branch of mathematics that permits the determination of network properties on the basis of the structure, geometry, or the interconnection of the network. It serves as a basis for the general method of node analysis that will be developed in subsequent sections of this chapter. The use of network topology does *not* depend on the types or the magnitude of the elements that constitute the network.

Every network element, such as the lumped admittances in Fig. 7.5*b*, may be replaced by a single line called a *branch*, which has two ends called *nodes*.

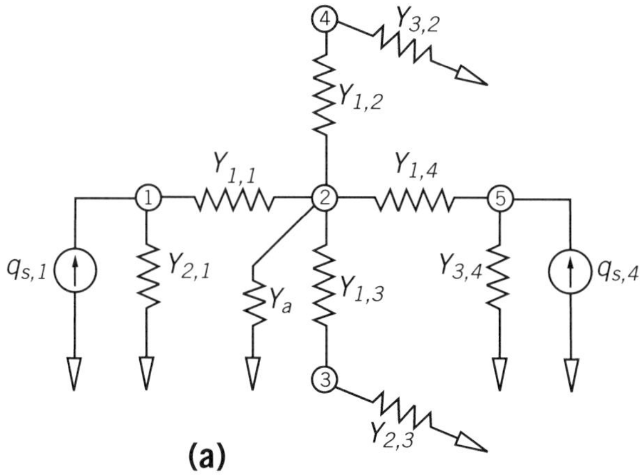

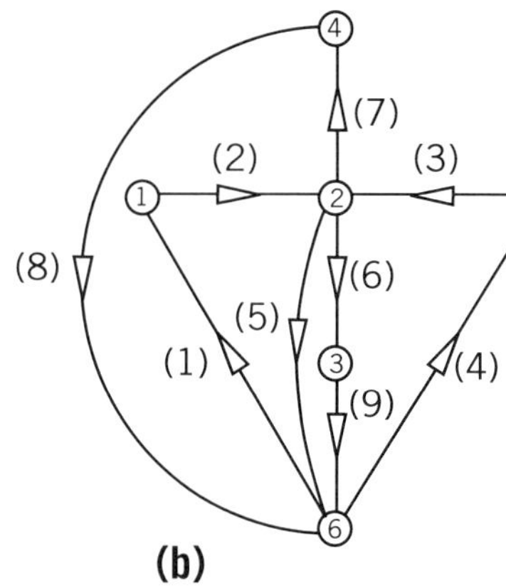

Figure 7.6 (*a*) A simplification of the network in Fig. 7.5*b* and (*b*) an oriented graph for the network of (*a*). The oriented graph contains six nodes and nine branches. Node-6 represents the environment.

A *linear graph* is a collection of branches and nodes that is arranged to exactly represent the geometry of the network. In the context of finned array analysis, the linear graph may be referred to as the *array graph,* and it is customary to represent the branches of the array graph by numerals within parentheses and the nodes by numerals within circles. When each branch of an array graph carries an arrow to indicate its orientation, the array graph is then said to be an *oriented array graph*.

A simplification of the arrangement of admittances in Fig. 7.5*b* is shown in Fig. 7.6*a* where

$$Y_a = Y_{3,1} + Y_{2,2} + Y_{3,3} + Y_{2,4}$$

and Fig. 7.6*b* shows an oriented graph of the network in Fig. 7.6*a*. Again, note

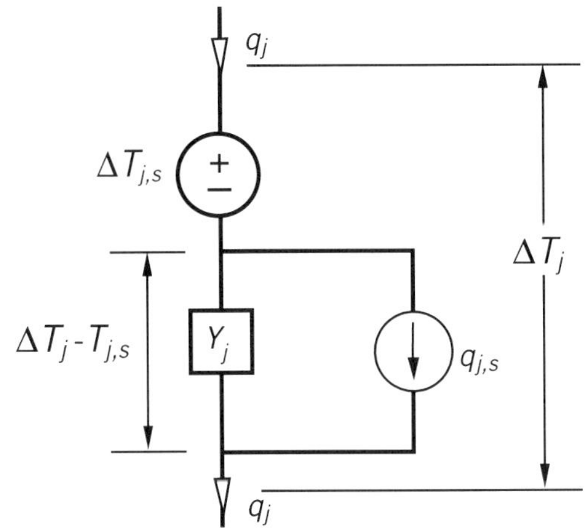

Figure 7.7 A general branch containing an admittance, a heat source, and a temperature source.

the use of circles to designate nodes and numbers in parentheses to designate branches. Observe that branches containing heat sources must also contain an admittance. The heat sources are included in branches 1 and 4, and the branch orientations are discretionary except for the branches containing the heat sources where the positive orientation must correspond to the direction of the heat input. The objective is to obtain all node temperature excesses in a computationally efficient manner and then, knowing the temperature excesses and heat flows at nodes 1 and 5, to determine the input admittances at nodes 1 and 5.

7.4.2 The General Branch

The most general case of the jth branch is the case, shown in Fig. 7.7 where a heat source, q_{js}, and a temperature source, ΔT_{js}, is present. The branch must also contain one of the thermal admittances.

Because the total temperature drop across each branch must be the sum of component temperature drops, it is observed that the temperature drop across the parallel combination of the branch heat source, q_{js}, and the branch admittance, Y_j, must be $\Delta T_j - \Delta T_{js}$, where ΔT_{js} is the temperature source in branch j. Continuity, then, dictates that the branch heat flow, q_j, is represented, in matrix form, for all j branches, by

$$\mathbf{Q} = \mathbf{Q}_s + \mathbf{Y}(\Delta\mathbf{T} - \Delta\mathbf{T}_s) \tag{7.25}$$

Here, with b branches,

$\mathbf{Q}$ is a $b \times 1$ heat flow vector

$\mathbf{Q}_s$ is a $b \times 1$ heat source vector (all nodes may possess heat sources)

$\mathbf{Y}$ is a $b \times b$ branch admittance matrix that is diagonal

$\Delta\mathbf{T}_s$ is a $b \times 1$ temperature source vector that may be (and most often is) null

$\Delta\mathbf{T}$ is a $b \times 1$ temperature vector

7.4.3 The Node Branch Incidence Matrix

For a graph containing n_t nodes and b branches, a matrix that exactly represents the graph may be formulated. This matrix, which is called the *augmented node-branch incidence matrix,* will have n_t rows and b columns and will contain elements $a_{jk} = +1, -1$, or 0 in accordance with the scheme

$$a_{jk} = \begin{cases} +1 & \text{if branch } j \text{ leaves node } k \\ -1 & \text{if branch } j \text{ enters node } k \\ 0 & \text{if branch } j \text{ does not touch node } k \end{cases} \tag{7.26}$$

For example, the augmented node-branch incidence matrix for the oriented graph in Fig. 7.6*b* will be $n_t = 6 \times b = 9$ or 6×9:

$$\mathbf{A}_a = \begin{bmatrix} -1 & 1 & 0 & 0 & 0 & 0 & 0 & 0 & 0 \\ 0 & -1 & -1 & 0 & 1 & 1 & 1 & 0 & 0 \\ 0 & 0 & 0 & 0 & 0 & -1 & 0 & 0 & 1 \\ 0 & 0 & 0 & 0 & 0 & 0 & -1 & 1 & 0 \\ 0 & 0 & 1 & -1 & 0 & 0 & 0 & 0 & 0 \\ 1 & 0 & 0 & 1 & -1 & 0 & 0 & -1 & -1 \end{bmatrix} \tag{7.27}$$

It can be noted that every column contains a single +1 and a single −1 and that a summation of all elements in each column yields a zero.

The node-branch incidence matrix contains $n_t - 1$ rows and b branches. It is obtained from $\mathbf{A}_a$ by merely deleting the row that represents the surrounding environment. In Fig. 7.6*b*, node 6 represents the environment and when it is deleted,

$$\mathbf{A} = \begin{bmatrix} -1 & 1 & 0 & 0 & 0 & 0 & 0 & 0 & 0 \\ 0 & -1 & -1 & 0 & 1 & 1 & 1 & 0 & 0 \\ 0 & 0 & 0 & 0 & 0 & -1 & 0 & 0 & 1 \\ 0 & 0 & 0 & 0 & 0 & 0 & -1 & 1 & 0 \\ 0 & 0 & 1 & -1 & 0 & 0 & 0 & 0 & 0 \end{bmatrix} \tag{7.28}$$

Because the row representing the environment has been deleted, the analysis becomes a node-to-datum analysis, with the environment as the datum and all node temperatures to be considered as temperature excesses. Thus, $n = n_t - 1$ may now be used to represent the number of nodes.

7.4.4 Continuity

The product **AQ**, which postmultiplies an $n \times b$ matrix by a $b \times 1$ column vector, represents a statement of continuity at each node:

$$\mathbf{AQ} = \begin{bmatrix} -1 & 1 & 0 & 0 & 0 & 0 & 0 & 0 & 0 \\ 0 & -1 & -1 & 0 & 1 & 1 & 1 & 0 & 0 \\ 0 & 0 & 0 & 0 & 0 & -1 & 0 & 0 & 1 \\ 0 & 0 & 0 & 0 & 0 & 0 & -1 & 1 & 0 \\ 0 & 0 & 1 & -1 & 0 & 0 & 0 & 0 & 0 \end{bmatrix} \begin{bmatrix} q_1 \\ q_2 \\ q_3 \\ q_4 \\ q_5 \\ q_6 \\ q_7 \\ q_8 \\ q_9 \end{bmatrix}$$

$$= \begin{bmatrix} q_2 - q_1 \\ q_5 + q_6 + q_7 - q_2 - q_3 \\ q_9 - q_6 \\ q_8 - q_7 \\ q_3 - q_4 \end{bmatrix} = \begin{bmatrix} 0 \\ 0 \\ 0 \\ 0 \\ 0 \end{bmatrix}$$

and this can be confirmed by an inspection of the oriented graph in Fig. 7.6*b*. Thus, the equation

$$\mathbf{AQ} = \mathbf{0} \tag{7.29}$$

is a matrix statement of the continuity at every node in the array.

A matrix that links the branch temperature drops, $\Delta\mathbf{T}$, with a node temperature excess vector, Θ, can also be found. Define the elements c_{jk} of the matrix **C** in

$$\Delta\mathbf{T} = \mathbf{C}\Theta \tag{7.30}$$

by

$$c_{jk} = \begin{cases} +1 & \text{if branch } j \text{ leaves node } k \\ -1 & \text{if branch } j \text{ enters node } k \\ 0 & \text{if branch } j \text{ does not touch node } k \end{cases} \tag{7.31}$$

The matrix **C** will contain b rows and n columns, and for the $b \times n$ matrix for the oriented graph of Fig. 7.6*b* (without node 6) will be

$$\mathbf{C} = \begin{bmatrix} -1 & 0 & 0 & 0 & 0 \\ 1 & -1 & 0 & 0 & 0 \\ 0 & -1 & 0 & 0 & 1 \\ 0 & 0 & 0 & 0 & -1 \\ 0 & 1 & 0 & 0 & 0 \\ 0 & 1 & -1 & 0 & 0 \\ 0 & 1 & 0 & -1 & 0 \\ 0 & 0 & 0 & 1 & 0 \\ 0 & 0 & 1 & 0 & 0 \end{bmatrix} \tag{7.32}$$

There are two kinds of branches. Those that touch two nodes, say nodes r and s, will posses a temperature drop, $\Delta T = \theta_r - \theta_s$. Those that touch a node and the datum node will have a temperature drop equal to the temperature excess for that node. In both cases, the branch temperature drops depend on the node temperature excesses. If the matrix of Eq. (7.32) is put into Eqs. (7.30), the result is

$$\mathbf{C\Theta} = \begin{bmatrix} -1 & 0 & 0 & 0 & 0 \\ 1 & -1 & 0 & 0 & 0 \\ 0 & -1 & 0 & 0 & 1 \\ 0 & 0 & 0 & 0 & -1 \\ 0 & 1 & 0 & 0 & 0 \\ 0 & 1 & -1 & 0 & 0 \\ 0 & 1 & 0 & -1 & 0 \\ 0 & 0 & 0 & 1 & 0 \\ 0 & 0 & 1 & 0 & 0 \end{bmatrix} \begin{bmatrix} \theta_1 \\ \theta_2 \\ \theta_3 \\ \theta_4 \\ \theta_5 \end{bmatrix} = \begin{bmatrix} -\theta_1 \\ \theta_1 - \theta_2 \\ \theta_5 - \theta_2 \\ -\theta_5 \\ \theta_2 \\ \theta_2 - \theta_3 \\ \theta_2 - \theta_4 \\ \theta_4 \\ \theta_3 \end{bmatrix}$$

which is entirely correct. Moreover, a comparison of Eqs. (7.28) and (7.32) shows that $a_{jk} = c_{kj}$, which indicates that

$$\mathbf{C} = \mathbf{A}^{\mathbf{T}}$$

and that

$$\mathbf{\Delta T} = \mathbf{A}^{\mathbf{T}}\mathbf{\Theta} \tag{7.33}$$

7.4.5 Node-to-Datum Analysis of Finned Arrays

The method of node-to-datum analysis is based on the branch equations, which is a form of Eqs. (7.25),

$$\mathbf{Q} = \mathbf{Q}_s - \mathbf{Y\Delta T}_s + \mathbf{Y\Delta T} \tag{7.34}$$

the expression of continuity

$$\mathbf{AQ} = \mathbf{0} \tag{7.29}$$

and the relationship between the branch temperature drops and the node temperature excesses

$$\mathbf{\Delta T} = \mathbf{A}^{\mathbf{T}}\mathbf{\Theta} \tag{7.33}$$

If Eqs. (7.34) are premultiplied by **A** and set equal to zero in accordance with Eqs. (7.29),

$$\mathbf{AQ} = \mathbf{0} = \mathbf{AQ}_s - \mathbf{AY\Delta T}_s + \mathbf{AY\Delta T}$$

or

$$\mathbf{AY\Delta T} = \mathbf{AY\Delta T}_s - \mathbf{AQ}_s \tag{7.35}$$

Then, a node heat source vector may be defined

$$\tilde{\mathbf{Q}} \equiv \mathbf{AY\Delta T}_s - \mathbf{AQ}_s \tag{7.36}$$

so that

$$\mathbf{AY\Delta T} = \tilde{\mathbf{Q}}$$

If Eqs. (7.33) are inserted here,

$$\mathbf{AYA}^{\mathbf{T}}\Theta = \tilde{\mathbf{Q}}$$

The node equations result

$$\mathbf{Y}_n\Theta = \tilde{\mathbf{Q}} \tag{7.37}$$

where the node admittance matrix, $\mathbf{Y}_n$, is defined by

$$\mathbf{Y}_n \equiv \mathbf{AYA}^{\mathbf{T}} \tag{7.38}$$

Solution of Eqs. (7.37)

$$\Theta = \mathbf{Y}_n^{-1}\tilde{\mathbf{Q}} \tag{7.39}$$

yields the temperature excess at each node, and, if desired, input admittances can then be evaluated at each point where heat is injected into the array.

7.5 TWO EXAMPLES

Example 7.3: In the finned array shown in Fig. 7.8, all fins are identical with $b = 4$ cm, $\delta = 0.250$ cm, and $L = 20$ cm. If $k = 200$ W/m-K and $h = 100$ W/m^2-K, find the temperature excesses at all points where heat is injected and at the fin intersection point. Heat inputs of 40, 80, and 20 W are indicated. The nomenclature, $q_{b,1}$, $q_{b,2}$, and $q_{b,3}$ suggests that these heat inputs might just as well be placed at nodes 1, 2, and 3.

SOLUTION All fins are identical with

$$m = \left(\frac{2h}{k\delta}\right)^{1/2} = \left[\frac{2(100)}{200(0.0025}\right]^{1/2} = 20.0000\,\text{m}^{-1}$$

$$mb = 20.0000(0.0400) = 0.8000$$

and

$$Y_o = (2hk\delta)^{1/2}L = [2(100)(200)(0.0025)]^{1/2}(0.20) = 2.0000\,\text{W/K}$$

All have identical $\mathbf{Y}$-matrices:

$$\begin{aligned} y_{11} &= Y_o \coth mb \\ &= (2.0000)(1.5059) \end{aligned}$$

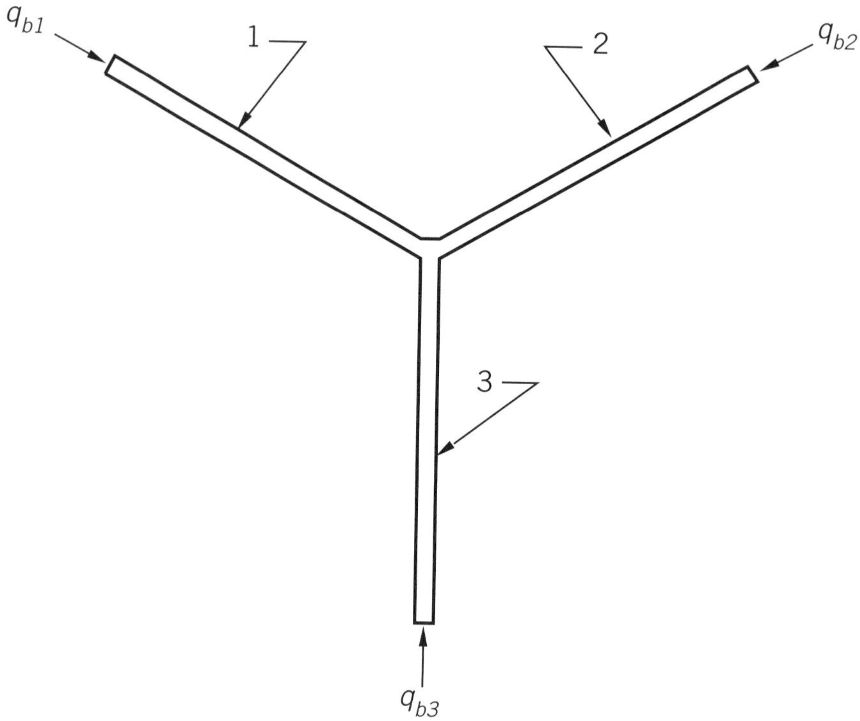

Figure 7.8 Finned array for Ex. 7.3.

$$
\begin{aligned}
&= 3.0119\,\mathrm{W/K} \\
y_{12} &= -Y_o \operatorname{csch} mb \\
&= -(2.0000)(1.1260) \\
&= -2.2520\,\mathrm{W/K} \\
y_{21} &= Y_o \operatorname{csch} mb \\
&= 2.2520\,\mathrm{W/K}
\end{aligned}
$$

and

$$
\begin{aligned}
y_{22} &= -Y_o \coth mb \\
&= -3.0119\ \mathrm{W/K}
\end{aligned}
$$

Then, by Eqs. (7.24),

$$Y_1 = -y_{12} = y_{21} = 2.2520\,\mathrm{W/K}$$

$$Y_2 = y_{11} + y_{21} = 3.0119 + (-2.2520) = 0.7599\,\mathrm{W/K}$$

and

$$Y_3 = y_{12} - y_{22} = -2.2520 - (-3.0119) = 0.7599\,\mathrm{W/K}$$

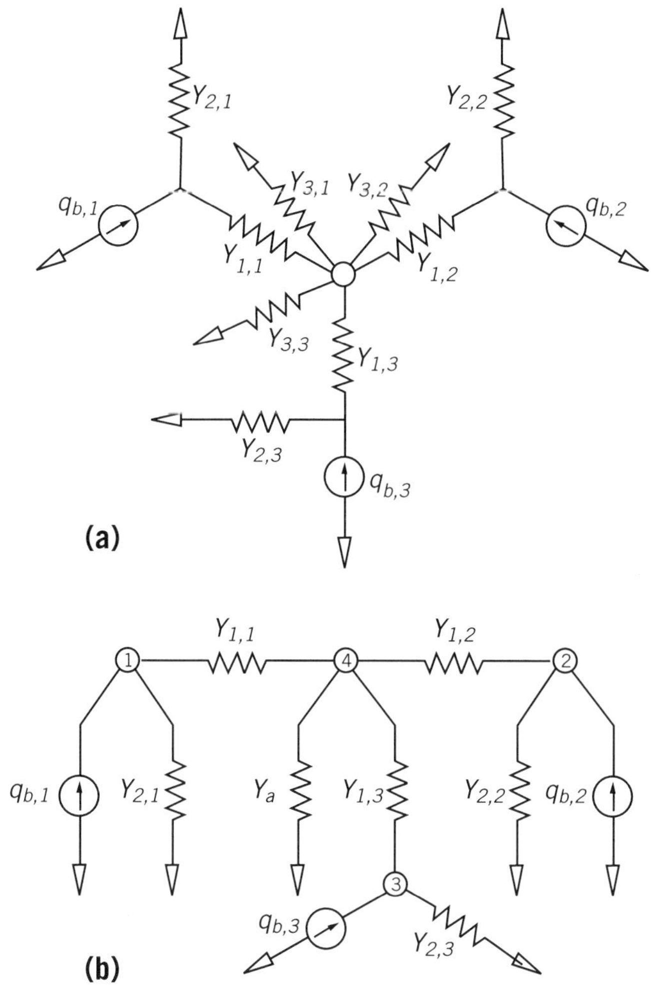

Figure 7.9 Equivalent network for the finned array of Fig. 6.8 and (*b*) a simplification of (*a*). Here, the heat source is designated by the subscript, b.

The equivalent pi-network is displayed in Fig. 7.9*a* and a simplification using Y_a is shown in Fig. 7.9*b*. Here

$$Y_a = Y_{3,1} + Y_{3,2} + Y_{3,3}$$

and because all of these particular Y's have the same value of 0.7599 W/K, then

$$Y_a = 3(0.7599) = 2.2797\ \mathrm{W/K}$$

An oriented array graph is shown in Fig. 7.10, and it may be noted that the graph contains seven branches and five nodes with node 5 as the datum node.

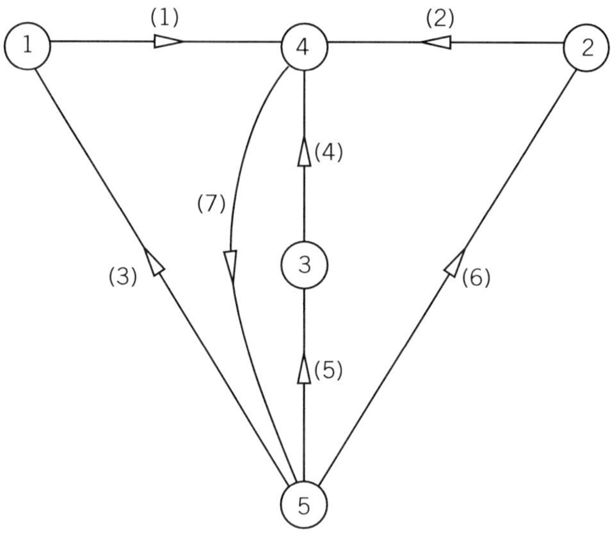

Figure 7.10 Oriented array graph for the network in Fig. 7.9*b*.

The augmented node branch incidence matrix is 5×7:

$$\mathbf{A}_a = \begin{bmatrix} 1 & 0 & -1 & 0 & 0 & 0 & 0 \\ 0 & 1 & 0 & 0 & 0 & -1 & 0 \\ 0 & 0 & 0 & 1 & -1 & 0 & 0 \\ -1 & -1 & 0 & -1 & 0 & 0 & 1 \\ 0 & 0 & 1 & 0 & 1 & 1 & -1 \end{bmatrix}$$

The node branch incidence matrix is 4×7 and is obtained by removing the datum node (row 5),

$$\mathbf{A} = \begin{bmatrix} 1 & 0 & -1 & 0 & 0 & 0 & 0 \\ 0 & 1 & 0 & 0 & 0 & -1 & 0 \\ 0 & 0 & 0 & 1 & -1 & 0 & 0 \\ -1 & -1 & 0 & -1 & 0 & 0 & 1 \end{bmatrix}$$

and this has a transpose which is 7×4:

$$\mathbf{A}^T = \begin{bmatrix} 1 & 0 & 0 & -1 \\ 0 & 1 & 0 & -1 \\ -1 & 0 & 0 & 0 \\ 0 & 0 & 1 & -1 \\ 0 & 0 & -1 & 0 \\ 0 & -1 & 0 & 0 \\ 0 & 0 & 0 & 1 \end{bmatrix}$$

The branch admittance matrix is 7 × 7 and is diagonal

$$\mathbf{Y} = \begin{bmatrix} 2.2520 & 0 & 0 & 0 & 0 & 0 & 0 \\ 0 & 2.2520 & 0 & 0 & 0 & 0 & 0 \\ 0 & 0 & 0.7599 & 0 & 0 & 0 & 0 \\ 0 & 0 & 0 & 2.2520 & 0 & 0 & 0 \\ 0 & 0 & 0 & 0 & 0.7599 & 0 & 0 \\ 0 & 0 & 0 & 0 & 0 & 0.7599 & 0 \\ 0 & 0 & 0 & 0 & 0 & 0 & 2.2797 \end{bmatrix}$$

There will be a heat source vector that is 7 × 1

$$\mathbf{Q} = \begin{bmatrix} 0 \\ 0 \\ 40 \\ 0 \\ 20 \\ 80 \\ 0 \end{bmatrix}$$

and the temperature source vector that is 7 × 1 is null

$$\Delta\mathbf{T}_s = \mathbf{0}$$

The node admittance matrix, $\mathbf{Y}_n$, is defined by Eq. (7.38):

$$\mathbf{Y}_n = \mathbf{AYA}^{\mathbf{T}} \tag{7.38}$$

Thus

$$\mathbf{Y}_n = \begin{bmatrix} 1 & 0 & -1 & 0 & 0 & 0 & 0 \\ 0 & 1 & 0 & 0 & 0 & -1 & 0 \\ 0 & 0 & 0 & 1 & -1 & 0 & 0 \\ -1 & -1 & 0 & -1 & 0 & 0 & 1 \end{bmatrix} \bullet$$

$$\begin{bmatrix} 2.2520 & 0 & 0 & 0 & 0 & 0 & 0 \\ 0 & 2.2520 & 0 & 0 & 0 & 0 & 0 \\ 0 & 0 & 0.7599 & 0 & 0 & 0 & 0 \\ 0 & 0 & 0 & 2.2520 & 0 & 0 & 0 \\ 0 & 0 & 0 & 0 & 0.7599 & 0 & 0 \\ 0 & 0 & 0 & 0 & 0 & 0.7599 & 0 \\ 0 & 0 & 0 & 0 & 0 & 0 & 2.2797 \end{bmatrix} \bullet$$

$$\begin{bmatrix} 1 & 0 & 0 & -1 \\ 0 & 1 & 0 & -1 \\ -1 & 0 & 0 & 0 \\ 0 & 0 & 1 & -1 \\ 0 & 0 & -1 & 0 \\ 0 & -1 & 0 & 0 \\ 0 & 0 & 0 & 1 \end{bmatrix}$$

$$= \begin{bmatrix} 3.0119 & 0 & 0 & -2.2520 \\ 0 & 3.0119 & 0 & -2.2520 \\ 0 & 0 & 3.0119 & -2.2520 \\ -2.2520 & -2.2520 & -2.2520 & 9.0357 \end{bmatrix}$$

The node heat source vector is 4×1 is given by Eq. (7.36):

$$\tilde{\mathbf{Q}} = \mathbf{AY}\Delta\mathbf{T}_s - \mathbf{AQ}_s \tag{7.36}$$

and because $\Delta\mathbf{T}_s$ is null

$$\tilde{\mathbf{Q}} = -\mathbf{AQ}_s$$

$$= -\begin{bmatrix} 1 & 0 & -1 & 0 & 0 & 0 & 0 \\ 0 & 1 & 0 & 0 & 0 & -1 & 0 \\ 0 & 0 & 0 & 1 & -1 & 0 & 0 \\ -1 & -1 & 0 & -1 & 0 & 0 & 1 \end{bmatrix} \begin{bmatrix} 0 \\ 0 \\ 40 \\ 0 \\ 20 \\ 80 \\ 0 \end{bmatrix} = \begin{bmatrix} 40 \\ 80 \\ 20 \\ 0 \end{bmatrix}$$

Thus, the node equations are in accordance with Eqs. (7.37),

$$\mathbf{Y}_n \boldsymbol{\theta} = \tilde{\mathbf{Q}}$$

$$\begin{bmatrix} 3.0119 & 0 & 0 & -2.2520 \\ 0 & 3.0119 & 0 & -2.2520 \\ 0 & 0 & 3.0119 & -2.2520 \\ -2.2520 & -2.2520 & -2.2520 & 9.0357 \end{bmatrix} \begin{bmatrix} \theta_1 \\ \theta_2 \\ \theta_3 \\ \theta_4 \end{bmatrix} = \begin{bmatrix} 40 \\ 80 \\ 20 \\ 0 \end{bmatrix}$$

and solution by a matrix inversion gives

$$\begin{bmatrix} \theta_1 \\ \theta_2 \\ \theta_3 \\ \theta_4 \end{bmatrix} = \begin{bmatrix} 32.93 \\ 46.21 \\ 26.28 \\ 26.27 \end{bmatrix}$$

Example 7.4: Confirm the result of Ex. 7.1 using node analysis. The configuration is given in Fig. 7.2, which is repeated here for convenience as Fig. 7.11*a*. Pertinent dimensions and thermal parameters can be obtained by referring to Ex. 7.1. Recall that all fins may be treated as longitudinal fins of rectangular profile and that $|q_{b,1}| = 20$ W and $|q_{a,4}| = 16$ W.

SOLUTION This configuration was employed in the development of the node analysis method, and much of the work for the setup of the node equations appears elsewhere throughout this chapter. To begin, refer to Fig. 7.11*a*, and recall from Ex. 7.1 that for fins 2 and 3,

$$m_2 b_2 = m_3 b_3 = 44.3678(0.001144) = 0.0508$$

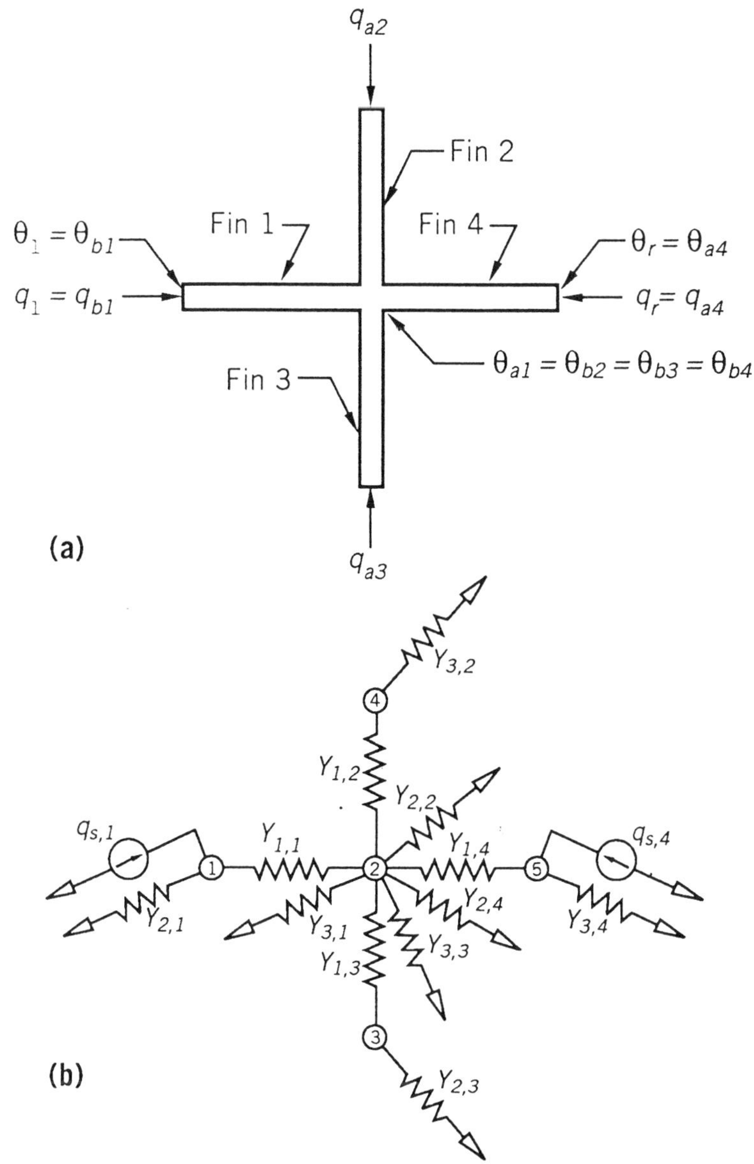

Figure 7.11 (*a*) An array of four fins with heat input at opposite ends and (*b*) an equivalent network composed of four pi-networks to represent the array. The finned array in (*a*) was considered in Ex. 7.1.

and

$$\begin{aligned} Y_{o,2} &= Y_{o,3} = (2hk\delta)^{1/2}L \\ &= [2(50)(200)(0.000254)]^{1/2}(0.4000) = 0.9016\,\text{W/K} \end{aligned}$$

For fins 1 and 4

$$m_1b_1 = m_4b_4 = 57.2786(0.00635) = 0.3637$$

and

$$\begin{aligned} Y_{o,1} &= Y_{o,4} = (2hk\delta)^{1/2}L \\ &= [2(50)(200)(0.0001524)]^{1/2}(0.4000) = 0.6983\,\text{W/K} \end{aligned}$$

The **Y**-matrix for fins 2 and 3, which possess identical dimensions, is obtained from Eqs. (3.11):

$$\begin{aligned} y_{11,2} = y_{11,3} &= Y_{o,2}\coth m_2b_2 \\ &= 0.9016(19.7187) \\ &= 17.7775\,\text{W/K} \\ y_{12,2} = y_{12,3} &= -Y_{o,2}\,\text{csch}\, m_2b_2 \\ &= -0.9016(19.6933) \\ &= -17.7546\,\text{W/K} \\ y_{21,2} = y_{21,3} &= Y_{o,2}\,\text{csch}\, m_2b_2 \\ &= 17.7546\,\text{W/K} \end{aligned}$$

and

$$\begin{aligned} y_{22,2} = y_{22,3} &= -Y_{o,2}\coth m_2b_2 \\ &= -17.7775\,\text{W/K} \end{aligned}$$

Then, by Eqs. (7.24),

$$\begin{aligned} Y_{1,2} = Y_{1,3} &= -y_{12,2} = y_{21,2} \\ &= 17.7546\,\text{W/K} \\ Y_{2,2} = Y_{2,3} &= y_{11,2} - y_{21,2} \\ &= 17.7753 + (-17.7546) \\ &= 0.0229\,\text{W/K} \end{aligned}$$

and

$$\begin{aligned} Y_{3,2} = Y_{3,3} &= y_{12,2} - y_{22,2} \\ &= -17.7546 - (-17.7775) \\ &= 0.0229\,\text{W/K} \end{aligned}$$

The $\mathbf{Y}$-matrix for fins 1 and 4 which also possess identical dimensions, is obtained from Eqs. (3.11):

$$\begin{aligned} y_{11,1} = y_{11,4} &= Y_{o,1} \coth m_1 b_1 \\ &= 0.6983(2.8696) \\ &= 2.0039 \text{ W/K} \\ y_{12,1} = y_{12,4} &= -Y_{o,1} \operatorname{csch} m_1 b_1 \\ &= -0.6983(2.6897) \\ &= -1.8783 \text{ W/K} \\ y_{21,1} = y_{21,4} &= Y_{o,1} \operatorname{csch} m_1 b_1 \\ &= 1.8783 \text{ W/K} \end{aligned}$$

and

$$\begin{aligned} y_{22,1} = y_{22,4} &= -Y_{o,1} \coth m_1 b_1 \\ &= -2.0039 \text{ W/K} \end{aligned}$$

Then, by Eqs. (7.24),

$$\begin{aligned} Y_{1,1} = Y_{1,4} &= -y_{12,1} = y_{21,1} \\ &= 1.8783 \text{ W/K} \\ Y_{2,1} = Y_{2,4} &= y_{11,1} + y_{21,1} \\ &= 2.0039 + (-1.8783) \\ &= 0.1256 \text{ W/K} \end{aligned}$$

and

$$\begin{aligned} Y_{3,1} = Y_{3,4} &= y_{12,1} - y_{22,1} \\ &= -1.8783 - (-2.0039) \\ &= 0.1256 \text{ W/K} \end{aligned}$$

The equivalent pi-network for the four fins is displayed in Fig. 7.11*b*, and a simplification of the network is shown in Fig. 7.12*a*. The oriented array graph is shown in Fig. 7.12*b*, and it is to be noted that $q_{s1} = 20$ W is in branch 1, and $q_{s4} = 16$ W is in branch 4. Here, the value of Y_a is seen to be

$$\begin{aligned} Y_a &= Y_{3,1} + Y_{2,2} + Y_{3,3} + Y_{2,4} \\ &= 0.1256 + 0.0229 + 0.0229 + 0.1256 \\ &= 0.2979 \text{ W/K} \end{aligned}$$

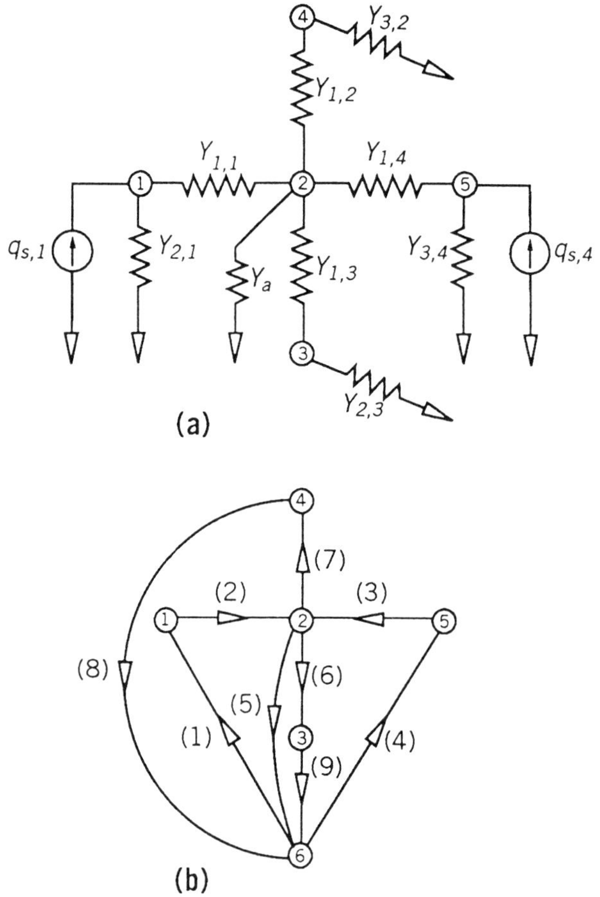

Figure 7.12 (*a*) A simplification of the network in Fig. 7.11*b* and (*b*) the oriented fin array graph.

The augmented node-branch incidence matrix and the node-branch incidence matrix have already been developed in the discussion in Sec. 7.4.3:

$$\mathbf{A}_a = \begin{bmatrix} -1 & 1 & 0 & 0 & 0 & 0 & 0 & 0 & 0 \\ 0 & -1 & -1 & 0 & 1 & 1 & 1 & 0 & 0 \\ 0 & 0 & 0 & 0 & 0 & -1 & 0 & 0 & 1 \\ 0 & 0 & 0 & 0 & 0 & 0 & -1 & 1 & 0 \\ 0 & 0 & 1 & -1 & 0 & 0 & 0 & 0 & 0 \\ 1 & 0 & 0 & 1 & -1 & 0 & 0 & -1 & -1 \end{bmatrix} \tag{7.27}$$

and with node 6 as the datum node

$$\mathbf{A} = \begin{bmatrix} -1 & 1 & 0 & 0 & 0 & 0 & 0 & 0 & 0 \\ 0 & -1 & -1 & 0 & 1 & 1 & 1 & 0 & 0 \\ 0 & 0 & 0 & 0 & 0 & -1 & 0 & 0 & 1 \\ 0 & 0 & 0 & 0 & 0 & 0 & -1 & 1 & 0 \\ 0 & 0 & 1 & -1 & 0 & 0 & 0 & 0 & 0 \end{bmatrix} \tag{7.28}$$

The branch admittance matrix is 9×9 and is diagonal. Inspection of Fig. 7.12 shows that

$$\mathbf{Y} = \begin{bmatrix} Y_{2,1} & 0 & 0 & 0 & 0 & 0 & 0 & 0 & 0 \\ 0 & Y_{1,1} & 0 & 0 & 0 & 0 & 0 & 0 & 0 \\ 0 & 0 & Y_{1,4} & 0 & 0 & 0 & 0 & 0 & 0 \\ 0 & 0 & 0 & Y_{3,4} & 0 & 0 & 0 & 0 & 0 \\ 0 & 0 & 0 & 0 & Y_a & 0 & 0 & 0 & 0 \\ 0 & 0 & 0 & 0 & 0 & Y_{1,3} & 0 & 0 & 0 \\ 0 & 0 & 0 & 0 & 0 & 0 & Y_{1,2} & 0 & 0 \\ 0 & 0 & 0 & 0 & 0 & 0 & 0 & Y_{3,2} & 0 \\ 0 & 0 & 0 & 0 & 0 & 0 & 0 & 0 & Y_{2,3} \end{bmatrix}$$

There will be no branch temperature source vector because there are no temperature sources. But because branches 1 and 4 contain heat sources, the branch heat source vector will be

$$\mathbf{Q}_s = \begin{bmatrix} 20 \\ 0 \\ 0 \\ 16 \\ 0 \\ 0 \\ 0 \\ 0 \\ 0 \end{bmatrix}$$

The solution may now proceed in accordance with the procedure in Sec. 7.4.5. First, a node admittance matrix is formulated using Eq. (7.38):

$$\mathbf{Y}_n = \mathbf{AYA}^{\mathbf{T}} \tag{7.38}$$

The reader may verify that

$$\mathbf{AY} = \begin{bmatrix} -Y_{2,1} & Y_{1,1} & 0 & 0 & 0 & 0 & 0 & 0 & 0 \\ 0 & -Y_{1,1} & -Y_{1,4} & 0 & Y_a & Y_{1,3} & Y_{1,2} & 0 & 0 \\ 0 & 0 & 0 & 0 & 0 & -Y_{1,3} & 0 & 0 & Y_{2,3} \\ 0 & 0 & 0 & 0 & 0 & 0 & -Y_{1,2} & Y_{3,2} & 0 \\ 0 & 0 & Y_{1,4} & -Y_{3,4} & 0 & 0 & 0 & 0 & 0 \end{bmatrix}$$

and then that

$$\mathbf{Y}_n = \mathbf{AYA}^T = \begin{bmatrix} Y_A & -Y_{1,1} & 0 & 0 & 0 \\ -Y_{1,1} & Y_B & -Y_{1,3} & -Y_{1,2} & -Y_{1,4} \\ 0 & -Y_{1,3} & Y_C & 0 & 0 \\ 0 & -Y_{1,2} & 0 & Y_D & 0 \\ 0 & -Y_{1,4} & 0 & 0 & Y_E \end{bmatrix}$$

where

$$\begin{aligned} Y_A &= Y_{2,1} + Y_{1,1} \\ Y_B &= Y_{1,1} + Y_{1,4} + Y_a + Y_{1,3} + Y_{1,2} \\ Y_C &= Y_{1,3} + Y_{2,3} \\ Y_D &= Y_{1,2} + Y_{3,2} \end{aligned}$$

and

$$Y_E = Y_{1,4} + Y_{3,4}$$

Use of the admittance values already developed leads to numerical values for Y_A through Y_E:

$$\begin{aligned} Y_A &= Y_{2,1} + Y_{1,1} \\ &= 0.1256 + 1.8783 \\ &= 2.0039\,\text{W/K} \end{aligned}$$

$$\begin{aligned} Y_B &= Y_{1,1} + Y_{1,4} + Y_a + Y_{1,3} + Y_{1,2} \\ &= 1.8783 + 1.8783 + 0.2970 + 17.7546 + 17.7546 \\ &= 39.5583\,\text{W/K} \end{aligned}$$

$$\begin{aligned} Y_C &= Y_{1,3} + Y_{2,3} \\ &= 17.7546 + 0.0229 \\ &= 17.7775\,\text{W/K} \end{aligned}$$

$$\begin{aligned} Y_D &= Y_{1,3} + Y_{2,3} \\ &= 17.7546 + 0.0229 \\ &= 17.7775\,\text{W/K} \end{aligned}$$

and

$$\begin{aligned} Y_E &= Y_{1,4} + Y_{3,4} \\ &= 1.8783 + 0.1256 \\ &= 2.0039\,\text{W/K} \end{aligned}$$

With all of the admittance values inserted

$$\mathbf{Y}_n = \begin{bmatrix} 2.0039 & -1.8783 & 0 & 0 & 0 \\ -1.8783 & 39.5628 & -17.7546 & -17.7546 & -1.8783 \\ 0 & -17.7546 & 17.7775 & 0 & 0 \\ 0 & -17.7546 & 0 & 17.7775 & 0 \\ 0 & -1.8783 & 0 & 0 & 2.0039 \end{bmatrix} \mathrm{K/W}$$

The node heat source vector is 4×1 is given by Eq. (7.36)

$$\tilde{\mathbf{Q}} \equiv \mathbf{AY}\Delta\mathbf{T}_s - \mathbf{AQ}_s \tag{7.36}$$

and because $\Delta\mathbf{T}_s$ is null

$$\tilde{\mathbf{Q}} = -\mathbf{AQ}_s =$$

$$-\begin{bmatrix} -1 & 1 & 0 & 0 & 0 & 0 & 0 & 0 & 0 \\ 0 & -1 & -1 & 0 & 1 & 1 & 1 & 0 & 0 \\ 0 & 0 & 0 & 0 & 0 & -1 & 0 & 0 & 1 \\ 0 & 0 & 0 & 0 & 0 & 0 & -1 & 1 & 0 \\ 0 & 0 & 1 & -1 & 0 & 0 & 0 & 0 & 0 \end{bmatrix} \begin{bmatrix} 20 \\ 0 \\ 0 \\ 16 \\ 0 \\ 0 \\ 0 \\ 0 \\ 0 \end{bmatrix} = \begin{bmatrix} 20 \\ 0 \\ 0 \\ 0 \\ 16 \end{bmatrix}$$

The formulation of the node equations is Eq. (7.37),

$$\mathbf{Y_n}\Theta = \tilde{\mathbf{Q}} \tag{7.37}$$

and the solution for the temperature excess vector is obtained from

$$\Theta = \mathbf{Y_n^{-1}}\tilde{\mathbf{Q}} \tag{7.39}$$

or $\Theta =$

$$\begin{bmatrix} 2.0039 & -1.8783 & 0 & 0 & 0 \\ -1.8783 & 39.5628 & -17.7546 & -17.7546 & -1.8783 \\ 0 & -17.7546 & 17.7775 & 0 & 0 \\ 0 & -17.7546 & 0 & 17.7775 & 0 \\ 0 & -1.8783 & 0 & 0 & 2.0039 \end{bmatrix}^{-1} \begin{bmatrix} 20 \\ 0 \\ 0 \\ 0 \\ 16 \end{bmatrix}$$

and finally

$$\begin{bmatrix} \theta_1 \\ \theta_2 \\ \theta_3 \\ \theta_1 \\ \theta_5 \end{bmatrix} = \begin{bmatrix} 64.7 \\ 58.4 \\ 58.3 \\ 58.3 \\ 62.7 \end{bmatrix} \ ^\circ\mathrm{C}$$

This matches the solution in Ex. 7.1 and is particularly gratifying in that the observation can be made that a more general method is available to back up the a priori approach illustrated in Ex. 7.1.

Of course, thought could be given to the use of a general electrical network analyzer such as *PSPICE* for the solution of more comprehensive networks that represent more complicated finned arrays. Such analyzers have the capability of handling both temperature sources (as represented by voltage sources) or heat sources (analogued as current sources). Printouts can be arranged to provide the heat flow from base to tip of every fin and the temperature at each node point. Further discussion of this important real-world attribute would need to delve into more specific cases that are beyond the scope of this chapter.

Chapter

8

A GENERAL ARRAY METHOD

8.1 INTRODUCTION

The method of node analysis discussed in Chap. 7 is solidly based on a foundation of network topology. In this chapter, a very general algorithm that also utilizes network topology will be described. This general method employs the $\mathbf{Y}$-matrix for every fin in the array and reduces the analysis of a complicated array, regardless of its size, to a formulation that does not require the generation of the equivalent pi-network required by a nodal analysis. Hence, it appears more general and does not offer the mental block that some may have for "*things that look electrical*."

8.2 THE GENERAL ARRAY ALGORITHM

To analyze an arbitrary fin configuration, with or without the preprocessing that can be provided by the cascade, cluster, and parallel algorithms discussed in Chap. 4, one first represents the array by a labeled, oriented graph with its nodes corresponding to the individual fin connecting points and with a branch for each fin in the array. The orientation arrows point from *base* to *tip* of each individual fin in the array. Thus, the graph of the array shown in Fig. 8.1 is displayed in Fig. 8.2. Note well that, unlike the graphs used in the method of node analysis, *no account is taken of the heat sources by including them in a particular branch.*

Consider the mathematical problem of solving for the e (e for *branch*)[1]

[1] e is used to avoid confusion with b, which will represent the base of each fin. The usage is based on the fact that, in graph theory, branches of a graph are frequently referred to as the *edges*.

Figure 8.1 A finned array between two dissipating surfaces.

heat flows, and temperature excesses. To obtain a rough count, disregard any singular fins. Each regular fin carries four unknowns. These are the base and tip temperature excesses and the base and tip heat flows and for e fins there are $4e$ unknowns. The **Y**-matrices account for $2e$ relations because they are linear transformations that relate two of the unknowns to the other two. At each node that represents a connection of at least two fins, there is one continuity relationship or heat balance. If ℓ_i fins adjoin the ith node, there will be $\ell_i - 1$ temperature excess balance equations. Moreover, it is observed that $\sum \ell_i = 2e$ because every regular fin adjoins exactly two nodes. Thus for n nodes, the total number of equations is therefore

$$2e + \sum_{i=1}^{n}(1 + \ell_i - 1) = 2e + \sum_{i=1}^{n} l_i = 2e + 2e = 4e$$

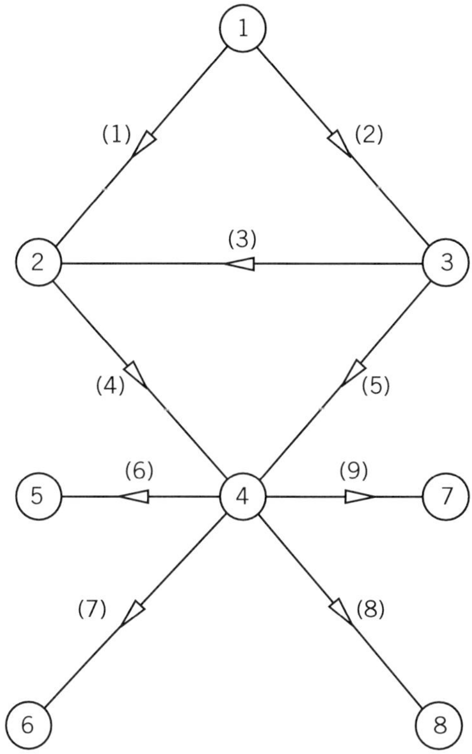

Figure 8.2 Oriented array graph for the array of Fig. 8.1. Numerals within circles designate nodes, and numerals within parentheses designate branches.

The classical strategy is to work from these $4e$ equations in $4e$ unknowns. Because $e = 7$ in Fig. 8.1 (disregarding fins 6 and 9, which may be represented by their input admittances Y_{i6} and Y_{i9}), the reader may appreciate the efficiency of the general array algorithm, which will now be developed.

To express the continuity equations systematically and efficiently, one defines an $n \times 2e$ incidence matrix containing *one* row for each node and *two* columns for each branch in the graph. The $n \times e$ incidence matrix is designated by $\mathbf{\Omega}$ with elements ω with subscript. The first column corresponding to a given branch indicates the node to which the base of that branch (that fin) is attached. The second column identifies the node to which the tip is attached. Formally, the elements of $\mathbf{\Omega}$ are

$$\begin{aligned} \omega_{i,2j-1} &= 1 \text{ if the } j\text{th base adjoins the } i\text{th node} \\ \omega_{i,2j} &= 1 \text{ if the } j\text{th tip adjoins the } i\text{th node} \\ \omega_{rs} &= 0 \text{ otherwise} \end{aligned} \tag{8.1}$$

For example. for the graph of Fig. 8.2, which represents the array of Fig. 8.1,

$\boldsymbol{\Omega}$ is given by

$$\begin{bmatrix} 1 & 0 & 1 & 0 & 0 & 0 & 0 & 0 & 0 & 0 & 0 & 0 & 0 & 0 & 0 & 0 & 0 & 0 \\ 0 & 1 & 0 & 0 & 0 & 1 & 1 & 0 & 0 & 0 & 0 & 0 & 0 & 0 & 0 & 0 & 0 & 0 \\ 0 & 0 & 0 & 1 & 1 & 0 & 0 & 0 & 1 & 0 & 0 & 0 & 0 & 0 & 0 & 0 & 0 & 0 \\ 0 & 0 & 0 & 0 & 0 & 0 & 0 & 1 & 0 & 1 & 1 & 0 & 1 & 0 & 1 & 0 & 1 & 0 \\ 0 & 0 & 0 & 0 & 0 & 0 & 0 & 0 & 0 & 0 & 0 & 1 & 0 & 0 & 0 & 0 & 0 & 0 \\ 0 & 0 & 0 & 0 & 0 & 0 & 0 & 0 & 0 & 0 & 0 & 0 & 0 & 1 & 0 & 0 & 0 & 0 \\ 0 & 0 & 0 & 0 & 0 & 0 & 0 & 0 & 0 & 0 & 0 & 0 & 0 & 0 & 0 & 0 & 0 & 1 \\ 0 & 0 & 0 & 0 & 0 & 0 & 0 & 0 & 0 & 0 & 0 & 0 & 0 & 0 & 0 & 1 & 0 & 0 \end{bmatrix}$$

Now, the continuity equations for the heat flow state that, at every node, the heat leaving each incident fin base must be balanced by the heat entering from each incident fin tip *plus* the heat input from the sources, if any. Thus at the ith node

$$\sum q_{b,j} - \sum q_{a,j} = q_{s,i} \tag{8.2}$$

where the first sum extends over the fins with bases incident to node i, and the second extends over those with tips incident to node i. The term $q_{s,i}$ represents the source heat injection at node i.

One can read off from the $+1$ entries of the ith row of $\boldsymbol{\Omega}$ precisely which terms come into the sum of Eq. (8.2). If the heat flows are arranged in a column vector for e branches as

$$\hat{\mathbf{Q}} = \begin{bmatrix} q_{b,1} \\ -q_{a,1} \\ q_{b,2} \\ -q_{a,2} \\ \vdots \\ q_{b,e} \\ -q_{a,e} \end{bmatrix} \tag{8.3}$$

then the continuity equations for heat flow can be expressed as

$$\boldsymbol{\Omega}\hat{\mathbf{Q}} = \mathbf{Q}_s \tag{8.4}$$

with $\mathbf{Q}_s$ a column vector of heat sources for the n nodes:

$$\mathbf{Q}_s = \begin{bmatrix} q_{s,1} \\ q_{s,2} \\ q_{s,3} \\ \vdots \\ q_{s,n} \end{bmatrix} \tag{8.5}$$

Compatability equations for the temperature excesses simply state that the temperature excesses at a given node are equal. Forming the vector

$$\Theta = \begin{bmatrix} \theta_{b,1} \\ \theta_{a,1} \\ \theta_{b,2} \\ \theta_{a,2} \\ \vdots \\ \theta_{b,e} \\ \theta_{a,e} \end{bmatrix} \tag{8.6}$$

one observes that the node to which a given base or tip is attached can be found by looking for a plus-one (+ 1) entry in the corresponding column of $\mathbf{\Omega}$. Thus, if one constructs a column vector, $\hat{\Theta}$, whose ith component is the common value of the temperature excess at node i,

$$\hat{\Theta} = \begin{bmatrix} \theta_1 \\ \theta_2 \\ \theta_3 \\ \vdots \\ \theta_n \end{bmatrix} \tag{8.7}$$

then the compatability for the temperature excesses can be expressed as

$$\Theta = \mathbf{\Omega}^T \hat{\Theta} \tag{8.8}$$

Finally, one can express all of the thermal admittance equations by means of the $\mathbf{Y}$-matrix for a single fin given by Eq. (2.13), repeated here as Eq. (8.9):

$$\begin{bmatrix} q_b \\ q_a \end{bmatrix} = \mathbf{Y} \begin{bmatrix} \theta_b \\ \theta_a \end{bmatrix} = \begin{bmatrix} y_{11} & y_{12} \\ y_{21} & y_{22} \end{bmatrix} \begin{bmatrix} \theta_b \\ \theta_a \end{bmatrix} \tag{8.9}$$

via a single partitioned array graph incidence matrix, $\mathbf{Y}_g$,

$$\mathbf{Y}_g = \begin{bmatrix} \mathbf{Y}_1 & \mathbf{0} & \mathbf{0} & \cdots & \mathbf{0} \\ \mathbf{0} & \mathbf{Y}_2 & \mathbf{0} & \cdots & \mathbf{0} \\ \mathbf{0} & \mathbf{0} & \mathbf{Y}_3 & \cdots & \mathbf{0} \\ \cdots & \cdots & \cdots & \cdots & \cdots \\ \mathbf{0} & \mathbf{0} & \mathbf{0} & \cdots & \mathbf{Y}_n \end{bmatrix} \tag{8.10}$$

where each $\mathbf{Y}$-matrix in $\mathbf{Y}_g$ is 2×2.

The formulation for the entire array in terms of $\mathbf{Y}_g$ and the column vector of heat flows

$$\mathbf{Q} = \begin{bmatrix} q_{b,1} \\ q_{a,1} \\ q_{b,2} \\ q_{a,2} \\ \vdots \\ q_{b,e} \\ q_{a,e} \end{bmatrix} \tag{8.11}$$

is similar to the formulation for a single fin given by Eq. (8.9).[2] Here, Eqs. (8.6), (8.8), and (8.10) are used to yield

$$\mathbf{Q} = \mathbf{Y}_g \Theta \tag{8.12}$$

Then, a combination of Eqs. (8.4), (8.8), and (8.12) results in the node equations for the n temperature excesses, one at each node

$$\mathbf{K}\hat{\Theta} = \mathbf{Q} \tag{8.13}$$

where

$$\mathbf{K} = \Omega \mathbf{S} \mathbf{Y}_g \Omega^T \tag{8.14}$$

and where $\mathbf{S}$ is a *sign-correcting* matrix

$$\mathbf{S} = \begin{bmatrix} 1 & 0 & 0 & 0 & \cdots & 0 & 0 \\ 0 & -1 & 0 & 0 & \cdots & 0 & 0 \\ 0 & 0 & 1 & 0 & \cdots & 0 & 0 \\ 0 & 0 & 0 & -1 & \cdots & 0 & 0 \\ \cdots & \cdots & \cdots & \cdots & \cdots & \cdots & \cdots \\ 0 & 0 & 0 & 0 & \cdots & 1 & 0 \\ 0 & 0 & 0 & 0 & \cdots & 0 & -1 \end{bmatrix} \tag{8.15}$$

relating $\hat{\mathbf{Q}}$ and $\mathbf{Q}$ via

$$\hat{\mathbf{Q}} = \mathbf{S}\mathbf{Q} \tag{8.16}$$

The awkward appearance of $\mathbf{S}$ in the node equations of Eq. (8.13) could be eliminated by reversing the sign convention for q_a in the original formulation for the linear transformations. This would cause difficulty elsewhere and would practically invalidate the cascade algorithm.

The node equations can be constructed from the fin graph, and the number is obviously one for each node. However, preprocessing by the cascade, cluster, and parallel algorithms may reduce this number, and by a special technique that is applicable to nodes occurring at isolated fin tips, such as nodes 5 and 7 in Fig. 8.2, the number may be reduced still further. This technique will be described in the next section.

8.3 A TECHNIQUE FOR NODE REDUCTION

If the tip of a particular fin has no other fins or sources connected to it (such as fins 6 and 9 in Fig. 8.2), the corresponding nodes (nodes 5 and 7) can be eliminated. Observe that, inasmuch as such fins are effectively singular because the tips are adiabatic, their contribution to the array graph admittance matrix, $\mathbf{Y}_g$, produces a row of zeroes, because for singular fins

$$\begin{bmatrix} q_b \\ q_a \end{bmatrix} = \begin{bmatrix} Y_{in} & 0 \\ 0 & 0 \end{bmatrix} \begin{bmatrix} \theta_b \\ \theta_a \end{bmatrix} \tag{8.17}$$

[2] The difference between Eqs. (8.3) and (8.11) is readily apparent.

where Y_{in} is the fin input admittance. Thus, the equation is redundant, and the associated node temperature excess does not appear in the system represented by Eq. (8.13).

Of course, there is no loss of information because the performance (the input admittance) of such a fin does not depend on its tip temperature, as is demonstrated by the definition of the input admittance

$$Y_{in} \equiv \frac{q_b}{\theta_b}$$

To take advantage of this, one may delete the fin from further consideration, replacing it in the scheme of things by an equivalent *negative* heat source at its *base* node, injecting heat *into* the node at a rate

$$q_{in} = -Y_{in}\theta_i \tag{8.18}$$

This substitution produces no new unknowns and preserves the linearity of the node equations.

8.4 THREE EXAMPLES

Three examples will now be considered. The first considers the fin configuration in Fig. 8.1 whose array graph is displayed in Fig. 8.2. The second and third are reworks of Exs. 6.3 and 6.4.

Example 8.1: Apply the techniques discussed in this chapter to determine the total heat dissipated by the array of Fig. 8.2 if heat is injected at the base of fin 1 and the tips of fins 7 and 8 with $\theta_1 = \theta_6 = \theta_8$ (the temperature excesses at nodes 1, 6, and 8) are equal.

SOLUTION Refer to Fig. 8.2 and observe that fins 6 and 9 form a cluster. Thus, by the cluster algorithm of Eq. (5.10), their Y_{in}'s may be added. In addition, fins 7 and 8 are connected in parallel so that, in accordance with the parallel algorithm of Eq. (5.16), their $\mathbf{Y}$-matrices may be added. These preliminary simplifications are examples of what has heretofore been referred to as *preprocessing*, and their consideration permits the array graph to be reduced to the graph shown in Fig. 8.3, with fins 6 and 7 appropriately modified. The location of the heat sources in accordance with the problem statement is also shown in Fig. 8.3.

As mentioned in the previous section, node 5 and fin 6 can be replaced by an effective heat source feeding node 4 at the rate of

$$q_4 = -Y_{i6}\theta_4$$

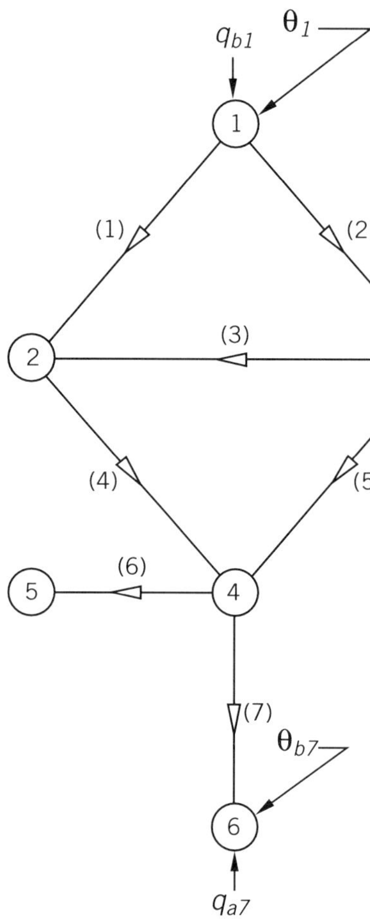

Figure 8.3 Array graph of Fig. 8.2 with preliminary simplifications.

The array graph incidence matrix for Fig. 8.3 then becomes

$$\boldsymbol{\Omega} = \left[\begin{array}{cc|cccccccccccc} \mathrm{N} & 1 & 1 & 0 & 1 & 0 & 0 & 0 & 0 & 0 & 0 & 0 & 0 & 0 \\ \mathrm{O} & 2 & 0 & 1 & 0 & 0 & 0 & 1 & 1 & 0 & 0 & 0 & 0 & 0 \\ \mathrm{D} & 3 & 0 & 0 & 0 & 1 & 1 & 0 & 0 & 0 & 1 & 0 & 0 & 0 \\ \mathrm{E} & 4 & 0 & 0 & 0 & 0 & 0 & 0 & 0 & 1 & 0 & 1 & 1 & 0 \\ \mathrm{S} & 6 & 0 & 0 & 0 & 0 & 0 & 0 & 0 & 0 & 0 & 0 & 0 & 1 \end{array}\right]$$

There will be five node temperature excesses for nodes 1, 2, 3, 4, and 6

$$\Theta = \begin{bmatrix} \theta_1 \\ \theta_2 \\ \theta_3 \\ \theta_4 \\ \theta_6 \end{bmatrix}$$

and five heat sources (some of which are zero) for nodes 1, 2, 3, 4, and 6:

$$\mathbf{Q}_s = \begin{bmatrix} q_{b,1} \\ 0 \\ 0 \\ -Y_{in,6}\theta_4 \\ q_{a,7} \end{bmatrix}$$

The matrix $\mathbf{Y}_g$ is given by

$$\mathbf{Y}_g = \begin{bmatrix} \mathbf{Y}_1 & \mathbf{0} & \mathbf{0} & \mathbf{0} & \mathbf{0} & \mathbf{0} \\ \mathbf{0} & \mathbf{Y}_2 & \mathbf{0} & \mathbf{0} & \mathbf{0} & \mathbf{0} \\ \mathbf{0} & \mathbf{0} & \mathbf{Y}_3 & \mathbf{0} & \mathbf{0} & \mathbf{0} \\ \mathbf{0} & \mathbf{0} & \mathbf{0} & \mathbf{Y}_4 & \mathbf{0} & \mathbf{0} \\ \mathbf{0} & \mathbf{0} & \mathbf{0} & \mathbf{0} & \mathbf{Y}_5 & \mathbf{0} \\ \mathbf{0} & \mathbf{0} & \mathbf{0} & \mathbf{0} & \mathbf{0} & \mathbf{Y}_7 \end{bmatrix} \tag{8.19}$$

The matrix $\mathbf{K}$ is first assembled in accordance with Eqs. (8.13),

$$\mathbf{K} = \mathbf{\Omega S Y}_g \mathbf{\Omega}^T \tag{8.13}$$

so that the node equations become

$$\mathbf{K} \begin{bmatrix} \theta_1 \\ \theta_2 \\ \theta_3 \\ \theta_4 \\ \theta_6 \end{bmatrix} = \begin{bmatrix} q_{b,1} \\ 0 \\ 0 \\ -Y_{in,6}\theta_4 \\ q_{a,7} \end{bmatrix} \tag{8.20}$$

Observe that if θ_1 and θ_6 are specified, Eq. (8.20) expresses five linear equations in the unknowns $\theta_2, \theta_3, \theta_4, \theta_{b,1}$, and $q_{a,7}$. On the other hand, if $q_{b,1}$ and $q_{a,7}$ are specified, then Eq. (8.20) expresses five linear equations for $\theta_1, \theta_2, \theta_3, \theta_4$, and θ_6, from which θ_1 and θ_6 are calculated. Unfortunately, θ_4 is an unknown and $-Y_{i6}\theta_4$ appears on the right-hand side of Eq. (8.20).

This is easily remedied by adding Y_{i6} to the (4,4) entry of $\mathbf{K}$ and calling the resulting matrix $\mathbf{J}$. When this is done, Eq. (8.20) becomes

$$\mathbf{J} \begin{bmatrix} \theta_1 \\ \theta_2 \\ \theta_3 \\ \theta_4 \\ \theta_6 \end{bmatrix} = \begin{bmatrix} q_{b,1} \\ 0 \\ 0 \\ 0 \\ q_{a,7} \end{bmatrix} \tag{8.21}$$

A matrix inversion then yields the unknown θ's:

$$\begin{bmatrix} \theta_1 \\ \theta_2 \\ \theta_3 \\ \theta_4 \\ \theta_6 \end{bmatrix} = \mathbf{J}^{-1} \begin{bmatrix} q_{b,1} \\ 0 \\ 0 \\ 0 \\ q_{a,7} \end{bmatrix} \tag{8.22}$$

Now, in the practical case where $q_{b,1}$ and $q_{a,7}$ are specified and it is desired to know what temperature excesses are required for their dissipation, a single performance index can be obtained when $\theta_1 = \theta_6$. It is the ratio of the total required heat dissipation to the temperature excess on either side,[3] namely,

$$Y_x = \frac{q_{b,1} + q_{b,7}}{\theta_1} \tag{8.23}$$

where the subscript, x, is intended to suggest *exchanger*.

Observe that in Eq. (8.22), only the corner elements of $\mathbf{J}^{-1}$ are important to this particular problem. If these are labeled $u, v, w,$ and z, then

$$\mathbf{J}^{-1} = \begin{bmatrix} u & \cdots & \cdots & \cdots & v \\ \vdots & & & & \vdots \\ \vdots & & & & \vdots \\ \vdots & & & & \vdots \\ w & \cdots & \cdots & \cdots & z \end{bmatrix}$$

and one finds that

$$\theta_1 = uq_{b,1} + vq_{a,7}$$

and

$$\theta_6 = wq_{b,1} + zq_{a,7}$$

With $\theta_1 = \theta_6$, a little algebra shows that

$$Y_x = \frac{q_{b,1} + q_{b,7}}{\theta_1} = \frac{u + z - v - w}{uz - vw}$$

and the required $\theta_1 = \theta_6$ to dissipate $q_{b,1} + q_{b,7}$ will be

$$\theta_1 = \theta_6 = Y_x(q_{b,1} + q_{b,7})$$

[3]As in a compact heat exchanger core.

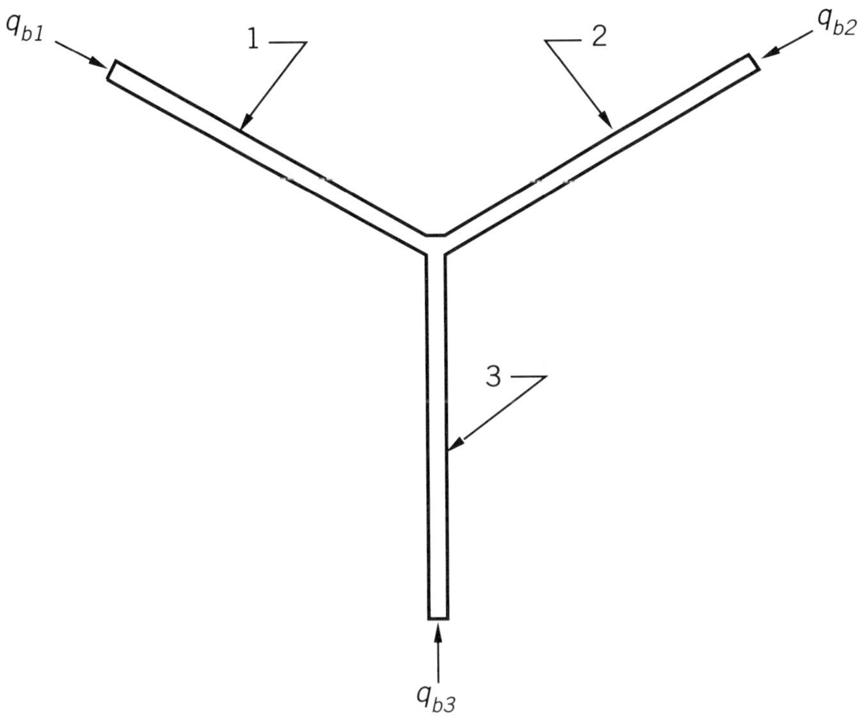

Figure 8.4 The array of Ex. 7.3. The three fins have identical dimensions, and all three have a heat input.

Example 8.3: Rework Ex. 7.3 by the general array method. The configuration is shown in Fig. 7.8 and is repeated here as Fig. 8.4. The tips of fins 1, 2, and 3 form node 4. The heat inputs to fins 1, 2, and 3 are 40, 80, and 20 W respectively. The objective is to prove that

$$\begin{bmatrix} \theta_1 \\ \theta_2 \\ \theta_3 \\ \theta_4 \end{bmatrix} = \begin{bmatrix} 32.9 \\ 46.2 \\ 26.3 \\ 26.3 \end{bmatrix} {}^\circ C$$

SOLUTION The **Y**-matrices for the three identical fins are obtained from Ex. 7.3:

$$\mathbf{Y} = \begin{bmatrix} 3.0119 & -2.2520 \\ 2.2520 & -3.0119 \end{bmatrix} \text{W/K}$$

The array graph for the configuration of Fig. 8.4 is shown in Fig. 8.5, and the array graph incidence matrix for Fig. 8.5 is 4×6 (four nodes and three

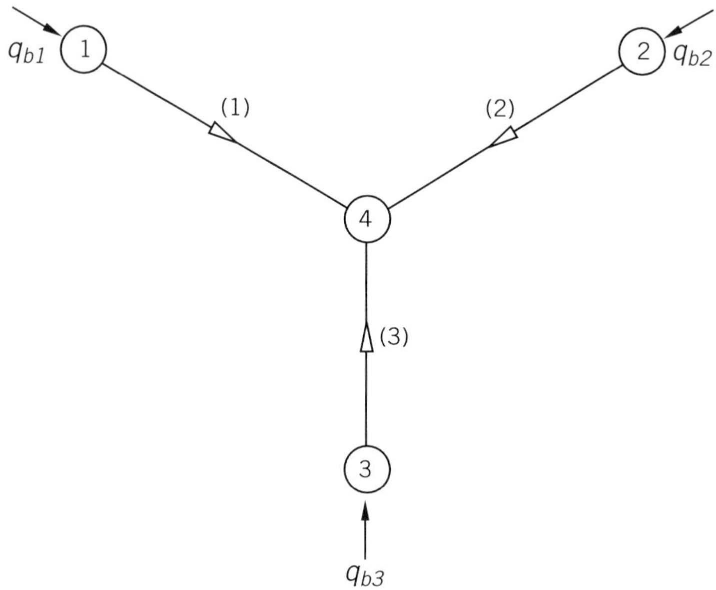

Figure 8.5 Oriented array graph for the array of Fig. 8.4.

branches),

$$\mathbf{\Omega} = \begin{bmatrix} 1 & 0 & 0 & 0 & 0 & 0 \\ 0 & 0 & 1 & 0 & 0 & 0 \\ 0 & 0 & 0 & 0 & 1 & 0 \\ 0 & 1 & 0 & 1 & 0 & 1 \end{bmatrix}$$

the temperature excess vector is 4×1,

$$\Theta = \begin{bmatrix} \theta_1 \\ \theta_2 \\ \theta_3 \\ \theta_4 \end{bmatrix}$$

and the heat source vector is 4×1,

$$\mathbf{Q}_s = \begin{bmatrix} q_{s,1} \\ q_{s,2} \\ q_{s,3} \\ q_{s,4} \end{bmatrix} = \begin{bmatrix} 40 \\ 80 \\ 20 \\ 0 \end{bmatrix}$$

The fin graph admittance matrix has the form

$$\mathbf{Y}_g = \begin{bmatrix} 3.0119 & -2.2520 & 0 & 0 & 0 & 0 \\ 2.2520 & -3.0119 & 0 & 0 & 0 & 0 \\ 0 & 0 & 3.0119 & -2.2520 & 0 & 0 \\ 0 & 0 & 2.2520 & -3.0119 & 0 & 0 \\ 0 & 0 & 0 & 0 & 3.0119 & -2.2520 \\ 0 & 0 & 0 & 0 & 2.2520 & -3.0119 \end{bmatrix} \text{W/K}$$

Then

$$\mathbf{K} = \mathbf{\Omega} \mathbf{S} \mathbf{Y}_g \mathbf{\Omega}^T$$

and with

$$\mathbf{\Omega S} = \begin{bmatrix} 1 & 0 & 0 & 0 & 0 & 0 \\ 0 & 0 & 1 & 0 & 0 & 0 \\ 0 & 0 & 0 & 0 & 1 & 0 \\ 0 & 1 & 0 & 1 & 0 & 1 \end{bmatrix} \begin{bmatrix} 1 & 0 & 0 & 0 & 0 & 0 \\ 0 & -1 & 0 & 0 & 0 & 0 \\ 0 & 0 & 1 & 0 & 0 & 0 \\ 0 & 0 & 0 & -1 & 0 & 0 \\ 0 & 0 & 0 & 0 & 1 & 0 \\ 0 & 0 & 0 & 0 & 0 & -1 \end{bmatrix}$$

$$= \begin{bmatrix} 1 & 0 & 0 & 0 & 0 & 0 \\ 0 & 0 & 1 & 0 & 0 & 0 \\ 0 & 0 & 0 & 0 & 1 & 0 \\ 0 & -1 & 0 & -1 & 0 & -1 \end{bmatrix}$$

and

$$\mathbf{Y}_g \mathbf{\Omega}^T = \begin{bmatrix} 3.0119 & -2.2520 & 0 & 0 & 0 & 0 \\ 2.2520 & -3.0119 & 0 & 0 & 0 & 0 \\ 0 & 0 & 3.0119 & -2.2520 & 0 & 0 \\ 0 & 0 & 2.2520 & -3.0119 & 0 & 0 \\ 0 & 0 & 0 & 0 & 3.0119 & -2.2520 \\ 0 & 0 & 0 & 0 & 2.2520 & -3.0119 \end{bmatrix} \bullet$$

$$\begin{bmatrix} 1 & 0 & 0 & 0 \\ 0 & 0 & 0 & 1 \\ 0 & 1 & 0 & 0 \\ 0 & 0 & 0 & 1 \\ 0 & 0 & 1 & 0 \\ 0 & 0 & 0 & 1 \end{bmatrix} =$$

$$\begin{bmatrix} 3.0119 & 0 & 0 & -2.2520 \\ 2.2520 & 0 & 0 & -3.0119 \\ 0 & 3.0119 & 0 & -2.2520 \\ 0 & 2.2520 & 0 & -3.0119 \\ 0 & 0 & 3.0119 & -2.2520 \\ 0 & 0 & 2.2520 & -3.0119 \end{bmatrix}$$

one obtains

$$\mathbf{K} = \boldsymbol{\Omega}\mathbf{S}\mathbf{Y}_g\boldsymbol{\Omega}^T = \begin{bmatrix} 1 & 0 & 0 & 0 & 0 & 0 \\ 0 & 0 & 1 & 0 & 0 & 0 \\ 0 & 0 & 0 & 0 & 1 & 0 \\ 0 & -1 & 0 & -1 & 0 & -1 \end{bmatrix} \bullet$$

$$\begin{bmatrix} 3.0119 & 0 & 0 & -2.2520 \\ 2.2520 & 0 & 0 & -3.0119 \\ 0 & 3.0119 & 0 & -2.2520 \\ 0 & 2.2520 & 0 & -3.0119 \\ 0 & 0 & 3.0119 & -2.2520 \\ 0 & 0 & 2.2520 & -3.0119 \end{bmatrix}$$

$$= \begin{bmatrix} 3.0119 & 0 & 0 & -2.2520 \\ 0 & 3.0119 & 0 & -2.2520 \\ 0 & 0 & 3.0119 & -2.2520 \\ -2.2520 & -2.2520 & -2.2520 & 9.0357 \end{bmatrix}$$

The formulation is

$$\mathbf{K}\Theta = \mathbf{Q}_s \begin{bmatrix} 3.0119 & 0 & 0 & -2.2520 \\ 0 & 3.0119 & 0 & -2.2520 \\ 0 & 0 & 3.0119 & -2.2520 \\ -2.2520 & -2.2520 & -2.2520 & 9.0357 \end{bmatrix} \begin{bmatrix} \theta_1 \\ \theta_2 \\ \theta_3 \\ \theta_4 \end{bmatrix} = \begin{bmatrix} 40 \\ 80 \\ 20 \\ 0 \end{bmatrix}$$

and a matrix inversion then provides the solution

$$\begin{bmatrix} \theta_1 \\ \theta_2 \\ \theta_3 \\ \theta_4 \end{bmatrix} = \begin{bmatrix} 32.93 \\ 46.21 \\ 26.28 \\ 26.27 \end{bmatrix} {}^\circ\mathrm{C}$$

which checks Ex. 7.3.

Example 8.4: Rework Ex. 7.4 by the general array method. The configuration is shown in Fig. 7.11 and is repeated here as Fig. 8.6. The tips of fins 2 and 4 are adiabatic, and their heat inputs are 20 W at the base of fin 1 and 16 W at the tip of fin 4. The objective is to prove that $\theta_{b,1} = 64.7°\mathrm{C}$, and $\theta_{a,4} = 62.7°\mathrm{C}$. Pertinent physical and thermal data may be found in Ex. 7.4.

SOLUTION The **Y**-matrices for all fins are obtained from Ex. 7.4,

$$\mathbf{Y}_1 = \mathbf{Y}_4 = \begin{bmatrix} 2.0039 & -1.8783 \\ 1.8783 & -2.0039 \end{bmatrix} \mathrm{W/K}$$

and

$$\mathbf{Y}_2 = \mathbf{Y}_3 = \begin{bmatrix} 17.7775 & -17.7546 \\ 17.7546 & -17.7775 \end{bmatrix} \mathrm{W/K}$$

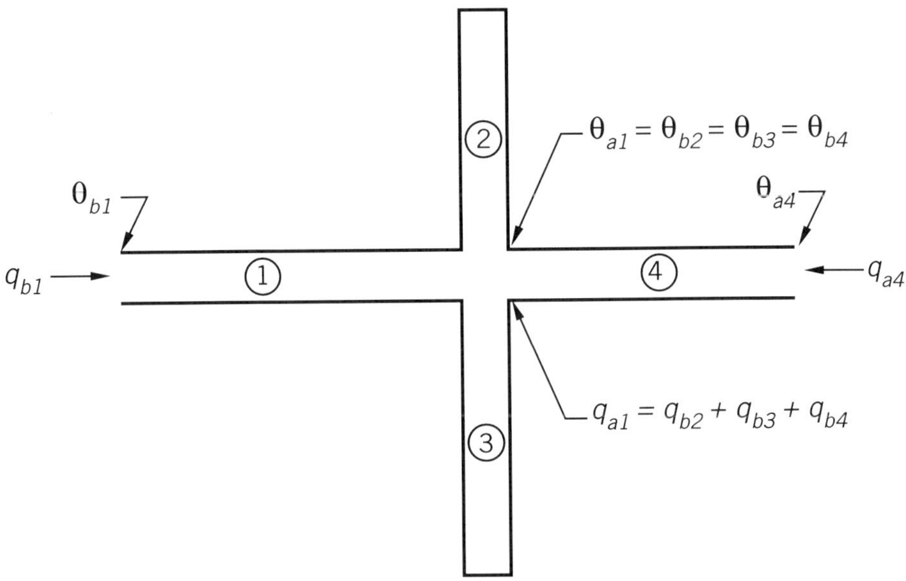

Figure 8.6 The array of Ex. 7.4. It represents a repeating section of a double contact heat exchanger stack subject to uneven heat loading. Heat enters at the left and right, and the tips of fin 2 and 3 are adiabatic.

with determinant

$$\det \mathbf{Y}_2 = \det \mathbf{Y}_3 = \begin{vmatrix} 17.7775 & -17.7546 \\ 17.7546 & -17.7775 \end{vmatrix} = -0.8137$$

The input admittance for fins 2 and 3 is obtained from data contained in Table 2.2. Because the tips of fins 2 and 3 are adiabatic, it is obtained from $\tau_{21,2} = \tau_{21,3}$ and $\tau_{11,2} = \tau_{11,3}$, using the reflection relationship of Eq. (5.6).

$$\tau_{21,2} = \tau_{21,3} = \frac{-\Delta_{Y2}}{y_{21,2}} = \frac{-(-0.8137)}{17.7546} = 0.0458$$

and

$$\tau_{11,2} = \tau_{11,3} = \frac{-y_{22,2}}{y_{21,2}} = \frac{-(-17.7775)}{17.7546} = 1.0013$$

Then from Eq. (4.6)

$$Y_{in,2} = Y_{in,3} = \frac{\tau_{21,2}}{\tau_{11,2}} = \frac{0.0458}{1.0013} = 0.0457\,\mathrm{W/K}$$

The array graph for the configuration of Fig. 8.6 is shown in Fig. 8.7*a*. It will be reduced to the graph shown in Fig. 8.7*b* because fins 2 and 3 in Fig. 8.6 with adiabatic tips form a cluster with

$$Y_{in} = Y_{in,2} + Y_{in,3} = 0.0457 + 0.0457 = 0.0914\,\mathrm{W/K}$$

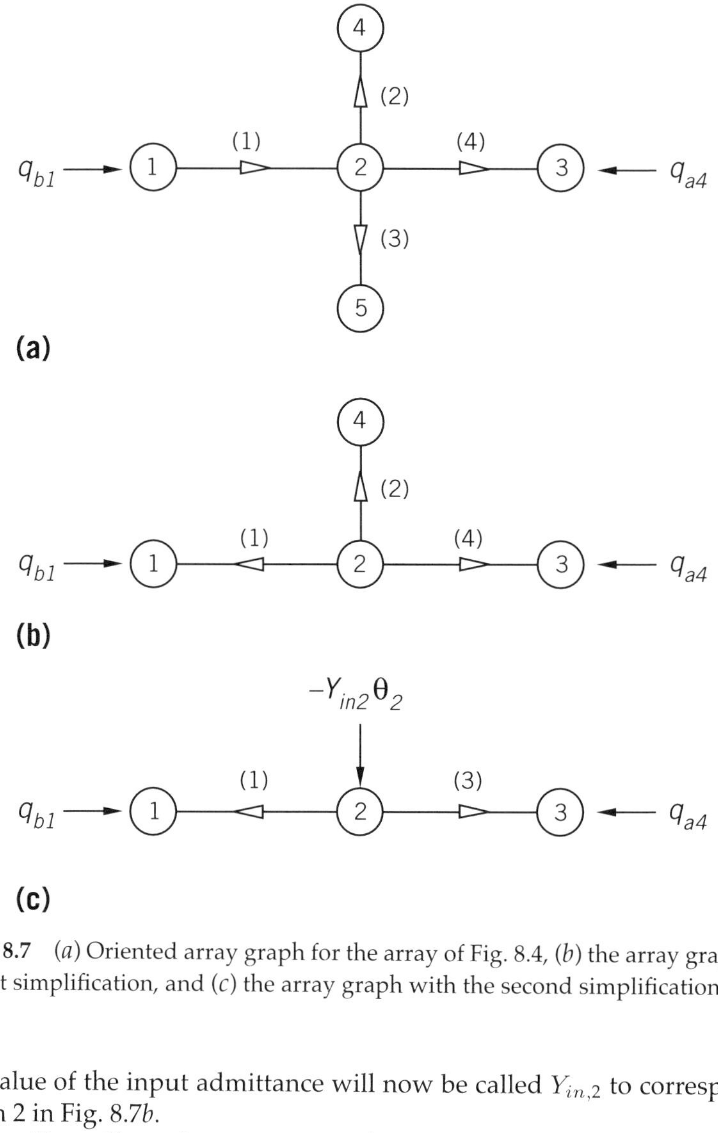

Figure 8.7 (*a*) Oriented array graph for the array of Fig. 8.4, (*b*) the array graph with the first simplification, and (*c*) the array graph with the second simplification.

This value of the input admittance will now be called $Y_{in,2}$ to correspond to branch 2 in Fig. 8.7*b*.

The effect of branch 2 can now be handled as shown in Fig. 8.7*c*, which shows two branches and three nodes. The heat source at node 2 is $Y_{i2}\theta_2$ to represent the effects of both fins 2 and 3 in Figs. 8.6 and Fig. 8.7*a*. Notice that branch 4 in Fig. 8.7*b* has been renumbered to be called branch 3.

The array graph incidence matrix for Fig. 8.7*c* is 3×4 (three nodes and two branches),

$$\boldsymbol{\Omega} = \begin{bmatrix} 1 & 0 & 0 & 0 \\ 0 & 1 & 1 & 0 \\ 0 & 0 & 0 & 1 \end{bmatrix}$$

the temperature excess vector is 3×1,

$$\Theta = \begin{bmatrix} \theta_1 \\ \theta_2 \\ \theta_3 \end{bmatrix}$$

and the heat source vector is 3×1

$$\mathbf{Q}_s = \begin{bmatrix} q_{b,1} \\ -Y_{in,2}\theta_2 \\ q_{a3} \end{bmatrix}$$

The fin graph admittance matrix has the form

$$\mathbf{Y}_g = \begin{bmatrix} 2.0039 & -1.8783 & 0 & 0 \\ 1.8783 & -2.0039 & 0 & 0 \\ 0 & 0 & 2.0039 & -1.8783 \\ 0 & 0 & 1.8783 & -2.0039 \end{bmatrix} \text{W/K}$$

Then

$$\mathbf{K} = \mathbf{\Omega S Y}_g \mathbf{\Omega}^T$$

and with

$$\mathbf{\Omega S} = \begin{bmatrix} 1 & 0 & 0 & 0 \\ 0 & 1 & 1 & 0 \\ 0 & 0 & 0 & 1 \end{bmatrix} \begin{bmatrix} 1 & 0 & 0 & 0 \\ 0 & -1 & 0 & 0 \\ 0 & 0 & 1 & 0 \\ 0 & 0 & 0 & -1 \end{bmatrix} = \begin{bmatrix} 1 & 0 & 0 & 0 \\ 0 & -1 & 1 & 0 \\ 0 & 0 & 0 & -1 \end{bmatrix}$$

and

$$\mathbf{Y}_g \mathbf{\Omega}^T = \begin{bmatrix} 2.0039 & -1.8783 & 0 & 0 \\ 1.8783 & -2.0039 & 0 & 0 \\ 0 & 0 & 2.0039 & -1.8783 \\ 0 & 0 & 1.8783 & -2.0039 \end{bmatrix} \begin{bmatrix} 1 & 0 & 0 \\ 0 & 1 & 0 \\ 0 & 1 & 0 \\ 0 & 0 & 1 \end{bmatrix}$$

$$= \begin{bmatrix} 2.0039 & -1.8783 & 0 \\ 1.8783 & -2.0039 & 0 \\ 0 & 2.0039 & -1.8783 \\ 0 & 1.8783 & -2.0039 \end{bmatrix}$$

one obtains

$$\mathbf{K} = \mathbf{\Omega S Y}_g \mathbf{\Omega}^T = \begin{bmatrix} 1 & 0 & 0 & 0 \\ 0 & -1 & 1 & 0 \\ 0 & 0 & 0 & -1 \end{bmatrix} \begin{bmatrix} 2.0039 & -1.8783 & 0 \\ 1.8783 & -2.0039 & 0 \\ 0 & 2.0039 & -1.8783 \\ 0 & 1.8783 & -2.0039 \end{bmatrix}$$

$$= \begin{bmatrix} 2.0039 & -1.8783 & 0 \\ 1.8783 & 4.0078 & -1.8783 \\ 0 & -1.8783 & 2.0039 \end{bmatrix}$$

The formulation

$$\mathbf{K}\Theta = \mathbf{Q}_s \left[\begin{array}{ccc} 2.0039 & -1.8783 & 0 \\ 1.8783 & 4.0078 & -1.8783 \\ 0 & -1.8783 & 2.0039 \end{array} \right] \left[\begin{array}{c} \theta_1 \\ \theta_2 \\ \theta_3 \end{array} \right] = \left[\begin{array}{c} 20 \\ -0.0914\theta_2 \\ 16 \end{array} \right]$$

can be modified to $\mathbf{J}\Theta = \mathbf{Q}_s$ by adding 0.0916 to the 2,2 element of $\mathbf{K}$

$$\begin{aligned} j_{22} &= k_{22} + 0.0914 \\ &= 4.0078 + 0.0914 \\ &= 4.0992 \end{aligned}$$

Thus,

$$\mathbf{J}\Theta = \mathbf{Q}_s$$

or

$$\left[\begin{array}{ccc} 2.0039 & -1.8783 & 0 \\ 1.8783 & 4.0992 & -1.8783 \\ 0 & -1.8783 & 2.0039 \end{array} \right] \left[\begin{array}{c} \theta_1 \\ \theta_2 \\ \theta_3 \end{array} \right] = \left[\begin{array}{c} 20 \\ 0 \\ 16 \end{array} \right]$$

and a matrix inversion then provides the solution

$$\left[\begin{array}{c} \theta_1 \\ \theta_2 \\ \theta_3 \end{array} \right] = \left[\begin{array}{ccc} 2.0039 & -1.8783 & 0 \\ 1.8783 & 4.0994 & -1.8783 \\ 0 & -1.8783 & 2.0039 \end{array} \right]^{-1} \left[\begin{array}{c} 20 \\ 0 \\ 16 \end{array} \right]$$

or

$$\left[\begin{array}{c} \theta_1 \\ \theta_2 \\ \theta_3 \end{array} \right] = \left[\begin{array}{c} 64.7 \\ 58.3 \\ 62.7 \end{array} \right] \ {}^\circ\mathrm{C}$$

which checks both Exs. 7.1 and 7.4.

8.5 CLOSURE

This concludes the discussion of the general array method. It may be compared to the method of node analysis, which is the alternate method of analysis when the array contains loops in the graph theoretical sense. Both appear to have advantages.

The method of node analysis requires an understanding of the equivalence provided by the equivalent pi-network. One may observe that there is no real need to consider this equivalence as an electric analog. Yet, the temptation to do so is quite formidable, and it seems only natural that one would seek either an electrical network analysis solver such as *PSPICE* or an equation solver to solve the $n \times n$ set of linear, algebraic equations that the user will generate by hand. As mentioned in a previous section, the use of the network equation solver method will work quite well.

On the other hand, the general network method has the advantage in that the network graph may be constructed directly from the configuration of the array without the formulation of the thermal (or electrical for those who are so inclined) resistances from the equivalent pi-networks. But for an array of only a moderate degree of complexity, this will result in the multiplication of large matrices which, computationally, will lead to problems if the matrices are ill conditioned.

Chapter

9

CONVECTIVE OPTIMIZATIONS

9.1 INTRODUCTION[1]

Extended surfaces, whether as individual fins, or as fins in combination in finned arrays or arrays of fins are employed in many thermal applications to enhance the heat transfer between the primary surface and the convective environment. Kraus (1988) traced the history of the development of extended surface technology and its applications and provided a brief review of the important aspects of the technology. One noteworthy aspect of particular importance is the optimization of extended surfaces that is not only of considerable theoretical interest but also, in some cases, may be critical in the design of heat sinks for electrical and electronic equipment. The purpose of this chapter is to focus on individual fin optimizations in the convective environment and to provide an in-depth review. Optimization of an array of fins is considered in Chap. 10.

Fin optimization problems can be classified into two types:

1. Given a fin shape (type and profile), determine the dimensions that lead to maximum heat transfer for a given fin volume. This may also be posed as given a fin shape and a desired heat dissipation, determine the dimensions so that the volume of material used is a minimum.

2. For a given heat dissipation, determine the shape (or profile) of the fin so that the volume of the material used is a minimum. The solutions

[1]The authors are indebted to Professor Abdul Aziz of Gonzaga University for his encouragement and for his permission to extract much of his article entitled *Optimum Dimensions of Extended Surfaces Operating in a Convective Environment,* which was published in the May 1992 *Applied Mechanics Reviews* (ASME, New York).

to this group of problems result in optimum fin shapes with curved surfaces. Such optimum shapes are difficult and expensive to fabricate, and, therefore, from a practical point of view, it is the first group of problems that is more relevant and important.

9.2 LONGITUDINAL FINS

The optimization problem for three longitudinal fin profiles are considered in this section. These are the longitudinal fins of rectangular, triangular, and concave parabolic profile. Two sets of expressions for optimum dimensions are provided:

1. Given the profile area, determine the aspect ratio, or alternatively, the fin height and the fin base thickness. Some expressions in this category were provided by Kern and Kraus (1972).
2. Given the heat dissipation, determine the aspect ratio, or alternatively, the fin height and the fin base thickness.

The section concludes with a design example and a comparison of the quantity of material required for the fins of optimum dimensions in each case. All analyses are subject to the Murray (1938) and Gardner (1945) assumptions delineated in Sec. 1.3.6.

9.2.1 The Longitudinal Fin of Rectangular Profile

Consider the longitudinal fin of rectangular profile shown in Fig. 9.1a. Its dimensions are length, L, height, b, and thickness, δ, and it has a thermal conductivity, k. Both of its faces are exposed to a convective environment at temperature, T_s, with heat-transfer coefficient, h.

Let the dissipation per unit length be $Q = q_b/L$ so that, with no heat loss from the fin tip, and with the base temperature held at T_b so that $\theta_b = T_b - T_s$, the heat dissipation per unit length, Q, is obtained from a modification of Eq. (2.3),

$$q_b = (2hk\delta)^{1/2} L\theta_b \tanh mb \tag{2.3}$$

where

$$m = \left(\frac{2h}{k\delta}\right)^{1/2}$$

as

$$Q = \frac{q_b}{L} = (2hk\delta)^{1/2}\theta_b \tanh mb$$

or, with m substituted,

$$Q = k\delta\left(\frac{2h}{k\delta}\right)^{1/2}\theta_b \tanh\left[\left(\frac{2h}{k\delta}\right)^{1/2} b\right] \tag{9.1}$$

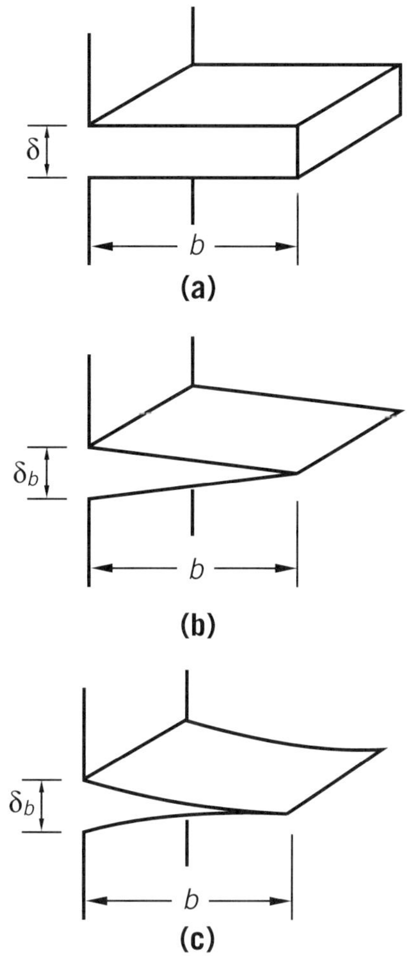

Figure 9.1 Longitudinal fin geometries: (*a*) rectangular profile, (*b*)triangular profile, and (*c*) concave parabolic profile.

The problem is to find b and δ such that, for a fixed profile area, $A_p = \delta b$, and hence, a fixed volume, Q, is maximized. In eliminating b in favor of δ via

$$b = \frac{A_p}{\delta}$$

Eq. (9.1) can be rearranged to

$$Q = (2hk)^{1/2}\theta_b\delta^{1/2}\tanh\beta \tag{9.2}$$

where

$$\beta = mb = A_p\left(\frac{2h}{k\delta^3}\right)^{1/2}$$

The customary procedure of finding where $dQ/d\delta = 0$ leads to the transcendental equation [Kern and Kraus (1972)],

$$3\beta \operatorname{sech}^2\beta = \tanh\beta \tag{9.3a}$$

and, as Aziz (1992) has shown, this can be modified through the use of the simple hyperbolic function identity

$$\sinh 2x = 2\sinh x\cosh x$$

to

$$\sinh 2\beta = 6\beta \tag{9.3b}$$

In either case, a trial-and-error solution gives $\beta = 1.4192$, and this value leads to the optimum values of fin thickness, δ_o, and fin height, b_o, given by

$$\delta_o = 0.9977\left(\frac{hA_p^2}{k}\right)^{1/3} \tag{9.4}$$

and

$$b_o = 1.0023\left(\frac{kA_p}{h}\right)^{1/3} \tag{9.5}$$

Equations (9.4) and (9.5) are to be used when A_p is specified. If, however, the fin is to be designed for a given q_b, then

$$\tanh\left[A_p\left(\frac{2h}{k\delta^3}\right)^{1/2}\right] = \tanh\beta = \tanh 1.4192 = 0.8894$$

is substituted into Eq. (9.2) to yield

$$Q = (2hk)^{1/2}\theta_b\delta^{1/2}(0.8894)$$

so that

$$\delta_o = \frac{0.6321}{hk}\left(\frac{Q}{\theta_b}\right)^2 \tag{9.6}$$

and

$$b_o = 0.7978\frac{Q}{h\theta_b} \tag{9.7}$$

9.2.2 The Longitudinal Fin of Triangular Profile

For the longitudinal fin of triangular profile, which is shown in Fig. 9.1b, an expression for the heat dissipation that ignores the fin taper angle, ϕ, can be developed from the governing differential equation for temperature excess (Kern and Kraus, 1972),

$$x\frac{d^2\theta}{dx^2} + \frac{d\theta}{dx} - m^2 b\theta = 0$$

where m is based on the base thickness

$$m = \left(\frac{2h}{k\delta_b}\right)^{1/2}$$

This differential equation possesses a general solution,

$$\theta(x) = C_1 I_0(2m\sqrt{bx}) + C_2 K_0(2m\sqrt{bx})$$

and it is observed that, in order to have a finite temperature excess at the fin tip [where $x = 0$ and $K_0(0)$ is unbounded], C_2 must equal 0. Thus, with $C_2 = 0$,

$$\theta(x) = C_1 I_0(2m\sqrt{bx})$$

and because at $x = b$ where $\theta(x = b) = \theta_b$

$$C_1 = \frac{\theta_b}{I_0(2mb)}$$

and

$$\theta(x) = \theta_b \frac{I_0(2m\sqrt{bx})}{I_0(2mb)}$$

The heat dissipation is the heat flow through the base, which may be obtained from $q = k\delta_b L(d\theta/dx)$ as

$$q_b = \frac{2hL\theta_b I_1(2mb)}{mI_0(2mb)}$$

and if it is noted that the fin profile area is

$$A_p = \frac{\delta_b b}{2}$$

a procedure that is identical to that used for the longitudinal fin of rectangular profile can be used to show that the heat flow per unit length is given by

$$Q = \frac{q_b}{L} = (2hk)^{1/2}\theta_b \delta_b^{1/2} \frac{I_1(\beta)}{I_0(\beta)} \tag{9.8}$$

where

$$\beta = 4A_p \left(\frac{2h}{k\delta_b^3}\right)^{1/2}$$

This time, the optimization procedure of finding where $dQ/d\delta_b = 0$ yields the transcendental equation

$$I_0^2(\beta) - \frac{4}{3}\frac{I_0(\beta)I_1(\beta)}{\beta} - I_1^2(\beta) = 0 \tag{9.9}$$

and a trial-and-error solution gives

$$\beta = 2.6188$$

The optimum base thickness and fin height for a given profile area then follow:

$$\delta_{bo} = 1.6710 \left(\frac{hA_p^2}{k} \right)^{1/3} \tag{9.10}$$

and

$$b_o = 1.1969 \left(\frac{kA_p}{h} \right)^{1/3} \tag{9.11}$$

The corresponding expressions in terms of a specified Q are

$$\delta_o = \frac{0.8273}{hk} \left(\frac{Q}{\theta_b} \right)^2 \tag{9.12}$$

and

$$b_o = 0.8422 \frac{Q}{h\theta_b} \tag{9.13}$$

9.2.3 The Longitudinal Fin of Concave Parabolic Profile

Figure 9.1*c* shows a fin of concave parabolic profile. The heat dissipation per unit length, in this case, can be determined from a modification of Eq. (4.12),

$$Q = \frac{q_b}{L} = \frac{\alpha k \delta_b \theta_b}{b}$$

where

$$\alpha = -\frac{1}{2} + \frac{(1 + 4m^2b^2)^{1/2}}{2}$$

and where, as in the case of the fin of triangular profile,

$$m = \left(\frac{2h}{k\delta_b} \right)^{1/2}$$

The profile area is

$$A_p = \frac{\delta_b b}{3}$$

and Q can be written terms of A_p

$$Q = \frac{k\delta_b\theta_b}{2b} \left[\left(1 + \frac{72A_p^2 h}{k\delta_b^3} \right)^{1/2} - 1 \right] \tag{9.14}$$

When b is expressed in terms of A_p and δ_b, Q may be rearranged as a function of δ_b. With

$$\beta = mb = 3A_p \left(\frac{2h}{k\delta_b^3} \right)^{1/2}$$

inserted into Eq. (9.14), the application of the optimization procedure of finding where $dq/d\delta_b = 0$ gives

$$\beta = 3A_p \left(\frac{2h}{k\delta_b^3} \right)^{1/2} = \sqrt{2}$$

The optimum base thickness and height for a given A_p then follow as

$$\delta_{bo} = 2.0801 \left(\frac{hA_p^2}{k} \right)^{1/3} \tag{9.15}$$

and

$$b_o = 1.4422 \left(\frac{kA_p}{h} \right)^{1/3} \tag{9.16}$$

The corresponding expressions in terms of a specified Q are

$$\delta_{bo} = \frac{Q^2}{hk\theta_b^2} \tag{9.17}$$

and

$$b_o = \frac{Q}{h\theta_b} \tag{9.18}$$

9.2.4 Comparison of Optimum Fins

It is enlightening to compare the profile areas for rectangular, triangular, and concave parabolic longitudinal fins for the same heat dissipation. The optimum profile area, A_{po}, for each geometry can be obtained by using the expression [2] for δ_{bo} or b_o in terms of Q. The result of such manipulation gives

rectangular profile $$A_{po} = \frac{0.5044Q^3}{h^2k\theta_b^3} \tag{9.19}$$

triangular profile $$A_{po} = \frac{0.3484Q^3}{h^2k\theta_b^3} \tag{9.20}$$

concave parabolic profile $$A_{po} = \frac{0.3333Q^3}{h^2k\theta_b^3} \tag{9.21}$$

[2]For the longitudinal fin of rectangular profile, δ_{bo} becomes δ_o.

A comparison of Eqs. (9.19) through (9.21) shows that a concave parabolic fin requires only 66 percent as much material as the rectangular profile fin for the same Q and the same operating conditions. However, the triangular fin requires only about 5 percent more material than the concave parabolic fin. It is obvious that the concave parabolic fin uses the least amount of material and is consequently the lightest of the three shapes. However, a concave parabolic fin is more difficult to fabricate, and a triangular fin may be a preferable alternative to a concave parabolic fin without a significant penalty in weight.

Another interesting point is that A_p (and hence the weight) is proportional to the cube of Q. Thus, if Q is to be doubled, the increase in A_p is eightfold. Because of this, it is advantageous to employ two identical fins rather than to attempt to achieve twice the dissipation using a single bulky fin. Indeed, a larger number of shorter and lighter fins leads to a better design than fewer longer and heavier fins.

The preceding discussion assumed that the fins were all made of the same material. If a choice of materials is available, then Eqs. (9.19) through (9.21) show that A_p or the volume, V, of the fin is inversely proportional to k. Because $m = \rho V$, the mass, m, is proportional to ρ/k. The material with the lowest ρ/k value gives the lightest fin. For carbon steel, aluminum, and copper, these values are 129.8, 11.4, and 22.3 kg K/Wm, respectively. Thus, for the same volume, a copper fin is 1.96 times and a carbon steel fin 11.4 times heavier than an aluminum fin.

Example 9.1: The rear panel of a rack of electronics is one meter high and is to be equipped externally with 16 AISI 1010 vertical longitudinal steel fins (k = 58.3 W/m-K). The density of steel is 7832 kg/m. It is expected that the fins will operate in a convective environment with a heat-transfer coefficient of 100 W/m^2-K. The total mass of the material to be used is 20.387 kg. Determine the optimum dimensions for (*a*) rectangular, (*b*) triangular, and (*c*) concave parabolic profiles. and then calculate the heat dissipated from these fins if the temperature difference between the panel surface and the surroundings is $\theta_b = 80°$C.

SOLUTION The permitted profile area of each fin is

$$A_p = \frac{20.387}{16(7832)(1)} = 1.627 \times 10^{-4}\ \text{m}^2$$

Rectangular Profile: Using Eq. (9.4),

$$\begin{aligned} \delta_o &= 0.9977 \left(\frac{hA_p^2}{k} \right)^{1/3} \\ &= 0.9977 \left[\frac{100(1.627 \times 10^{-4})^2}{58.3} \right]^{1/3} \end{aligned}$$

$$= 3.56 \times 10^{-3}\,\text{m or } 3.56\,\text{mm}\ \ (0.140\,\text{in})$$

and then

$$\begin{aligned} b_o &= \frac{A_p}{\delta_o} \\ &= \frac{1.627 \times 10^{-4}}{3.56 \times 10^{-3}} \\ &= 0.0457\,\text{m or } 45.7\,\text{mm}\ \ (1.799\,\text{in}) \end{aligned}$$

Using a rearrangement of Eq. (9.7),

$$\begin{aligned} Q &= \frac{b_o h \theta_b}{0.7978} \\ &= \frac{0.0457(100)(80)}{0.7978} \\ &= 458.4\,\text{W/m} \end{aligned}$$

Triangular Profile: Using Eq. (9.10),

$$\begin{aligned} \delta_{bo} &= 1.6710 \left(\frac{hA_p^2}{k} \right)^{1/3} \\ &= 1.6710 \left[\frac{100(1.627 \times 10^{-4})^2}{58.3} \right]^{1/3} \\ &= 5.96 \times 10^{-3}\,\text{m or } 5.96\,\text{mm}\ \ (0.235\,\text{in}) \end{aligned}$$

and then

$$\begin{aligned} b_o &= \frac{2A_p}{\delta_{bo}} \\ &= \frac{2(1.627 \times 10^{-4})}{5.96 \times 10^{-3}} \\ &= 0.0546\,\text{m or } 54.6\,\text{mm}\ \ (2.150\,\text{in}) \end{aligned}$$

Using a rearrangement of Eq. (9.13),

$$\begin{aligned} Q &= \frac{b_o h \theta_b}{0.8422} \\ &= \frac{0.0546(100)(80)}{0.8422} \\ &= 518.5\,\text{W/m} \end{aligned}$$

Concave Parabolic Profile: Using Eq. (9.15),

$$\begin{aligned}\delta_{bo} &= 2.0801\left(\frac{hA_p^2}{k}\right)^{1/3} \\ &= 2.0801\left[\frac{100(1.627\times10^{-4})^2}{58.3}\right]^{1/3} \\ &= 7.42\times10^{-3}\,\text{m or } 7.42\,\text{mm}\ \ (0.292\,\text{in})\end{aligned}$$

and then

$$\begin{aligned}b_o &= \frac{3A_p}{\delta_{bo}} \\ &= \frac{3(1.627\times10^{-4})}{7.42\times10^{-3}} \\ &= 0.0658\,\text{m or } 65.8\,\text{mm}\ \ (2.591\,\text{in})\end{aligned}$$

Using a rearrangement of Eq. (9.18),

$$\begin{aligned}Q &= b_o h\theta_b \\ &= 0.0658(100)(80) \\ &= 526.2\,\text{W/m}\end{aligned}$$

The foregoing results show that the concave parabolic fin provides the highest heat removal. However, its performance is closely rivaled by the triangular fin. If a choice of shape was to be made between the two, one would settle in favor of the triangular profile because it is relatively easy to fabricate. However, it must be recognized that the actual fabrication process might result in a trapezoidal fin rather than the true triangular fin.

9.3 SPINES

The analysis for the optimum dimensions of convecting spines parallels that of longitudinal fins and is based on the same assumptions. Here, results are presented for three spine shapes or profiles: cylindrical, conical, and concave parabolic. The optimization results for a cylindrical spine were first given by Sonn and Bar-Cohen (1980), but a more general analysis for a convective spine of arbitrary shape was presented by Razelos (1983). The heat dissipation expressions for the three geometries are also given by Kern and Kraus (1972).

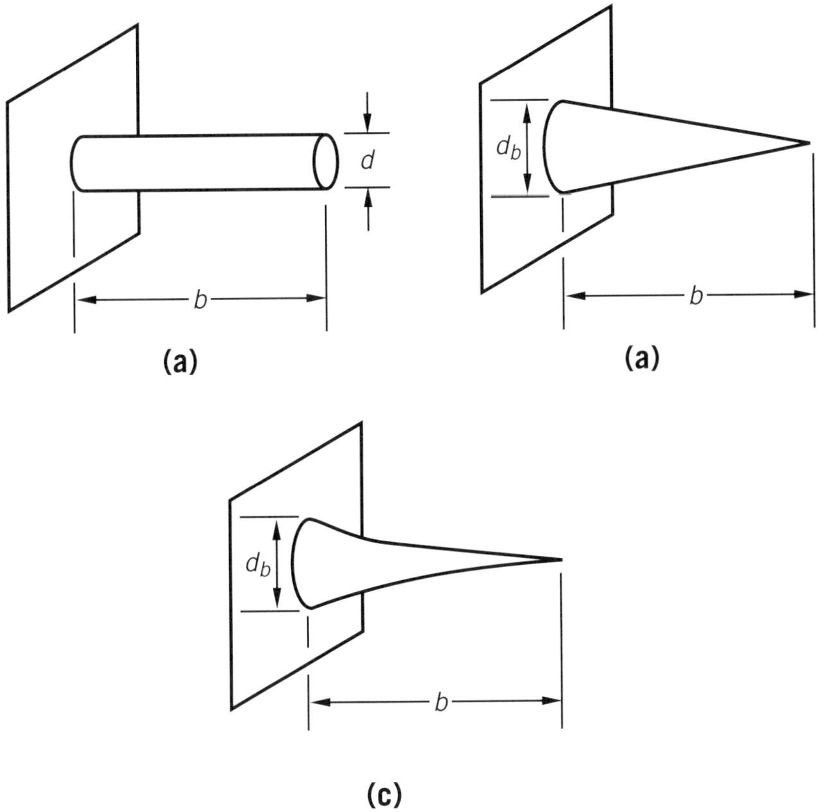

Figure 9.2 Spine geometries: (*a*) cylindrical, (*b*) conical, and (*c*) concave parabolic.

9.3.1 The Cylindrical Spine

Figure 9.2*a* shows a cylindrical spine of diameter, d, and height, b, made of a material with thermal conductivity, k. The spine is operating in a convective environment with a surrounding uniform temperature, T_s, with heat-transfer coefficient, h. The boundary conditions are those of constant base temperature, T_b (so that $\theta_b = T_b - T_s$), and negligible heat loss from the tip. The heat dissipation, q_b, as shown in Sec. 3.9, is given by

$$q_b = \frac{1}{2}(\pi^2 hkd^3)^{1/2}\theta_b \tanh mb$$

where

$$m = \left(\frac{4h}{kd}\right)^{1/2}$$

Let

$$\beta = mb = \left(\frac{4h}{kd}\right)^{1/2} b$$

so that the heat dissipation can be written as

$$q_b = \frac{\pi d^2 k\theta_b \beta \tanh \beta}{4b} \tag{9.22}$$

The problem is to find d and b such that, for a fixed volume,

$$V = \frac{\pi}{4} d^2 b$$

q_b is a maximum. Eliminating b in favor of d via the spine volume so that

$$b = \frac{4V}{\pi d^2}$$

q_b can be expressed as a function of d alone. Setting $dq_b/dd = 0$ and simplifying, the transcendental equation

$$10\beta = 3\sinh 2\beta \tag{9.23}$$

is obtained.

The trial-and-error solution of Eq. (9.23) gives $\beta = 0.9193$. The optimum diameter, d_o, and the optimum height, b_o, are given by

$$d_o = 1.5031 \left(\frac{hV^2}{k}\right)^{1/5} \tag{9.24}$$

and

$$b_o = 0.5636 \left(\frac{Vk^2}{h^2}\right)^{1/5} \tag{9.25}$$

The corresponding expressions in terms of q_b are

$$d_o = 0.9165 \left(\frac{q_b^2}{hk\theta_b^2}\right)^{1/3} \tag{9.26}$$

and

$$b_o = 0.4400 \left(\frac{q_b k}{h^2\theta_b}\right)^{1/3} \tag{9.27}$$

9.3.2 The Conical Spine

Figure 9.2*b* shows a conical spine having a base diameter, d_b, and height, b. The heat dissipation can be obtained from a modification of Eq. (4.14),

$$q_b = \frac{k\pi n r_b^2 \theta_b}{b^{1/2}} \frac{I_2(w_b)}{I_1(w_b)} \tag{4.14}$$

where

$$n = \left[\frac{4h^2}{r_b^2 k^2}(b^2 + r_b^2) \right]^{1/4}$$

and

$$w_b = 2nb^{1/2}$$

Aziz (1992) has considered the case of $b \gg r_b$ and has recast the expression for the heat dissipation to the form

$$q_b = \frac{\pi d_b^2 k \theta_b}{4b^{1/2}} \frac{I_2(\beta)}{I_1(\beta)} \tag{9.28}$$

where

$$\beta = w_b = 4\left(\frac{hb^2}{kd_b} \right)^{1/2}$$

Noting that the volume, V, in this case is given by

$$V = \frac{\pi d_b^2 b}{12}$$

and following the same procedure as that used for the cylindrical spine, the expressions for the optimum base diameter, d_{bo}, and the optimum spine height, b_o, take the form

$$d_{bo} = 1.9536 \left(\frac{hV^2}{k} \right)^{1/5} \tag{9.29}$$

and

$$b_o = 1.0008 \left(\frac{k^2 V}{h^2} \right)^{1/5} \tag{9.30}$$

The corresponding equations in terms of q are

$$d_{bo} = 1.0988 \left(\frac{q_b^2}{hk\theta_b^2} \right)^{1/3} \tag{9.31}$$

and

$$b_o = 0.7505 \left(\frac{q_b k}{h^2 \theta_b} \right)^{1/3} \tag{9.32}$$

which coincide with the results given by Razelos (1983).

9.3.3 The Concave Parabolic Spine

A concave parabolic spine of base diameter, d_b, and height, b, is shown in Fig. 9.2*c*. The heat flow can be obtained from a modification of Eq. (4.16). For the same thermal conditions as in the previous geometries,

$$q = \frac{k\pi r_b^2 \alpha \theta_b}{b}$$

where

$$\alpha = -\frac{3}{2} + \frac{(1 + 4m^2b^2)^{1/2}}{2}$$

The optimization provided by Aziz (1992) is based on an alternative representation,

$$q_b = \frac{\pi k d_b^2 \theta_b}{8b} \left[\left(9 + \frac{16hb^2}{kd_b} \right)^{1/2} - 3 \right] \tag{9.33}$$

and, as in previous discussions for other profiles, d_{bo}, and b_o, are found to give a maximum q for a fixed volume,

$$V = \frac{\pi \delta_b^2 b}{20}$$

The resulting equations for d_{bo} and b_o are

$$d_{bo} = 2.0968 \left(\frac{hV^2}{k} \right)^{1/5} \tag{9.34}$$

and

$$b_o = 1.4481 \left(\frac{k^2 V}{h} \right)^{1/5} \tag{9.35}$$

In terms of q

$$d_{bo} = 1.1746 \left(\frac{q_b^2}{hk\theta_b^2} \right)^{1/3} \tag{9.36}$$

and

$$b_o = 1.0838 \left(\frac{q_b k}{h^2 \theta_b} \right)^{1/3} \tag{9.37}$$

Example 9.2: It is desired to dissipate 10 W from a primary surface by convection at 90°C by attaching a spine. Determine the optimum dimensions for the cylindrical, conical, and concave parabolic spines. The spines are to be fabricated of a material having a thermal conductivity of 65 W/m-K. The environment temperature is 30°C and the heat-transfer coefficient is expected to be 80 W/m^2-K.

SOLUTION For all spines

$$\theta_b = 90 - 30 = 60°\text{C}$$

Cylindrical Spine : Using Eq. (9.26),

$$\begin{aligned} d_o &= 0.9165\left(\frac{q_b^2}{hk\theta_b^2}\right)^{1/3} \\ &= 0.9165\left[\frac{(10)^2}{80(65)(60)^2}\right]^{1/3} \\ &= 0.0160\,\text{m or } 16.0\,\text{mm}\ \ (0.630\,\text{in}) \end{aligned}$$

and then using Eq. (9.27),

$$\begin{aligned} b_o &= 0.4400\left(\frac{q_b k}{h^2\theta_b}\right)^{1/3} \\ &= 0.4400\left[\frac{10(65)}{(80)^2(60)}\right]^{1/3} \\ &= 0.0524\,\text{m or } 52.4\,\text{mm}\ \ (2.063\,\text{in}) \end{aligned}$$

Conical Spine: Using Eq. (9.31),

$$\begin{aligned} d_{bo} &= 1.0988\left(\frac{q_b^2}{hk\theta_b^2}\right)^{1/3} \\ &= 1.0988\left[\frac{(10)^2}{80(65)(60)^2}\right]^{1/3} \\ &= 0.0192\,\text{m or } 19.2\,\text{mm}\ \ (0.756\,\text{in}) \end{aligned}$$

Using Eq. (9.32),

$$\begin{aligned} b_o &= 0.7505\left(\frac{q_b k}{h^2\theta_b}\right)^{1/3} \\ &= 0.7505\left[\frac{10(65)}{(80)^2(60)}\right]^{1/3} \\ &= 0.0895\,\text{m or } 89.5\,\text{mm}\ \ (3.524\,\text{in}) \end{aligned}$$

Concave Parabolic Spine: Using Eq. (9.36),

$$\begin{aligned} d_{bo} &= 1.1746\left(\frac{q_b^2}{hk\theta_b^2}\right)^{1/3} \\ &= 1.1746\left[\frac{(10)^2}{80(65)(60)^2}\right]^{1/3} \\ &= 0.0205\,\text{m or } 20.5\,\text{mm}\ \ (0.807\,\text{in}) \end{aligned}$$

and then using Eq. (9.37),

$$\begin{aligned} b_o &= 1.0838 \left(\frac{q_b k}{h^2 \theta_b} \right)^{1/3} \\ &= 1.0838 \left[\frac{10(65)}{(80)^2(60)} \right]^{1/3} \\ &= 0.1292\,\text{m or } 129.2\,\text{mm} \;\; (5.087\,\text{in}) \end{aligned}$$

Comments: For a single spine, the optimum design yields "chubby" spines. However, if two spines are used, each dissipating 5 W, the dimensions for the cylindrical spine are $d_b = 10.1$ mm and $b = 41.6$ mm. The spine is shorter and slimmer. The volume of material used with a single spine is 48,985 mm^3, but with two spines, almost one-seventh of the volume is needed (6665 mm^3). If one uses four spines, each dissipating 2.5 W, the optimum dimensions of a cylindrical spine are $d_b = 6.4$ mm and $b = 33.0$ mm. The volume of material needed for four spines is 4146 mm^3, which is about one-twelfth of the volume of a single cylindrical spine. The four-spine design might be preferred.

One could consider the conical spine for a further reduction in volume. If four such spines are employed, the optimum dimensions of each would be $d_b = 7.6$ mm and $b = 56.3$ mm, and the total volume of four spines would be 3405 mm^3.

9.4 RADIAL FINS

The problem of the optimum dimensions of radial fins has been studied by Brown (1965), Mikk (1980), Razelos and Imre (1980), and Ullmann and Kalman (1989). While Brown has considered fins of rectangular profile, Mikk has considered both rectangular and triangular profiles. The work of Razelos and Imre considers three profiles: rectangular, triangular, and trapezoidal, and the Ullmann and Kalman analysis includes four shapes: rectangular, triangular, hyperbolic, and parabolic.

9.4.1 The Radial Fin of Rectangular Profile

Consider a radial fin of uniform thickness δ (rectangular profile) operating in a convective environment at temperature, T_s, with heat-transfer coefficient h, as shown in Fig. 9.3a. Let the base and tip radii be r_b and r_a, respectively. The thermal conductivity of the fin is k. Assuming a constant temperature, T_b, at the base, and no heat flow through the tip, the heat dissipation that is the heat flow through the base can be obtained from the development in Sec. 3.7 as

$$q_b = 2\pi r_b \delta k m \theta_b \left[\frac{I_1(mr_a)K_1(mr_b) - I_1(mr_b)K_1(mr_a)}{I_0(mr_b)K_1(mr_a) + I_1(mr_a)K_0(mr_b)} \right] \qquad (9.38)$$

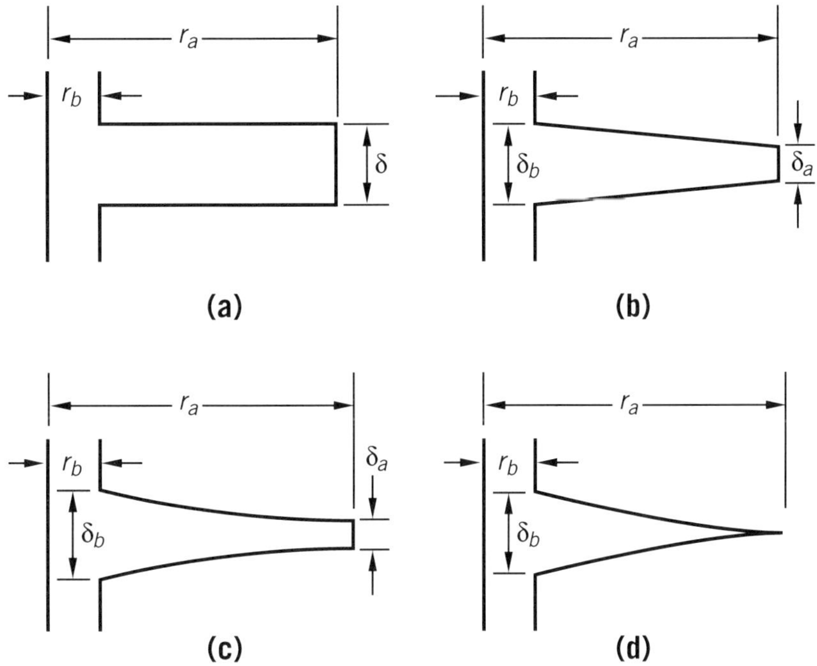

Figure 9.3 Radial fin geometries: (*a*) rectangular profile, (*b*) trapezoidal profile, (*c*) hyperbolic profile, and (*d*) parabolic profile.

where

$$m = \left(\frac{2h}{k\delta}\right)^{1/2}$$

Using the fin volume

$$V = \pi\delta(r_a^2 - r_b^2)$$

Eq. (9.38) may be written as

$$q_b = 2\pi k r_b \theta_b \left(\frac{2h}{k\delta}\right)^{1/2} \left[\frac{I_1(Z_a)K_1(Z_b) - I_1(Z_b)K_1(Z_a)}{I_0(Z_b)K_1(Z_a) + I_1(Z_a)K_0(Z_b)}\right] \tag{9.39}$$

where

$$Z_b = mr_b = \left(\frac{2hr_b^2}{k\delta}\right)^{1/2}$$

and

$$Z_a = \left[\frac{2hr_b^2(1 + V/\pi\delta r_b^2)}{k\delta}\right]^{1/2}$$

Equation (9.39) represents the heat flow as a function of fin thickness, δ. Keeping h, k, θ_b, r_b, and V constant, the heat dissipated by the fin will be a maximum when $dq_b/d\delta$ vanishes. Differentiating Eq. (9.39) with respect to δ, setting the result equal to zero, and rearranging gives

$$2 + Z_b\mathbf{\Lambda}_1 + Z_a C\mathbf{\Lambda}_2\mathbf{\Lambda}_3 = 0 \tag{9.40}$$

where

$$\mathbf{\Lambda}_1 = \left[\frac{I_1(Z_b)K_1(Z_a) - I_1(Z_a)K_1(Z_b)}{I_0(Z_b)K_1(Z_a) + I_1(Z_a)K_0(Z_b)} - \frac{I_0(Z_b)K_1(Z_a) + I_1(Z_a)K_0(Z_b)}{I_1(Z_b)K_1(Z_a) - I_1(Z_a)K_1(Z_b)}\right]$$

$$\mathbf{\Lambda}_2 = \frac{I_0(Z_b)K_1(Z_b) + I_1(Z_b)K_0(Z_b)}{I_0(Z_b)K_1(Z_a) + I_1(Z_a)K_0(Z_b)}$$

$$\mathbf{\Lambda}_3 = \frac{I_0(Z_a)K_1(Z_a) + I_1(Z_a)K_0(Z_a)}{I_1(Z_b)K_1(Z_a) - I_1(Z_a)K_1(Z_b)}$$

and

$$C = 1 + \frac{Z_b^2 V}{\pi r_b b Z_a^2}$$

The optimum dimensions for a desired q_b can be obtained by solving Eqs. (9.39) and (9.40). Such a solution has been reported by Brown (1965), and involves four parameters:

$$u = \left(\frac{2hr_b}{k}\right)^{1/2}$$

$$s = \frac{r_b}{\delta}$$

$$v = \frac{V}{\pi r_b^3}$$

and

$$Q = \frac{q_b}{2\pi k r_b \theta_b}$$

The solution is displayed in Fig. 9.4 where s (dashed lines) and v (solid lines) are plotted against Q for different values of u. For a given q_b and r_b, Fig. 9.4 allows the determination of δ via s and V via v. Once V is found, r_a can be obtained from the volume, $V = \pi\delta(r_a^2 - r_b^2)$. Mikk (1980) has also considered the optimization of a radial fin of rectangular profile. His solution

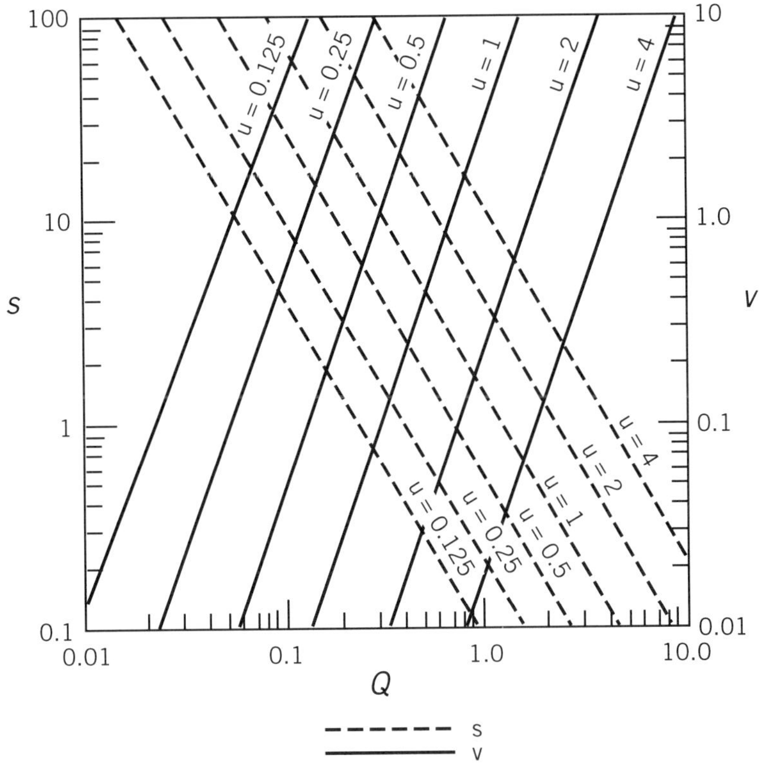

Figure 9.4 Optimum dimensions of a radial fin of rectangular profile (adapted from Brown, 1965).

is in terms of three dimensionless parameters:

$$\kappa = \frac{r_a}{r_b}$$

$$\sigma = \frac{h(r_a - r_b)^2}{k\delta}$$

and

$$\nu = 8\pi^3 r_b^2 (r_a^2 - r_b^2) h^2 k\delta \left(\frac{\theta_b}{q_b}\right)^3$$

The optimum values of these parameters are shown in Table 9.1 which also gives the efficiency of the fin, η, defined as

$$\eta \equiv \frac{q_b}{2\pi(r_a^2 - r_b^2)h\theta_b}$$

TABLE 9.1 Optimum Values of κ, σ and v for the Radial Fin of Rectangular Profile.

κ	σ	v	η
1.0	1.0071	1.0085	0.6267
1.5	0.7657	0.8115	0.6362
2.0	0.6328	0.6618	0.6426
2.5	0.5507	0.5493	0.6462
3.0	0.4944	0.4628	0.6488
4.0	0.4208	0.3426	0.6522
5.0	0.3741	0.2646	0.6546
6.0	0.3410	0.2116	0.6564
∞	0.0000	0.0000	0.6667

For a given q_b and r_b, Mikk's results require trial and error to determine r_a and δ. A value of r_a can be assumed, and this fixes the value of κ. Table 9.1 then gives σ and ν. The former gives δ and the latter yields q_b. If the calculated q_b does not match the given q_b, a new value of r_a must be assumed and the calculation repeated with a new value of r_a.

Another study of the optimization of the radial fin of rectangular profile is that of Razelos and Imre (1980), who employed a Runge-Kutta scheme in conjunction with a quasi-Newton algorithm to locate the optimum. Their results are plotted in Fig. 9.5. The parameters v, $U^{1/2}$, and κ are plotted against Q where

$$v = r_b \left(\frac{2h}{k\delta} \right)^{1/2}$$

$$U^{1/2} = \left(\frac{kV}{4\pi r_b^4 h} \right)^{1/2}$$

$$\kappa = \frac{r_a}{r_b}$$

and

$$Q = \frac{q_b}{4\pi h r_b^2 \theta_b}$$

The calculation procedure involves a given dissipation, q_b, and given base radius, r_b. The first step is to compute Q and to then use Fig. 9.5 to find κ and v. The value of κ leads to r_a, and the value of v provides δ. Note that it is not necessary to use $U^{1/2}$ to find V because r_b, r_a and δ can be used to find V directly.

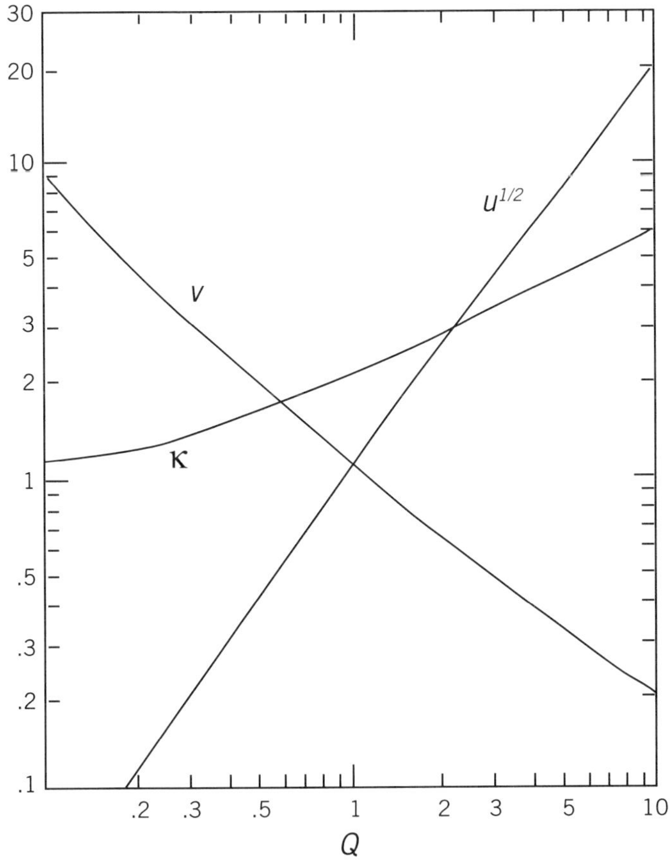

Figure 9.5 Optimum dimensions of a radial fin of rectangular profile (adapted from Razelos and Imre, 1980).

9.4.2 Radial Fins of Trapezoidal and Triangular Profiles

Figure 9.3*b* shows a radial fin of trapezoidal profile having a base thickness δ_b and a tip thickness δ_a operating under the same thermal conditions (h, k, T_b, and T_s) as in Sec. 9.4.1. Based on the work of Razelos and Imre (1980) and Netrakanti and Huang (1983), Fig 9.6 shows the effect of the fin profile in terms of the taper ratio

$$\lambda = \frac{\delta_a}{\delta_b}$$

on the optimum dimensions. Here, the variation of three parameters, δ_b/δ_c, V/V_c, and $Q = q_b/4\pi h r_b^2 \theta_b$, are plotted as a function of λ with the subscript c intended to refer to the fin of constant thickness (the radial fin of rectangular profile). The curves for $\lambda = 0$ correspond to a fin of triangular profile.

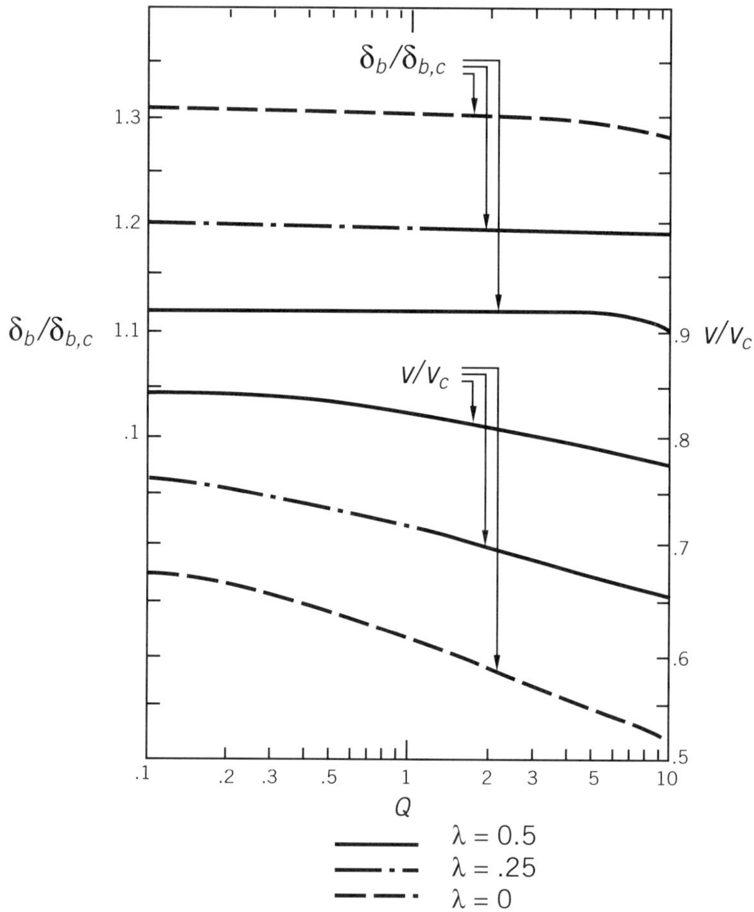

Figure 9.6 Optimum dimensions of radial fins of trapezoidal and triangular profiles (adapted from Razelos and Imre, 1980, and Netrakanti and Huang, 1985).

Figure 9.6 shows that for a given Q, the optimum base thickness increases as λ decreases but the volume decreases. Although not shown, the decrease in λ also causes a slight increase in $\kappa = r_a/r_b$, and this increases the fin height, $b = r_a - r_b$. However, the increase in κ as a function of λ does not exceed 3 percent (compared to the constant thickness fin) even for the limiting case of the triangular fin where $\lambda = 0$.

Mikk (1980) has also provided the solution for the dimensions of the optimum radial fin of triangular profile. His results are provided in Table 9.2. The parameters are defined as

$$\kappa = \frac{r_a}{r_b} \tag{9.41}$$

TABLE 9.2 Optimum Values of κ, σ and v for the Radial Fin of Triangular Profile.

κ	σ	v	η
1.0	0.8574	0.7037	0.5917
1.5	0.6143	0.5312	0.6116
2.0	0.4941	0.4143	0.6225
2.5	0.4232	0.3312	0.6296
3.0	0.3758	0.2708	0.6349
4.0	0.3169	0.1911	0.6416
5.0	0.2814	0.1426	0.6457
6.0	0.2571	0.1109	0.6384
∞	0.0000	0.0000	0.6667

TABLE 9.3 Optimum Dimensions of Radial Fins of Rectangular Profile.

Material	v	$U^{1/2}$	κ	$\delta \times 10^3$ (m)	$V \times 10^4$ (m^3	r_a (m)
Aluminum	1.2	0.97	2	3.29	1.01	0.12
Copper	1.2	0.97	2	1.96	0.60	0.12
Carbon steel	1.2	0.97	2	15.96	4.90	0.12

$$\sigma = \frac{h(r_a - r_b)^2}{k\delta_b} \tag{9.42}$$

(9.43)

and

$$v = 8\pi^3 r_b^2 h^2 k\delta_b (r_a^2 - r_b^2)\left(\frac{\theta_b}{q_b}\right)^3 \tag{9.44}$$

As mentioned in Sec. 9.4.1, Mikk's results require a trial-and-error approach if q_b and r_b are specified and r_a and δ_b are to be found.

Example 9.3: Design radial fins with optimum dimensions for (a) rectangular, (b) triangular ($\lambda = 0$), and (c) trapezoidal ($\lambda = 0.5$) profiles to dissipate 600 W. The base of the fin has a radius of 6 cm and the temperature at the fin base is 130°C, while the coolant temperature is 30°C. The average convective heat-transfer coefficient is 150 W/m^2-K. There is a choice of three materials: aluminum ($k = 228$ W/m-K), copper ($k = 382$ W/m-K), and carbon steel ($k = 47$ W/m-K).

SOLUTION

Rectangular Profile: The solution for an aluminum fin is presented. For aluminum with $k = 228$ W/m-K and with $\theta_b = 130 - 30 = 100°$C

$$Q = \frac{q_b}{4\pi h r_b^2 \theta_b} = \frac{600}{4\pi(150)(0.06)^2(100)} = 0.8842\,\text{W}$$

Then $v, U^{1/2}$ and κ are read from Fig. 9.5

$$v = 1.2$$
$$U^{1/2} = 0.97$$

and

$$\kappa = 2$$

Thus

$$\delta_b = \frac{2hr_b^2}{kv^2} = \frac{2(150)(0.06)^2}{228(1.2)^2} = 3.289 \times 10^{-3} \text{ or } 3.29\,\text{mm } (0.130\,\text{in})$$

$$U = (U^{1/2})^2 = (0.97)^2 = 0.941$$

$$V = \frac{4\pi r_b^4 hU}{k} = \frac{4\pi(0.06)^4(150)(0.941)}{228} = 1.008 \times 10^{-4}\,\text{m}^3$$

and

$$r_a = \kappa r_b = 2(0.06) = 0.12\,\text{m } (4.72\,\text{in})$$

The foregoing computations may be repeated for copper and carbon steel fins and the results of all computations may be assembled as in Table 9.3.

TABLE 9.4 Optimum Dimensions of Radial Fins of Triangular Profile.

Material	δ/δ_c	V/V_c	$\delta_b \times 10^3$ (m)	$V \times 10^5$ (m^3)	r_a (m)
Aluminum	1.3	0.61	4.28	6.15	0.12
Copper	1.3	0.61	2.55	3.67	0.12
Carbon steel	1.3	0.61	20.75	29.83	0.12

Triangular Profile: For the triangular fin ($\lambda = 0$), Fig. 9.6 may be used to read the values of δ_b/δ_c and V/V_c. For $Q = 0.8842$,

$$\delta_b/\delta_c = 1.30$$
$$V/V_c = 0.61$$

and

$$\kappa = 2$$

is the same as for the radial fin of rectangular profile. Thus, for an aluminum fin,

$$\begin{aligned}\delta_b &= 1.30\delta_c \\ &= 1.30(3.289 \times 10^{-3}) \\ &= 4.276 \times 10^{-3}\,\text{m or } 4.28\,\text{mm } (0.168\,\text{in}) \\ V &= 0.61V_c \\ &= 0.61(1.008 \times 10^{-4}) \\ &= 6.149 \times 10^{-5}\,\text{m}^3\end{aligned}$$

and

$$\begin{aligned}r_a &= \kappa r_b \\ &= 2(0.06) \\ &= 0.12\,\text{m } (4.72\,\text{in})\end{aligned}$$

Complete results for the radial fin of triangular profile, which include the optimizations for the copper and the carbon steel fins, are displayed in Table 9.4. Data for the copper and carbon steel fins were obtained in a procedure identical to that used in the foegoing computations for the aluminum fin.

Trapezoidal Profile: For the trapezoidal fin with $\lambda = 0.5$ and $Q = 0.8842$, Fig. 9.6 is read to give

$$\delta_b/\delta_c = 1.12$$

TABLE 9.5 Optimum Dimensions of Radial Fins of Trapezoidal Profile.

material	δ_b/δ_c	V/V_c	$\delta_b \times 10^3$ (m)	$V \times 10^5$ (m^3)	r_a (m)
Aluminum	1.12	0.83	3.68	8.37	0.12
Copper	1.12	0.83	2.20	4.99	0.12
Carbon steel	1.12	0.83	17.88	40.59	0.12

$$V/V_c = 0.83$$

and

$$\kappa = 2$$

is the same as for the radial fin of rectangular profile. Thus, for an aluminum fin,

$$\begin{aligned} \delta_b &= 1.12\delta_c \\ &= 1.12(3.289 \times 10^{-3}) \\ &= 3.684 \times 10^{-3}\,\text{m or } 3.68\,\text{mm } (0.145\,\text{in}) \\ V &= 0.83V_c \\ &= 0.83(1.008 \times 10^{-4}) \\ &= 8.366 \times 10^{-5}\,\text{m}^3 \end{aligned}$$

and

$$\begin{aligned} r_a &= \kappa r_b \\ &= 2(0.06) \\ &= 0.12\,\text{m } (4.72\,\text{in}) \end{aligned}$$

The results are summarized in Table 9.5.

Observations

1. For any given profile, the volume for a specified q is inversely proportional to the thermal conductivity of the the material. For example, the rectangular profile carbon steel fin requires about 8 times the volume required for the corresponding copper fin, and about 5 times that required for the aluminum fin (see Table 9.3).

2. As the taper ratio, λ, decreases, the base thickness, δ_b, increases. Comparing the three aluminum fins, the base thicknesses are 3.29 mm for $\lambda = 1$ (rectangular), 3.68 mm for $\lambda = 0.5$ (trapezoidal), and 4.28 mm for $\lambda = 0$ (triangular).

3. The volume decreases as λ decreases. Comparison of the tabulated values shows that for aluminum fins, $V = 101\ \text{cm}^3$ for $\lambda = 1$ (rectangular), $V = 83.7\ \text{cm}^3$ for $\lambda = 0.5$ (trapezoidal), and $V = 61.5\ \text{cm}^3$ for $\lambda = 0$ (triangular). Thus for the same q, a triangular aluminum fin consumes about 38 percent less material than the corresponding rectangular fin.

4. The height of the fin, $r_a - r_b$, is virtually independent of the fin shape. More accurate calculations of Razelos and Imre (1980) and Netrakanti and Huang (1983) reveal an increase in height of about 3 percent from the rectangular to the triangular shape.

5. For the rectangular geometry, Brown's solution (Fig. 9.4) or Mikk's solution (Table 9.1) could have been used. Moreover, for the triangular geometry, one can alternatively use Table 9.2. For the trapezoidal geometry, the results of Netrakanti and Huang (1983) provide an alternative to Fig. 9.6. Such calculations have revealed that the final results are essentially the same.

9.4.3 Radial Fins of Parabolic and Hyperbolic Profiles

Ullman and Kalman (1989) have presented solutions for the optimum dimensions of radial fins of four different profiles: rectangular, triangular, hyperbolic (Fig. 9.3*c*), and parabolic (Fig. 9.3*d*). Their results are shown in Fig. 9.7, where the parameters

$$\psi = \frac{k\delta_b}{2hr_b^2}$$

$$Q = \frac{q_b(n+1)(n+2)}{2\pi^2 k\delta_b\theta_b[\kappa - 1][\kappa + n + 1]} \qquad (n = 0, 1, 2)$$

which, for $n = 0$, reduces to

$$Q = \frac{q_b}{\pi^2 k\delta_b\theta_b[\kappa^2 - 1]} \qquad (n = 0)$$

and for $n = -1$,

$$Q = \frac{q_b}{2\pi^2 k\delta_b\theta_b[\kappa - 1]} \qquad (n = -1)$$

are plotted against

$$\kappa = \frac{r_a}{r_b}$$

for n = 0 (rectangular profile), $n = 1$ (triangular profile), $n = -1$ (hyperbolic profile), and $n = 2$ (parabolic profile). If the fin is to be designed for a given q_b, then the procedure is one of trial and error, which evolves around the selection of a trial value of δ_b, which fixes the parameter $\psi = k\delta_b/2hr_b^2$. Figure 9.7 may then be read to obtain Q and κ. With κ known, Q may be used to establish q_b for comparison with the given q. The procedure must be repeated until a match is obtained.

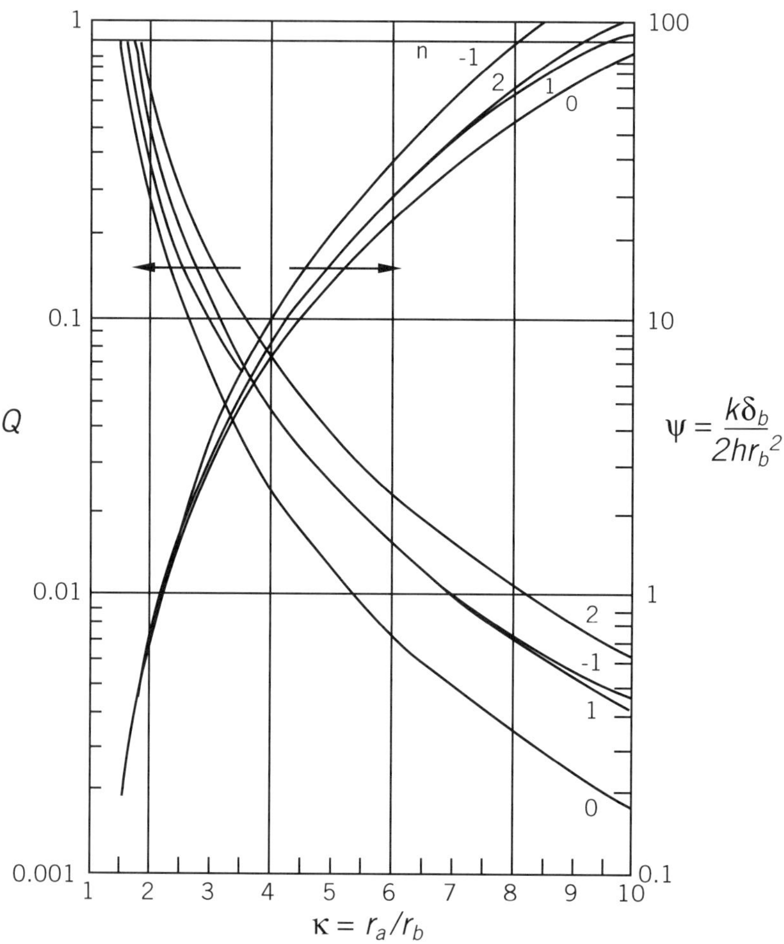

Figure 9.7 Optimum dimensions of radial fins: $n = 0$ (rectangular profile), $n = 1$ (triangular profile), $n = -1$ (hyperbolic profile), and $n = 2$ (parabolic profile) (adapted from Ullman and Kalman, 1989).

Example 9.4: Design radial fins with optimum dimensions for (*a*) rectangular, (*b*) triangular, (*c*) hyperbolic, and (*d*) parabolic profiles to dissipate 600 W. The base of the fin has a radius of 6 cm and the temperature at the fin base is 130°C, while the coolant temperature is 30°C. The average convective heat-transfer coefficient is 150 W/m^2-K. The fin material is aluminum, with $k = 228$ W/m-K, and the design is to be based on Fig. 9.7.

SOLUTION

Rectangular Profile: Here, $n = 0$, and as a first trial, select $\delta_b = 3\,\text{mm}$. Then

$$\psi = \frac{k\delta_b}{2hr_b^2} = \frac{228(0.003)}{2(150)(0.06)^2} = 0.633$$

Use of Fig. 9.7 yields $\kappa = 1.9$ and $Q = 0.28$. Thus

$$r_a = \kappa r_b = 1.9(0.06) = 0.114\,\text{m}\,(4.49\,\text{in})$$

and

$$\begin{aligned} q_b &= \pi^2 k\delta_b\theta_b[\kappa^2 - 1]Q \\ &= \pi^2(228)(0.003)(100)[(1.9)^2 - 1](0.28) \\ &= 493.3\,\text{W} \end{aligned}$$

This does not match the required $q = 600\,W$.

As a second trial, choose $\delta_b = 3.5\,\text{mm}$. Then

$$\psi = \frac{k\delta_b}{2hr_b^2} = \frac{228(0.0035)}{2(150)(0.06)^2} = 0.739$$

and Fig. 9.7 gives $\kappa = 2.0$ and $Q = 0.25$. Thus

$$r_a = \kappa r_b = 2.0(0.06) = 0.120\,\text{m}\,(4.72\,\text{in})$$

and

$$\begin{aligned} q_b &= \pi^2 k\delta_b\theta_b[\kappa^2 - 1]Q \\ &= \pi^2(228)(0.0035)(100)[(2.00)^2 - 1](0.25) \\ &= 590.7\,\text{W} \end{aligned}$$

This value of q_b is close to the given value ($591 \approx 600\,\text{W}$). Because of the inaccuracy inherent in reading in Fig. 9.7, one may deem these two values close enough to merit acceptance so that further refinements need not be attempted.

The results of this example match those of Ex. 9.3. The value of 3.5 mm (0.14 in.) can be compared with the value of 3.3 mm (0.130 in.) obtained in

Ex. 9.3. Such close agreement vouches for the consistency of the results of Brown (1965), Mikk (1980), Razelos and Imre (1980) and Ullman and Kalman (1989).

Triangular Profile: Here, n = 1, and as a first trial, select $\delta_b = 3$ mm. Then

$$\psi = \frac{k\delta_b}{2hr_b^2} = \frac{228(0.003)}{2(150)(0.06)^2} = 0.633$$

Use of Fig. 9.7 yields $\kappa = 1.90$ and $Q = 0.39$. Thus

$$r_a = \kappa r_b = 1.90(0.06) = 0.114\,\text{m}\,(4.49\,\text{in})$$

and

$$q_b = \frac{2\pi^2 k\delta_b\theta_b[\kappa - 1][\kappa + n + 1]Q}{(n+1)(n+2)} = \frac{2\pi^2(228)(0.003)(100)[1.90 - 1][1.90 + 1 + 1](0.39)}{(1+1)(1+2)} = 308.0\,\text{W}$$

This value of $q = 308$ W falls far short of the required value of 600 W.

For the second trial, choose $\delta_b = 4$ mm. Then

$$\psi = \frac{k\delta_b}{2hr_b^2} = \frac{228(0.004)}{2(150)(0.06)^2} = 0.844$$

and Fig. 9.7 gives $\kappa = 2.10$ and $Q = 0.38$. Thus

$$r_a = \kappa r_b = 2.10(0.06) = 0.126\,\text{m}\,(4.96\,\text{in})$$

and

$$q = \frac{2\pi^2 k\delta_b\theta_b[\kappa - 1][\kappa + n + 1]Q}{(n+1)(n+2)} = \frac{2\pi^2(228)(0.004)(100)[2.1 - 1][2.1 + 1 + 1](0.38)}{(1+1)(1+2)} = 514.1\,\text{W}$$

This value of q is closer to the required value ($q = 600$ W), but not close enough. A third trial is required, and when the selection is $\delta_b = 4.3$ mm (0.169 in), then $\psi = 0.907$ and Fig. 9.7 gives $\kappa = 2.20$ and $Q = 0.37$ so that $r_a = 0.132$ m (5.20 in) and $q_b = 601$ W, which is very close to $q = 600$ W.

The value of $\delta_b = 4.3$ mm (0.169 in) is very close to the value of 4.28 mm (0.168 in) obtained in Ex. 9.3. However, the value of $r_a = 0.132$ mm is about 9 percent higher than the value of $r_a = 0.12$ mm obtained in Ex. 9.3.

Parabolic Profile: Here, n = 2, and as a first trial, choose $\delta_b = 4$ mm. Then

$$\psi = \frac{k\delta_b}{2hr_b^2} = \frac{228(0.004)}{2(150)(0.06)^2} = 0.844$$

Use of Fig. 9.7 yields $\kappa = 2.10$ and $Q = 0.54$. Thus

$$r_a = \kappa r_b = 2.10(0.06) = 0.126\ \text{m}\ (4.96\ \text{in})$$

and

$$q_b = \frac{2\pi^2 k\delta_b\theta_b[\kappa - 1][\kappa + n + 1]Q}{(n+1)(n+2)} = \frac{2\pi^2(228)(0.004)(100)[2.10 - 1][2.10 + 2 + 1](0.54)}{(2+1)(2+2)} = 454.5\ \text{W}$$

This value of $q_b = 454.4$ W is less than the given value of 600 W.

For the second trial, choose $\delta_b = 4.3$ mm. Then

$$\psi = \frac{k\delta_b}{2hr_b^2} = \frac{228(0.0043)}{2(150)(0.06)^2} = 0.907$$

and Fig. 9.7 gives $\kappa = 2.20$ and $Q = 0.50$. Thus

$$r_a = \kappa r_b = 2.2(0.06) = 0.132\ \text{m}\ (5.20\ \text{in})$$

and

$$q = \frac{2\pi^2 k\delta_b\theta_b[\kappa - 1][\kappa + n + 1]Q}{(n+1)(n+2)}$$

$$= \frac{2\pi^2(228)(0.0043)(100)[2.20-1][2.20+2+1](0.50)}{(2+1)(2+2)}$$

$$= 503.2\,\text{W}$$

This value of q is closer to the required value ($q = 600$ W) but not close enough. A third trial is required, and when the selection is $\delta_b = 5.8$ mm (0.228 in), then $\psi = 1.22$ and Fig. 9.7 gives $\kappa = 2.30$ and $Q = 0.40$, so that $r_a = 0.138$ m (5.43 in) and $q_b = 599$ W, which is very close to $q = 600$ W.

The values of δ_b =5.8 mm (0.228 in) cannot be checked against other optimization data because no such data appear to exist.

Hyperbolic Profile: Here, $n = -1$, and as a first trial, choose $\delta_b = 5$ mm. Then

$$\psi = \frac{k\delta_b}{2hr_b^2} = \frac{228(0.005)}{2(150)(0.06)^2} = 1.055$$

Use of Fig. 9.7 yields $\kappa = 2.20$ and $Q = 0.27$. Thus

$$r_a = \kappa r_b = 2.20(0.06) = 0.132\,\text{m (5.20 in)}$$

and

$$q_b = 2\pi^2 k\delta_b\theta_b[\kappa-1]Q = 2\pi^2(228)(0.005)(100)[2.20-1](0.27) = 729\,\text{W}$$

This value of $q = 729$ W exceeds the given value of 600 W.

For the second trial, choose $\delta_b = 4.5$ mm. Then

$$\psi = \frac{k\delta_b}{2hr_b^2} = \frac{228(0.0045)}{2(150)(0.06)^2} = 0.950$$

and Fig. 9.7 gives $\kappa = 2.10$ and $Q = 0.30$. Thus

$$r_a = \kappa r_b = 2.1(0.06) = 0.126\,\text{m (4.96 in)}$$

and

$$
\begin{aligned}
q_b &= 2\pi^2 k\delta_b\theta_b[\kappa - 1]Q \\
&= 2\pi^2(228)(0.0045)(100)[2.10 - 1](0.30) \\
&= 668\,\mathrm{W}
\end{aligned}
$$

This value of q is closer to the required value ($q = 600\,\mathrm{W}$) but still too high. A third trial is in order, and when the selection is $\delta_b = 4\,\mathrm{mm}$, then $\psi = 0.84$ and Fig. 9.7 gives $\kappa = 2.10$ and $Q = 0.30$ so that $r_a = 0.126\,\mathrm{m}$ (4.96 in) and $q = 594\,\mathrm{W}$, which is deemed close enough to the required value.

Here, too, no other optimization data appear to exist, and these values cannot be checked.

9.5 CLOSURE

Aziz (1992) has also presented convective optimizations for individual fins operating under real constraints. These include the effects of tip heat loss, variable thermal conductivity, and internal heat generation. While of general interest, these optimizations are not considered here, and attention now turns to convective correlations and to the optimization of an array of fins in Chap. 10.

Chapter 10

HEAT TRANSFER-PARALLEL PLATE HEAT SINKS

10.1 Introduction

While heat sink thermal optimization may lead to relatively complex fin geometries, consideration of manufacturing costs, availability, and maintainability often dictates the use of the ubiquitous plate-fin heat sink, shown in Fig. 10.1. This plate-fin heat sink, consisting of parallel, longitudinal fins of rectangular profile, attached to a common base, can efficiently transfer heat both in forced and natural convection, as well as by radiation. Frequently, the heat dissipation occurs under a combination of these basic heat-transfer modes from the entire heat sink area.

Parallel-plate channels, formed by pairs of adjacent plate-fins, constitute the basic thermal *"building block"* of such heat sinks. Consequently, an understanding of forced convection, natural convection, and radiation heat transfer from parallel-plate channels is a necessary prerequisite for the design and optimization of plate-fin heat sinks. However, it should be noted that precise prediction of heat sink performance may require the recognition and the quantification of the differences between the heat-transfer coefficient along the base plate and on the fins. These differences are generally negligible for widely spaced fins where there is little boundary layer interference, and are of minor importance for closely spaced fins where the ratio of base surface to finned surface is low. Unless stated otherwise, later developments in this chapter assume that a single heat-transfer coefficient can be used to characterize convective heat transfer from the entire heat sink surface.

Prior to embracing this reliance on a parallel-plate "building block," the reader is reminded that relatively complex flow patterns may be encountered adjacent to heat sinks mounted on electronic components. Commonly, the presence of open areas, above and along-side the plate-fin heat sink, can result

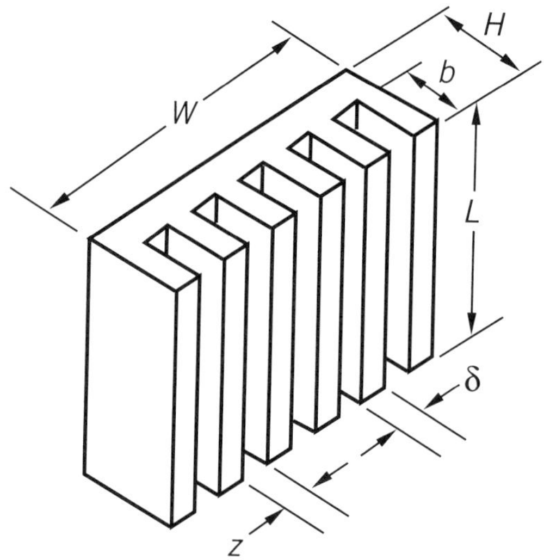

Figure 10.1 Typical plate-fin heat sink.

in a significant fraction of the flow bypassing the relatively high pressure-drop interfin channels. This phenomena was first studied numerically by Sparrow et al (1978), for fully-developed flow, and several years later by Ashiwake et al (1983) for actual chip packages with attached heat sinks. The latter study revealed that for the modest air velocities anticipated in this application (up to several meters per second), flow rate approximations based on consideration of hydraulic resistance in the parallel flow paths [Ellison (1993)], can yield accurate predictions for heat sink performance. Such analyses can be used to determine the appropriate, and perhaps even the optimum, fin spacing and height for a particular heat sink. It must also be noted that the use of appropriately shaped shrouds and flow diverters, placed strategically in the flow channel, can eliminate much of the "bypass" difficulty.

Succeeding sections of this chapter will review the fundamental relations for forced convection and natural convection in a parallel-plate channel. Attention will then turn to convective and radiative heat transfer from arrays of plate-fins and the thermal optimization of such heat sink structures.

10.2 Forced Convection in a Parallel-Plate Channel

The study of forced convection in parallel-plate channels follows a certain ordered pedagogy. Beginning with laminar flow and concluding with turbulent flow. Both of these flow regimes are treated in the subsections that now follow.

10.2.1 Thermofluid Analysis of Laminar Flow in a Channel

A study of forced convection in a parallel-plate channel requires consideration of phenomena associated with fluid mechanics, which lead to the development of a *velocity boundary layer*, and the temperature field within this boundary layer. The boundary layer discussion leads to consideration of the fully developed heat-transfer coefficient and then the analogy between heat transfer and fluid friction.

Attention turns first to a discussion of the fluid mechanics phenomena for laminar fluid flow in a parallel-plate channel.

Fluid Mechanics

Convective heat transfer is, by definition, concerned with heat exchange between a solid surface and a circulating viscous fluid. Determination of the temperature field and heat flow rate encountered in this thermal transport mode must therefore involve definition of the fluid behavior adjacent to the surface. The velocity of a fluid flowing past a stationary flat plate approaches zero near the surface, whereas the velocity of the fluid that is far removed from the surface asymptotically approaches the velocity of the unconstrained stream. In such a flow, frictionlike forces develop at the fluid-solid interface. The adhesion of fluid molecules to the confining surface and the resulting "*no-slip*" condition at an immersed surface is a basic postulate of fluid flow for all but rarefied gases. Although this postulate, expressed in terms of a monomolecular layer, may be justified by the action of intermolecular and surface tension forces, an additional mechanism is needed to explain the smooth transition from the wall to the "*free stream*" velocity and the generation, as well as the transmission, of surface forces into the fluid as a consequence of the fluid-solid interaction.

In liquids, at temperatures considerably above their freezing points, and in gases, this additional mechanism, the resistance to shearing offered by a flowing or contained fluid, is generally related to momentum exchange between the fluid molecules. Thermally excited, random molecular motion in the fluid causes the movement of a finite number of molecules from the wall layer to an adjacent layer. In this layer, collisions between fast, oncoming molecules and the nearly stationary wall layer molecules result in a momentum exchange and a reduction in the average velocity of the oncoming molecules. Repetition of this process in fluid layers further removed from the solid surface leads to the characteristic velocity profile shown in Fig. 10.2

Following Newton's law of motion, the change in momentum experienced by the fluid can be related to a force acting in a direction opposite to that of the flowing fluid, and it is this shear force that embodies the viscous resistance to the shearing deformation of the flowing fluid. In keeping with the foregoing model, the shear stress, or shear force per unit area, must be governed by the velocity gradient in the fluid and a fluid parameter related to the diffusion

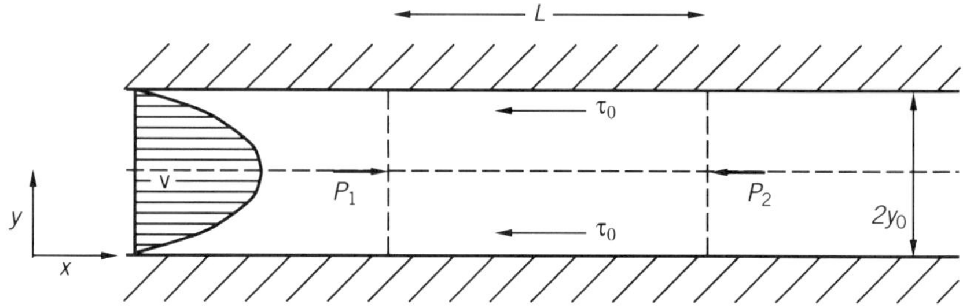

Figure 10.2 Geometric parameters and velocity profile in a parallel plate channel.

rate of molecules in the fluid. This fluid parameter is termed the *dynamic viscosity*, μ, and is extensively tabulated for most conventional fluids.

When the dynamic viscosity and velocity gradient are known, the shear stress, τ, is expressible as

$$\tau = -\mu \frac{dv}{dy} \tag{10.1}$$

where v is the velocity parallel to the surface and y is the coordinate normal to the surface. Equation (10.1) is known as *Newton's equation of viscosity* and fluids that obey this relation are known as *Newtonian fluids*. However, even in Newtonian fluids, Eq. (10.1) cannot be expected to apply in turbulent flow where the presence of swirling, eddy motion overshadows molecular-scale momentum exchange.

Example 10.1: Find the order of magnitude of the shear stress for air at 325 K flowing at 5 m/s in a pipe having a diameter of 1 cm.

SOLUTION To obtain a first estimate of the shear stress, use Eq. (10.1), with an estimated value of the velocity gradient $\Delta v/\Delta y$. At 325 K, the viscosity of air is approximately 2×10^{-5} N-s/m^2. With Δy taken as half the pipe diameter, the velocity gradient is of the order of

$$\frac{\Delta v}{\Delta y} = \frac{5\,\text{m/s}}{0.005\,\text{m}} = 1000\,\text{s}^{-1}$$

Then, the shear stress is

$$\tau = \mu \frac{\Delta v}{\Delta y} = (2 \times 10^{-5}\,\text{N-s/m}^2)(1000\,\text{s}^{-1}) = 0.02\,\text{N/m}^2 = 0.02\text{Pa}$$

Because of the frictionlike character of viscous shear stress, it is often convenient to relate the forces acting at a solid-fluid interface to a friction factor or coefficient. In channel flow this friction factor, f, can be simply related to the shear stress at the wall and the velocity head (or dynamic pressure) of the flow. According to Rohsenow and Choi (1961) and others,

$$f \equiv \frac{\tau_o}{\rho v^2/2} \tag{10.2}$$

where ρ is the fluid density and v is the average velocity. An identical relation is used to define the skin friction coefficient on a flat plate, C_f, for laminar and turbulent flow, with v_∞, the free stream velocity, replacing v.

Equation (10.1) suggests that viscous effects become progressively less significant as the velocity gradients decrease. In view of the characteristic velocity profile shown in Fig. 10.2, it might thus be expected that viscous effects dominate fluid behavior near solid surfaces, whereas inviscid flow prevails in regions far removed from these surfaces. This perception led Prandtl in 1904 to propose that the flow field, in fluid-solid interactions, be divided into two regions: a relatively thin viscous layer adjacent to the solid boundary, to be called the *boundary layer*, and an inviscid fluid core.

Prandtl's division implies that the boundary layer thickness, δ, be determined on the basis of a stated value of the velocity gradient. For convenience, however, δ is generally taken to equal the distance from the surface at which the fluid velocity (parallel to the wall) is within 1 percent of the unconstrained or free stream velocity. Other definitions of δ can also be used and are discussed in the relevant texts, including those of Rohsenow and Choi (1961), Eckert and Drake (1972), and Chapman (1987).

The temperature profile in a fluid flowing past a cold surface is generally similar to the fluid velocity profile. This similarity also exists in other configurations, and thus suggests that a thermal boundary layer, analogous to the velocity boundary layer, can be defined. External to the thermal boundary layer, temperature gradients and thermal conduction effects are considered negligible, whereas within the boundary layer thermal conduction is of significance. The thermal boundary-layer thickness, δ_T, is generally defined as the distance from the surface at which the fluid temperature is within 1 percent of the free stream temperature. It must be noted that although δ is generally not equal to δ_T, an equality ($\delta = \delta_T$) is often a satisfactory approximation for convective exchange between solids and gases.

The flow within the boundary layer near the leading edge of a flat surface is generally smooth and well ordered. The fluid appears to move past the surface in clearly defined sheets or *laminae* that flow parallel to the wall (as can be easily visualized by the addition of trace amounts of dye). This sheet-like behavior is termed *laminar flow*. However, as the flow path lengthens, minute disturbances in the flow field begin to grow, and a churning, eddying motion becomes superimposed on the laminar flow field. Somewhat further downstream, the laminar pattern is completely destroyed and replaced by large-eddy *turbulent flow*.

The transition to turbulent flow is dependent on fluid properties, the flow distance, and the level of turbulence in the free stream. The downstream distance at which transition occurs can be expressed in terms of the *Reynolds number*, Re_x, a nondimensional parameter that unifies the effects of flow, distance, and fluid properties. It is the ratio of the inertia forces to the viscous forces and is defined by

$$\text{Re}_x \equiv \frac{\rho v_\infty x}{\mu} \tag{10.3}$$

where x is the axial distance. The transition value of the Reynolds number varies from approximately 2×10^5 for flows with large disturbances to 10^6 for extremely smooth conditions. It is typically taken as 3×10^5 for both commercial and laboratory equipment.

Boundary layer flows within pipes and channels display a phenomenologically similar transition from laminar to turbulent flow. Transition in this case can also be related to a Reynolds number, the pipe Re_d, defined in terms of the pipe diameter rather than the flow distance

$$\text{Re}_d \equiv \frac{\rho v d}{\mu} \tag{10.4}$$

and 2300 is taken as a typical transition value.

Example 10.2: Determine the conditions under which the flow of air changes from the laminar to the turbulent mode by finding the (a) distance along a flat plate at which a 5 m/s flow of 325 K air becomes turbulent, (b) air velocity needed to achieve transition to turbulent flow at a 10-cm distance along a plate, and (c) air velocity needed to achieve turbulent flow in an infinitely long 2-cm diameter channel.

SOLUTION (a) Use Eq. (10.3) with Re = 3×10^5 and solve for x_{tr}. For air at 325 K, $\mu = 2 \times 10^{-5}$ N-s/m^2 and $\rho = 1$ kg/m^3. Then at $v = 5$ m/s,

$$x_{tr} = \frac{\text{Re}\mu}{\rho v} = \frac{3 \times 10^5 (2 \times 10^{-5})}{1(5)} = 1.2\,\text{m}$$

(b) Again, use Eq. (10.3) with Re = 3×10^5 and $x_{tr} = 0.1\,m$, and solve for v.

$$v = \frac{\text{Re}\mu}{\rho x_{tr}} = \frac{3 \times 10^5 (2 \times 10^{-5})}{1(0.1)} = 60\,\text{m/s}$$

(c) This time rearrange Eq. (10.4) at $\text{Re}_{tr} = 2300$, and solve for v_{tr}.

$$v_{tr} = \frac{\text{Re}_{tr}\mu}{\rho d} = \frac{2300(2 \times 10^{-5})}{1(0.02)} = 2.3\,\text{m/s}$$

The flow of a fluid between two heated parallel plates results in the formation and growth of momentum and temperature boundary layers along each plate surface. In laminar flow under steady-state conditions at a sufficiently downstream location to ensure that the respective boundary layers from opposite channel walls have merged, momentum conservation and flow continuity equations can be used to establish the relationship between the velocity and pressure fields in the channel (Rohsenow and Choi, 1961):

$$\frac{\partial^2 v}{\partial y^2} = \frac{1}{\mu}\frac{dp}{dx} \tag{10.5}$$

where p is the pressure. For this so-called *"fully developed"* flow, the velocity profile is symmetric about the channel centerline, and the velocity attains a zero value at the channel side walls. Using the symbols shown in Fig. 10.2, the velocity profile for this so-called *Poiseuille flow* can be obtained by integrating Eq. (10.5) and using the boundary conditions at the wall,

$$v(y = y_o) = 0 \tag{6a}$$

and at the centerline

$$\frac{\partial v}{\partial y}_{y=0} = 0 \tag{6b}$$

to evaluate the constants of integration. This yields

$$v = -\frac{1}{2\mu}\frac{dp}{dx}(y_o^2 - y^2) \tag{10.7}$$

With the average velocity, V, defined by

$$V \equiv \frac{1}{y_o}\int_0^{y_o} v\,dy \tag{10.8}$$

the pressure drop in the channel per unit length can be shown to be

$$\frac{\Delta p}{L} = \frac{3\mu V}{y_o^2} \tag{10.9}$$

As previously noted, in the analysis of flow in tubes and channels, it is common to define a friction factor, f, relating the pressure drop to the velocity head according to

$$f \equiv \frac{\Delta p/L}{4(\rho V^2/2)(1/d_e)} \tag{10.10}$$

Recalling the definition for the hydraulic diameter, d_e, for a parallel-plate channel,

$$d_e \equiv \frac{4A}{P} = 4y_o \tag{10.11}$$

and inserting this value and the pressure drop from Eq. (10.9) into Eq. (10.10), the friction factor for this parallel plate configuration is found to be

$$f = \frac{24\mu}{\rho V d_e} = \frac{24}{\text{Re}} \tag{10.12}$$

Observe that in round pipes with fully developed laminar flows, a similar definition of the friction factor yields f =16/Re. Thus, the hydraulic diameter is seen not to be a universal scaling parameter in laminar flow.

It is to be noted that other friction factor definitions exist. One of them omits the 4 in the denominator, yielding values of f that are four times greater than given by Eq. (10.10) and making f = 64/Re for fully developed laminar flow in tubes and pipes.

Example 10.3: Determine the pressure drop per unit length for air in laminar flow at 325 K and at a velocity of 2 m/s between two parallel plates spaced 1 cm apart.

SOLUTION Using Eq. (10.9), with $y_o = L/2 = 0.005\,\mathrm{m}$

$$\frac{\Delta p}{L} = \frac{3\mu V}{y_o^2} = \frac{3(2\times 10^{-5})(2)}{(0.005)^2} = 4.8\,(\mathrm{N/m^2})/\mathrm{m}$$

The Temperature Field

In a channel where both the velocity and temperature boundary layers have merged, the temperature profile is invariant in the axial direction. Thus, following Rohsenow and Choi (1961), with T as the temperature,

$$\frac{T_w - T}{T_w - T_m} = f(y) \tag{10.13}$$

and

$$\frac{\partial}{\partial x}\left(\frac{T_w - T}{T_w - T_m}\right) = 0 \tag{10.14}$$

where the mean fluid temperature, T_m, is defined by

$$T_m \equiv \frac{\int_0^{y_o} vT\,dy}{\int_0^{y_o} v\,dy} \tag{10.15}$$

with T_w as the wall temperature and y as the distance from the wall.

Although the temperature profile is presumed to be axially invariant, the individual temperatures (i.e., wall, local fluid, and mean fluid) may vary in the x-direction, subject to the constraints established by the expansion of Eq. (10.14):

$$\left(\frac{\partial T_w}{\partial x} - \frac{\partial T}{\partial x}\right) - \left(\frac{T_w - T}{T_w - T_m}\right)\left(\frac{\partial T_w}{\partial x} - \frac{\partial T_m}{\partial x}\right) = 0 \tag{10.16}$$

For fully developed, laminar flow at low or moderate velocities, the energy balance on a differential element yields (Rohsenow and Choi, 1961)

$$\frac{\partial^2 T}{\partial y^2} = \frac{v}{x}\frac{\partial T}{\partial x} \tag{10.17}$$

When the channel walls are isothermal, that is, $\partial T_w/\partial x = 0$, the fully developed axial temperature gradient is found via Eq. (10.16) to be

$$\left(\frac{\partial T}{\partial x}\right)_T = \frac{T_w - T}{T_w - T_m}\frac{\partial T_m}{\partial x} \neq f(x) \tag{10.18}$$

where the subscript T denotes the isothermal condition. When both channel walls dissipate an equal and uniform heat flux, Newton's cooling law with a uniform heat-transfer coefficient leads to a uniform wall-to-fluid mean temperature difference, $T_w - T_m$, along the channel. Thus, via Eqs. (10.14) and (10.16),

$$\left(\frac{\partial T}{\partial x}\right)_H = \frac{\partial T_w}{\partial x} = \frac{\partial T_m}{\partial x} \neq f(x) \tag{10.19}$$

where the subscript H denotes the isoflux condition.

The Fully Developed Heat-Transfer Coefficient

The convective heat transfer coefficient, h, defined by Newton's cooling law, is generally written as

$$h \equiv \frac{q''}{T_w - T_m} \tag{10.20}$$

where q'' is the surface heat flux. However, because heat must be conducted into the fluid at the wall-fluid interface, Fourier's law can be used to determine the heat flux as

$$q'' = -k\left(\frac{\partial T}{\partial y}\right)_w \tag{10.21}$$

where k is the thermal conductivity of the fluid and the subscript w defines the wall condition. Consequently, the heat-transfer coefficient is expressible in terms of the transverse fluid temperature gradient at the wall, that is,

$$h = \frac{-k(\partial T/\partial y)_w}{T_w - T_m} \tag{10.22}$$

Differentiation of the fully developed temperature profile of Eq. (10.13) shows the wall temperature gradient to be

$$\left(\frac{\partial T}{\partial y}\right)_w = -(T_w - T_m)[f'(y)]_w \tag{10.23}$$

A combinination of Eqs. (10.22) and (10.23) provides the fully developed heat transfer coefficient

$$h = k[f'(y)]_w \tag{10.24}$$

which is observed to be invariant in the axial direction.

Recognizing the importance of conduction in convective heat transfer, it is common to define a nondimensional heat-transfer coefficient for this configuration as the ratio of h to k/d_e to yield the *Nusselt number*, hd_e/k.

When Eq. (10.17) is integrated over the velocity profile of Eq. (10.7) and the axial gradient of Eq. (10.18) is used to provide the transverse fluid temperature gradient at an *isothermal* channel surface, the Nusselt number is found to be

$$\mathrm{Nu}_T \equiv \frac{hd_e}{k} = -d_e \frac{(\partial T/\partial y)_w}{T_w - T_m} = 7.6 \tag{10.25}$$

Similarly, for an *isoflux* channel with Eq. (10.19) for the axial gradient,

$$\mathrm{Nu}_H = 8.23 \tag{10.26}$$

Example 10.4: Determine the heat-transfer coefficient for fully-developed laminar flow of air at 325 K between uniformly heated, parallel plates spaced 5 mm apart.

SOLUTION Using Eq. (10.26),

$$\mathrm{Nu}_H = hd_e/k = 8.23$$

or

$$h = 8.23k/d_e$$

The thermal conductivity of air at 325 K is approximately 0.028 W/m-K and the hydraulic diameter for this parallel-plate channel is $d_e = 4y_o = 0.01\,\mathrm{m}$. Therefore, the heat-transfer coefficient is found to be

$$h = \frac{(8.23)(0.028)}{(0.01)} = 23.04\,\mathrm{W/m^2\text{-}K}$$

The Heat-Transfer/Friction Analogy

In analyzing transport phenomena, it is often convenient to utilize the *Stanton number*, defined as the ratio of the Nusselt number to the Reynolds number–Prandtl number (Re·Pr) product:

$$\mathrm{St} \equiv \frac{\mathrm{Nu}}{\mathrm{Re\,Pr}} \tag{10.27}$$

For the *isothermal* channel, the product of St and Pr thus equals

$$(\mathrm{StPr})_T = \frac{\mathrm{Nu}_T}{\mathrm{Re}} = \frac{7.6}{\mathrm{Re}} \tag{10.28}$$

whereas for the *isoflux* channel,

$$(\mathrm{StPr})_H = \frac{\mathrm{Nu}_H}{\mathrm{Re}} = \frac{8.23}{\mathrm{Re}} \tag{10.29}$$

The friction factor for this configuration was previously found to be equal to 24/Re. Consequently, in laminar, fully developed channel flow,

$$\mathrm{StPr} \approx \frac{f}{3} \tag{10.30}$$

It is significant to note that a similar calculation for flow in a circular tube yields

$$\mathrm{StPr} \approx \frac{f}{4}$$

10.2.2 Turbulent, Fully Developed, Smooth Channel Flow

This short section first considers fluid flow phenomena for fully developed, turbulent channel flow and concludes with a discussion of the heat-transfer coefficient.

Fluid Mechanics

In turbulent flow, eddy motion and velocity fluctuations increase the shear stress above the values associated with laminar flow. It is convenient to relate this increase to a so-called *momentum eddy diffusivity*, ϵ_m, and to thus express the relationship between the shear stress and the velocity gradient as

$$\frac{\tau}{\rho} = (\nu + \epsilon_m)\frac{dv}{dy} \tag{10.31}$$

where ν is the *dynamic viscosity* equal to the kinematic viscosity divided by the density.

A simple force balance on the control volume shown in Fig. 10.2 indicates that the wall shear stress is equal to

$$\tau_w = \frac{2y_o \Delta p}{2L} \tag{10.32}$$

Insertion of the friction factor as defined by Eq. (10.10) with $d_e = 4y_o$ yields

$$\tau_w = \left(\frac{f}{2}\right)\rho V^2 \tag{10.33}$$

The friction factor in turbulent pipe flow was found by Nikuradse (1930) to vary as

$$(f)^{-1/2} = 4\log(\mathrm{Re}\sqrt{f}) - 0.4 \tag{10.34}$$

In the midrange of the turbulent regime, $3 \times 10^4 < \mathrm{Re} < 5 \times 10^5$, Eq. (10.34) can be well approximated by

$$f = 0.046\mathrm{Re}^{-0.2} \tag{10.35}$$

Both Eqs. (10.34) and (10.35) can be applied as well to turbulent flow in channels formed by parallel plates when the Reynolds number is based on the hydraulic diameter d_e (Rohsenow and Choi, 1961).

The Heat Transfer Coefficient

The heat flux in the transverse (y) direction for turbulent flow in a tube or pipe can be expressed as

$$\frac{q''}{\rho c_p} = -(\alpha + \epsilon_h)\frac{dT}{dy} \tag{10.36}$$

where α is the thermal diffusivity and ϵ_h is the thermal eddy diffusivity and is analogous to ϵ_m introduced in Eq. (10.31). A division of this relation by the turbulent shear stress, provides a relation between q'' and τ:

$$\frac{q''}{c_p \tau} = \frac{\alpha + \epsilon_h}{v + \epsilon_m}\frac{\partial T}{\partial v} \tag{10.37}$$

For gases, where Pr is approximately unity, many investigators have assumed that $(\alpha + \epsilon_h)/(v + \epsilon_m)$ is also near unity and that the ratio of q'' to τ is independent of the transverse coordinate (Rohsenow and Choi, 1961). These two simplifications may be introduced into Eq. (10.37). Then, separation of the variables and subsequent integration from a point on the wall to the point in the channel where the mean velocity, V, and the mean temperature, T_m, are encountered, results in

$$\frac{q''_w}{c_p \tau_w}\int_w^m dv = -\int_w^m dT \tag{10.38}$$

Following the integration

$$\frac{q''_w V}{c_p \tau_w} = T_w - T_m \tag{10.39}$$

or, with $q''_w = h(T_w - T_m)$ and τ_w from Eq. (10.33)

$$\frac{h}{\rho c_p V} = \frac{f}{2} \tag{10.40}$$

It is interesting to note that the left side of this relation is identical to the Stanton number, St, which is found to equal $f/2$. It is thus apparent that the relationship between St and the friction factor in turbulent flow differs from that encountered in laminar flow, where, for $\mathrm{Pr} = 1$, $\mathrm{St} \approx f/3$.

Equation (10.40) was derived on the assumption of an equality between $\alpha + \epsilon_h$ and $v + \epsilon_m$. Experimental results, discussed by Rohsenow and Choi (1961), suggest that the ratio of the turbulent diffusivities, that is, $(\alpha + \epsilon_h)/(v + \epsilon_m)$, is, in fact, a function of the Prandtl number, which for conventional fluids may be approximated by $\mathrm{Pr}^{-2/3}$. Consequently, for Pr values other than unity, the relationship between the heat-transfer coefficient and the friction factor can be better stated as

$$\frac{h}{\rho c_p V}\mathrm{Pr}^{2/3} = \frac{f}{2} \tag{10.41a}$$

or

$$\mathrm{St}\,\mathrm{Pr}^{2/3} = \frac{f}{2} \tag{10.41b}$$

Introducing the definition of the Stanton number and the empirical relations for the friction factor stated in Eqs. (10.35) and (10.41), the Nusselt number for turbulent pipe flow is found to be equal to

$$\mathrm{Nu} = 0.023\mathrm{Re}^{0.8}\mathrm{Pr}^{0.33} \tag{10.42}$$

This equation is identical with the widely used Colburn (1933) correlation for pipe flow and differs from the familiar McAdams (1954) correlation by only a slightly lower (0.33 versus 0.4) Pr exponent.

Example 10.5: Determine the heat-transfer coefficient for the fully-developed turbulent flow of 325 K air, at a Reynolds number of 5000, flowing between isothermal, parallel plates spaced 5 mm apart.

SOLUTION Equation (10.42),

$$\mathrm{Nu} = 0.023\mathrm{Re}^{0.8}\mathrm{Pr}^{0.33} = \frac{hd_e}{k}$$

can be rearranged to yield h:

$$h = 0.023\frac{k}{d_e}\mathrm{Re}^{0.8}\mathrm{Pr}^{0.33}$$

The Prandtl number of air at 325 K is known to equal 0.70 and the thermal conductivity to equal to 0.028 W/m-K. Therefore, with d_e = 2(0.005 m) = 0.01 m,

$$\begin{aligned} h &= 0.023\frac{0.028}{0.01}(5000)^{0.8}(0.7)^{0.33} \\ &= (0.023)(2.8)(910.3)(0.89) \\ &= 52.06\,\mathrm{W/m^2\text{-}K} \end{aligned}$$

In turbulent, as opposed to laminar flow of conventional fluids, the boundary conditions exert only a very minor influence on the surface heat-transfer coefficient. This behavior can be related to the differences between the laminar and turbulent transverse temperature profile in the fluid. Thus, it is possible to use a single Nusselt number correlation for both isothermal and isoflux boundary conditions for all Pr greater than 0.7, with a maximum anticipated deviation of less than a few percent. The occurrence of the primary fluid thermal resistance in the region immediately adjacent to the surface also facilitates the application of Eq. (10.42) to most conduit geometries, as well as to turbulent flow between parallel plates. Of course, care must be taken to base both the Nusselt and Reynolds numbers on the hydraulic diameter d_e. The reader is reminded that the hydraulic diameter is not a universal scaling factor in laminar flow.

10.3 Natural Convection in a Parallel-Plate Channel

10.3.1 The Elenbaas Correlation

Elenbaas (1942) was apparently the first to document a detailed study of the thermofluid behavior of a vertical channel in natural convection. His experimental results for isothermal plates in air were later confirmed numerically by Bodoia and Osterle (1964) and shown to apply, as well, to constant heat flux conditions (Sobel et al., 1966). Aung (1972), Aung et al. (1972), Miyatake and Fujii (1972) and Miyatake et al. (1973) all extended the available results to include both asymmetric wall temperature and heat flux boundary conditions, including the case of the single insulated wall.

A unified picture of the thermal transport in such vertical channels emerged from these and complementary studies. Figure 10.3 shows that in the inlet region and in relatively short channels, individual momentum and thermal boundary layers are in evidence along each surface. Moreover, the heat-transfer rates approach those associated with laminar flow along isolated plates in infinite media. The development of these boundary layers progresses towards the fully developed condition at the channel outlet. Alternatively, for long narrow channels, the boundary layers merge near the entrance, and fully developed flow prevails along much of the channel.

In this fully developed regime, the local heat-transfer coefficient is constant (neglecting the temperature dependence of fluid properties) and equal to the well-documented forced convection values of Lundberg et al. (1963). However, because the local fluid temperature is not known explicitly, it is customary to reexpress the fully developed heat-transfer coefficient in terms of the wall-to-ambient or wall-to-inlet air temperature difference. The Nusselt number, Nu_o, for isothermal plates appropriate to this definition can be derived from the "*incompressible natural convection*" form of the Navier-Stokes equations. This was done semianalytically by Elenbaas (1942) and rederived via an integral formulation by Bar-Cohen and Rohsenow (1983).

The analytic Nusselt number relations for the fully developed flow region and the isolated-plate region, summarized in Table 10.1, can be expected to bound the Nusselt number values over the complete range of flow development. In the fully developed limit, the gap-based Nusselt number ($\text{Nu} = hz/k$), is seen to depend on a *gap-based Rayleigh number* with z taken as the spacing between plates

$$\text{Ra} = \frac{\rho^2 g \beta c_p z^4 (T_w - T_\infty)}{\mu k L}$$

for an isothermal channel and a *modified gap-based Rayleigh number*

$$\text{Ra}'' = \frac{\rho^2 g \beta c_p z^5 q''}{\mu k^2 L}$$

for an isoflux channel. In recognition of his seminal contributions to this field,

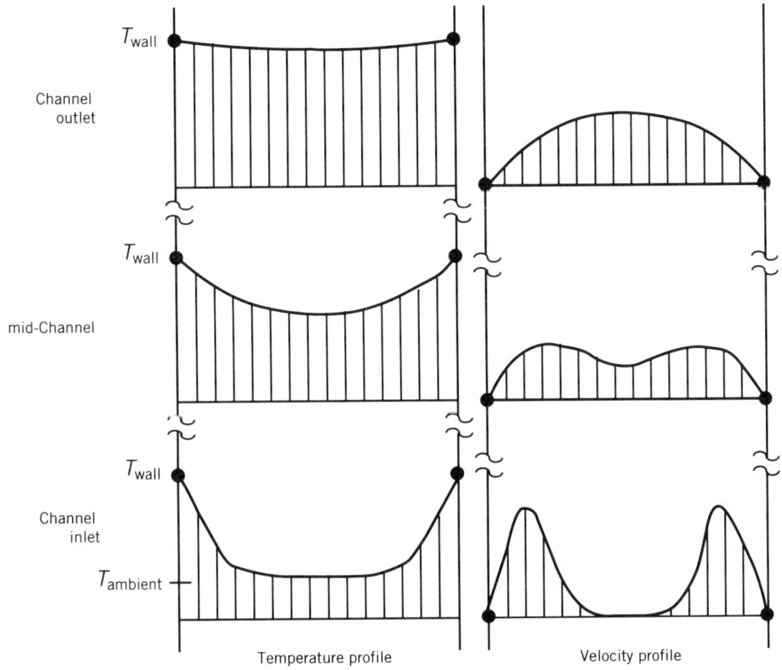

Figure 10.3 Developing temperature and velocity profiles for natural convection in a vertical channel.

these nondimensional groupings are, henceforth, referred to as the *Elenbaas number*, El, and the *modified Elenbaas number*, El$''$, respectively.

Elenbaas used his understanding of these bounding relations to establish the form of the relationship between the gap-based Nusselt number and the fluid properties over the entire range of channel geometries. Application of this form to the data for relatively short isothermal plates in air, yielded the well-known Elenbaas (1942) correlation:

$$\mathrm{Nu}_o = \frac{hz}{k} = \frac{\mathrm{El}}{24}\left(1 - e^{-35/\mathrm{El}}\right)^{3/4} \tag{10.43}$$

where Nu_o is the gap-based Nusselt number and z is the spacing between the plates.

10.3.2 Composite Relations

When a function is known to vary smoothly between two limiting expressions, which are themselves well defined, and when solutions for intermediate values of the function are either difficult to obtain or involve other tabulated functions, an approximate composite relation can be obtained by appropriately summing the two limiting expressions.

TABLE 10.1 Nusselt Numbers for Fully Developed and Isolated Plate Natural Convection in Two-Dimensional Channels

Boundary Conditions	Theoretical Expression	
	Fully Developed Limit	Isolated Plate Limit
Isothermal Plates:		
Symmetric - Both Plates at an Identical Temperature	$\mathrm{Nu}_{T,S} = \mathrm{El}/24$	
Asymmetric - One Plate Insulated	$\mathrm{Nu}_{T,A} = \mathrm{El}/12$	$\mathrm{Nu}_T = 0.59(\mathrm{El})^{1/4}$
Isoflux Plates - Based on Maximum Plate Temperature		
Symmetric - Both Plates at the Same Heat Flux	$\mathrm{Nu}_{H,S,L} = 0.144(\mathrm{El}'')^{1/2}$	
Asymmetric - One Plate Insulated	$\mathrm{Nu}_{H,A,L} = 0.204(\mathrm{El}'')^{1/2}$	$\mathrm{Nu}_H = 0.63(\mathrm{El})^{1/5}$
Isoflux Plates - Based on Mid-Height Temperature		
Symmetric - Both Plates at the Same Heat Flux	$\mathrm{Nu}_{H,S,L/2} = 0.289(\mathrm{El}'')^{1/2}$	$\mathrm{Nu}_H = 0.73(\mathrm{El})^{1/5}$
Asymmetric - One Plate Insulated	$\mathrm{Nu}_{H,A,L/2} = 0.41(\mathrm{El}'')^{1/2}$	
$\mathrm{Nu}_T \equiv [\, q''/(T_w - T_o)\,]\, z/k$	z = plate spacing	
$\mathrm{Nu}_{H,L} \equiv [\, q''/(T_{w,L} - T_o)\,]\, z/k$	$\mathrm{El} \equiv \rho^2 g \beta c_p (T_w - T_o) z^4/\mu k L$	
$\mathrm{Nu}_{H,L/2} \equiv [\, q''/(T_{w,L/2} - T_o)\,]\, z/k$	$\mathrm{El}'' \equiv \rho^2 g \beta c_p z^5 q''/\mu k^2 L$	

Fluid properties are based on the average fluid temperature.

This approach could be of particular benefit in predicting the behavior of air-cooled heat sinks in domains far from the Elenbaas parametric space or for heat sinks cooled by fluids other than air. Churchill and Usagi (1972) suggested that the frequently employed linear superposition be viewed as a special case of a more general summation of the form

$$y = [(Az^p)^n + (Bz^q)^n]^{1/n} \tag{10.44}$$

where

$$y \rightarrow Az^p \quad \text{as} \quad z \rightarrow 0$$
$$y \rightarrow Bz^q \quad \text{as} \quad z \rightarrow \infty$$

and

$$n > 0 \quad \text{if} \quad p < q$$
$$n < 0 \quad \text{if} \quad p > q$$

In using this approach for natural convection in a channel, care must be taken to express Nu in terms of the appropriately defined Elenbaas number for both the fully developed and isolated plate limits. This has been done in the isolated-plate correlations shown in Table 10.1. Multiplication of both sides of the conventional correlations by z/L easily converts those expressions to the relations shown between the gap Nusselt number, hz/k, and the Elenbaas number, El or El$''$ for isothermal and isoflux channels, respectively.

In applying Eq. (10.44) to natural convection in isothermal channels, it might thus be anticipated that Nu_o would vary according to

$$\text{Nu}_o = \{(A\,\text{El})^{-n} + [B(\text{El})^{1/4}]^{-n}\}^{-1/n} \tag{10.45}$$

10.3.3 Symmetric, Isothermal Plates

As suggested in Table 10.1, Nu for laminar-free convection on isothermal surfaces is dependent on $(\text{El})^{1/4}$, and configurational variations are generally reflected in different values of the coefficient. For moderately short vertical plates in air and $10^4 < \text{El} < 10^9$, McAdams (1954) reports the coefficient to equal 0.59. This expression, together with the fully developed Nu_o relation derived earlier, can be inserted into Eq. (10.45) to yield

$$\text{Nu}_o = \left[\left(\frac{1}{24}\text{El}\right)^{-n} + \left(0.59\text{El}^{1/4}\right)^{-n}\right]^{-1/n} \tag{10.46}$$

A comparison of Eqs. (10.43) and (10.46) shows that the choice of n = 2 offers very good agreement with the experimental results of Elenbaas. The composite equation for isothermal plates thus takes the form

$$\text{Nu}_o = \left[\frac{576}{(\text{El})^2} + \frac{2.87}{(\text{El})^{1/2}}\right]^{-1/2} \tag{10.47}$$

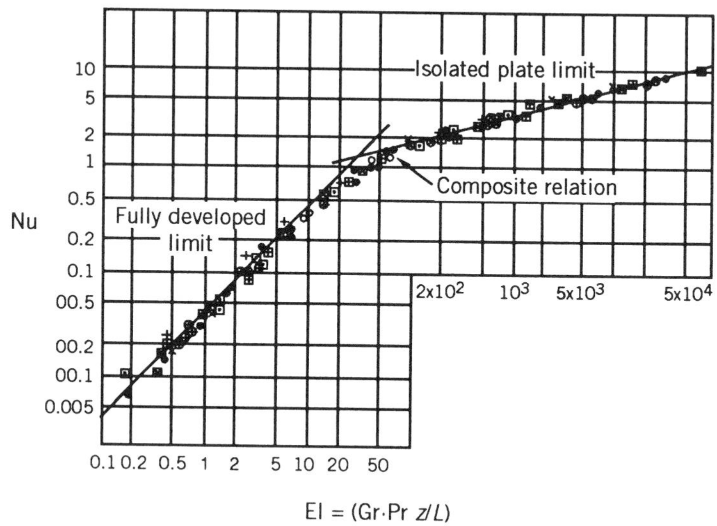

Figure 10.4 Nusselt number variation for symmetric isothermal plates (data points from Elenbaas, 1942).

where El and Nu are based on properties evaluated at the average fluid temperature. The validity of this relation, derived by Bar-Cohen and Rohsenow (1983), and the utility of the Churchill and Usagi (1972) approach, are borne out by the close proximity of the Elenbaas (1942) data points to the composite relation of Eq. (10.47) and the asymptotic equations at both limits, indicated in Fig. 10.4.

Example 10.6: Determine the heat-transfer coefficient for air in natural convection within a 15-cm long, parallel-plate channel. The ambient temperature is 300 K and the channel walls are maintained at 350 K. Three situations are to be considered, corresponding to plate spacings of 20 mm, 5 mm, and 2 mm.

SOLUTION Either Eqs. (10.43) or (10.47) (or Fig. 10.4) can be used here. Suppose Eq. (10.47) is employed. The Elenbaas number

$$\text{El} = \frac{g\rho^2 \beta c_p (T_w - T_o) z^4}{\mu k L}$$

is found first. The average air temperature is

$$T_{av} = \frac{300 + 350}{2} = 325\,\text{K}$$

and the fluid properties at this temperature are found to equal

$$\rho = 1\,\text{kg/m}^3$$

$$\begin{aligned} \beta &= 1/325 = 0.00308\,1/\text{K} \\ c_p &= 1000\,\text{J/kg-K} \\ \mu &= 2 \times 10^{-5}\,\text{N-s/m}^2 \end{aligned}$$

and

$$k = 0.028\,\text{W/m-K}$$

Thus with $T_w - T_o = 50\,\text{K}$ and $L = 0.15\,\text{m}$, El at 325 K is found to equal

$$\text{El} = 1.80 \times 10^{10} z^4$$

and thus

For the spacing = 20 mm, El = 2874	isolated-plate region
For the spacing = 5 mm, El = 11.23	intermediate region
For the spacing = 2 mm, El= 0.287	fully-developed region

Now a return to Eq. (10.47) or Fig. 10.4 will give the Nusselt numbers. For a spacing of 20 mm, Nu = 4.32 and

$$h = 4.32\frac{k}{z} = \frac{4.32(0.028)}{0.02} = 6.05\,\text{W/m}^2\text{-K}$$

For a spacing of 5 mm, Nu = 0.43 and

$$h = 0.43\frac{k}{z} = \frac{0.43(0.028)}{0.005} = 2.41\,\text{W/m}^2\text{-K}$$

and for a spacing of 2 mm, Nu = 0.012 and

$$h = 0.012\frac{k}{z} = \frac{0.012(0.028)}{0.002} = 0.168\,\text{W/m}^2\text{-K}$$

10.4 Thermal Radiation from a Parallel-Plate Channel

An isolated heated surface radiates energy to its surroundings in accordance with the Stefan-Boltzmann law introduced in Sec. 1.2.4, Eq. (1.27). This relationship applies as well to a finned surface, but care must be taken in formulating the radiant-exchange equations from the U-shaped channel, the "*building block*" of a typical plate-fin heat sink (Fig. 10.5).

In such a channel, only a fraction of the thermal radiation, originating from the area between the two parallel plates (surface 1), will flow directly to the surroundings (through surfaces 4, 5, and 6). The remainder of the radiation will be intercepted by the side walls of the channel and be reflected and/or absorbed by these surfaces. The side walls (surfaces 2 and 3) will themselves radiate to the surroundings, as well as back to the base surface (surface 1). Thus, correct determination of the net thermal radiation from a heat sink to

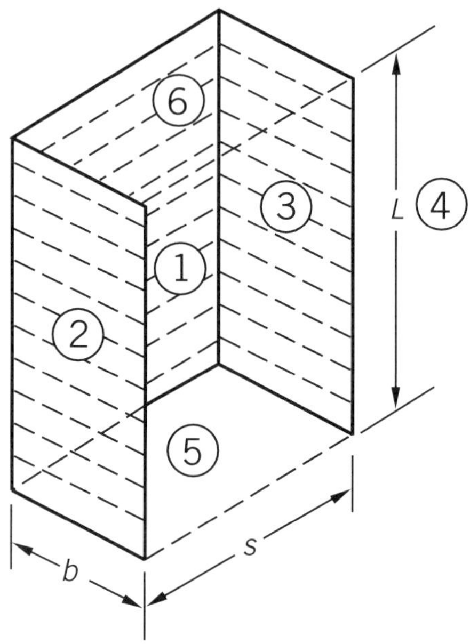

Figure 10.5 Geometric definition of U-shaped channels.

the ambient requires careful and detailed accounting for both emitted and reflected energy incident on each of the various surfaces. These geometric factors, are generally referred to as *radiation shape factors* in the thermal literature and can be obtained from algebraic equations and / or graphs appearing in the leading radiation texts (e.g., Siegel and Howell, 1992).

To analyze thermal radiation from a U-shaped channel, it is convenient to view the fin structure as an open-sided enclosure, with three opaque surfaces (1, 2, and 3) and three transparent surfaces (4, 5, and 6), providing an unobstructed view of the surroundings. To deal with multiple emitted and reflected radiation streams within such an enclosure, two new radiation terms are considered. The *radiosity*, J, is defined as the total radiant flux leaving each surface, and the *irradiation*, G, is defined as the total radiation flux incident on each surface. Then for the ith surface in the enclosure, the radiosity consists of the reflected irradiation and the surface emission,

$$J_i = \rho_i G_i + \epsilon_i E_{bi} \tag{10.48}$$

where ρ_i and ϵ_i are the surface reflectivity and emissivity, respectively, and E_{bi} is the black-body emissive power. If the surface temperature and irradiation over the surface of area, S_i, are uniform, then the net rate of heat loss (or gain) is

$$q_i = S_i(J_i - G_i) \tag{10.49}$$

If the surfaces are gray, then with α_i as the surface absorptivity of the ith surface,

$$\epsilon_i = \alpha_i$$

an important limiting assumption, and

$$\rho_i = 1 - \epsilon_i$$

Equation (10.48) can be rearranged to yield the irradiation, G_i,

$$G_i = \frac{1}{\rho_i}(J_i - \epsilon_i E_{bi})$$

After substituting for G_i in Eq. (10.49), the radiant heat loss can be expressed as

$$q_i = S_i \left[J_i - \frac{1}{\rho_i}(J_i - \epsilon_i E_{bi}) \right]$$

or

$$q_i = \frac{S_i}{\rho_i}[\epsilon_i E_{bi} + (\rho_i - 1)J_i] = \frac{S_i \epsilon_i}{1 - \epsilon_i}(E_{bi} - J_i) \tag{10.50}$$

If Eq. (10.50) is rewritten in the form

$$q_i = \frac{E_{bi} - J_i}{(1 - \epsilon_i)/S_i \epsilon_i} \tag{10.51}$$

$E_{bi} - J_i$ can be considered as a difference of potential and $(1 - \epsilon_i)/S_i \epsilon_i$ as the inherent radiative thermal resistance of the surface. This resistance governs the net flow of heat from the surface, based on the relationship between the "black-body" potential, $E_{bi} = \sigma T_i^4$, and the radiosity, J_i.

The direct radiative interchange between surface i and surface j is governed by the difference of the two radiosities, the surface, S_i, and the shape factor, F_{ij},

$$q_{ij} = \frac{J_i - J_j}{S_i F_{ij}} = \frac{J_i - J_j}{S_j F_{ji}} \tag{10.52}$$

where the *reciprocity relationship*, $S_i F_{ij} = S_j F_{ji}$, has been used. Here, the radiative thermal resistance is

$$\frac{1}{S_i F_{ij}} = \frac{1}{S_j F_{ji}}$$

This resistance governs the flow of heat between two node points or surfaces, each represented by a radiosity.

Equations (10.51) and (10.52) form the basis for a radiative resistance network, as in Fig. 10.6. The nodes marked E_{bi} and E_{bj} represent heat sources, one of which must be known. Each of these may be related to a temperature, $E_{bi} = \sigma T_i^4$. The resistances, $(1 - \epsilon_i)/S_i \epsilon_i$, transfer the E_{bi}'s to surface radiosities, J_i, thereby accounting for the fact that the surfaces are not black. The

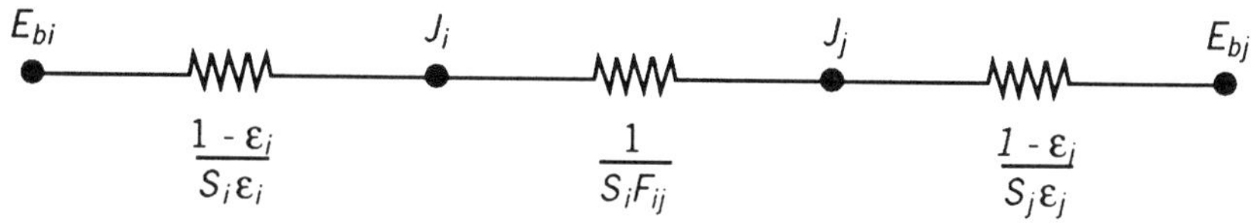

Figure 10.6 Radiative resistance network.

resistance, $1/S_iF_{ij} = 1/S_jF_{ji}$, links the radiosities and accounts for the actual radiative interchange between the ith and jth surfaces.

With this formulation and the assumption of diffuse radiation (Siegel and Howell, 1992), heat exchange within the enclosure of Fig. 10.5 can be evaluated from a *node* analysis of the thermal network of Fig 10.7. Observe in Fig. 10.7 that nodes 4, 5, and 6 are the nodes that represent the presumed black-body environment and that their node potentials are known,

$$J_4 = J_5 = J_6 = E_{bo} = \sigma T_o^4$$

where $T_o = 27 + 273 = 300$K.

In Fig. 10.7, the conductance values that are designated by K (for *konductance*) are:

$$K_a = \frac{S_1\epsilon_1}{1-\epsilon_1}$$

$$K_b = \frac{S_2\epsilon_2}{1-\epsilon_2}$$

$$K_c = \frac{S_3\epsilon_3}{1-\epsilon_3}$$

and

$$\begin{aligned}
K_d &= S_1F_{12} = S_2F_{21} \\
K_e &= S_2F_{23} = S_3F_{32} \\
K_f &= S_1F_{13} = S_3F_{31} \\
K_g &= S_1F_{14} = S_4F_{41} \\
K_h &= S_1F_{15} = S_5F_{51} \\
K_k &= S_1F_{16} = S_6F_{61} \\
K_\ell &= S_2F_{24} = S_4F_{42} \\
K_m &= S_2F_{25} = S_5F_{52} \\
K_n &= S_2F_{26} = S_6F_{62} \\
K_o &= S_3F_{34} = S_4F_{43} \\
K_p &= S_3F_{35} = S_5F_{53} \\
K_q &= S_3F_{36} = S_6F_{63}
\end{aligned}$$

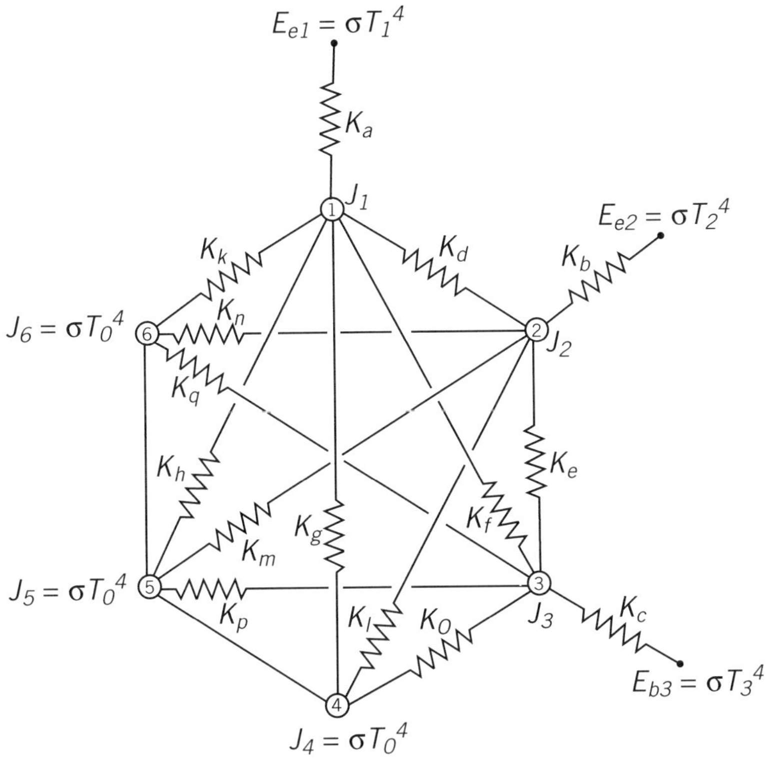

Figure 10.7 Radiation network for U-shaped channel (nodes J_4, J_5, and J_6 represent the surrounding environment. Hence, there is no resistance between them).

There is a certain symmetry in the configuration of Fig. 10.5 and its electrothermal analog of Fig. 10.7. Observe that nodes 4, 5, and 6 can be combined into one node (node 4) with a known potential, $E_o = \sigma T_o^4$ and that $S_2 = S_3, S_1 = S_4$, and $S_5 = S_6$. Moreover, with regard to node 1,

$$F_{12} = F_{13}$$

and

$$F_{15} = F_{16}$$

and for nodes 2 and 3

$$F_{24} = F_{34}$$

and

$$F_{25} = F_{35} = F_{26} = F_{36}$$

It may then be noted that

$$K_d = K_f$$

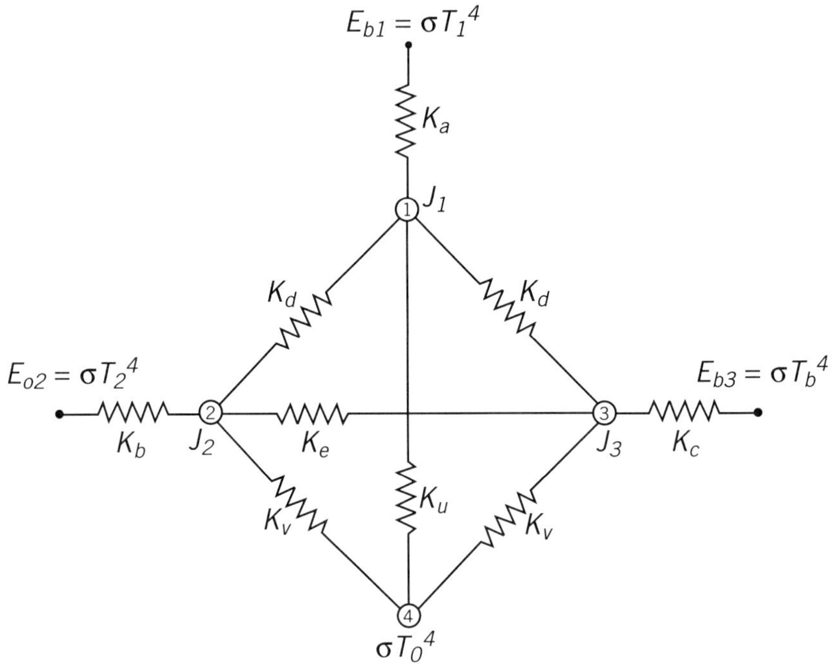

Figure 10.8 Simplification of the network in Fig. 10.7.

$$\begin{aligned} K_h &= K_k \\ K_m = K_n &= K_p = K_q \end{aligned}$$

and

$$K_o = K_\ell$$

With these equalities in hand and with the fact that nodes 4, 5, and 6, in Fig. 10.7 are all at the same potential, the network in Fig. 10.7 can then be simplified to that shown in Fig. 10.8 where K_a through K_e, remain unchanged but

$$\begin{aligned} K_u &= K_g + K_h + K_k \\ &= K_g + 2K_h \\ &= S_1 F_{14} + 2S_1 F_{15} \\ &= S_1(F_{14} + 2F_{15}) \end{aligned}$$

and

$$\begin{aligned} K_v &= K_\ell + K_m + K_n \\ &= K_\ell + 2K_n \end{aligned}$$

$$
\begin{aligned}
&= S_2 F_{24} + 2S_2 F_{25} \\
&= S_1 (F_{24} + 2F_{25})
\end{aligned}
$$

Now continuity may be used to write three *node equations*. With heat flows away from each node taken as positive, the algebraic sum of these heat flows must equal zero. Thus, for the three nodes in turn, the node equations are:

$$
\begin{aligned}
K_a(J_1 - \sigma T_1^4) + K_d(J_1 - J_2) + K_d(J_1 - J_3) + K_u(J_1 - \sigma T_o^4) &= 0 \\
K_b(J_2 - \sigma T_2^4) + K_d(J_2 - J_1) + K_e(J_2 - J_3) + K_v(J_2 - \sigma T_o^4) &= 0 \\
K_c(J_3 - \sigma T_3^4) + K_d(J_3 - J_1) + K_e(J_3 - J_2) + K_v(J_3 - \sigma T_o^4) &= 0
\end{aligned}
$$

or in the matrix form

$$
\mathbf{AJ} = \mathbf{E}
$$

$$
\begin{bmatrix} a_{11} & a_{12} & a_{13} \\ a_{21} & a_{22} & a_{33} \\ a_{31} & a_{22} & a_{33} \end{bmatrix}
\begin{bmatrix} J_1 \\ J_2 \\ J_3 \end{bmatrix}
=
\begin{bmatrix} \sigma(K_a T_1^4 + K_u T_0^4) \\ \sigma(K_b T_2^4 + K_v T_0^4) \\ \sigma(K_c T_3^4 + K_v T_0^4) \end{bmatrix}
\tag{10.53a}
$$

In Eq. (10.53a),

$$
\begin{aligned}
a_{11} &= K_a + 2K_d + K_u \\
&= S_1 \left[\frac{\epsilon_1}{1 - \epsilon_1} + 2F_{12} + F_{14} + 2F_{15} \right] \\
a_{22} &= K_b + K_d + K_e + K_v \\
&= S_2 \left[\frac{\epsilon_2}{1 - \epsilon_2} + F_{21} + F_{23} + F_{24} + 2F_{25} \right] \\
a_{33} &= K_c + K_d + K_e + K_v \\
&= S_3 \left[\frac{\epsilon_3}{1 - \epsilon_3} + F_{31} + F_{32} + F_{34} + 2F_{35} \right] \\
a_{12} = a_{21} &= -S_1 F_{12} = -S_2 F_{21} \\
a_{13} = a_{31} &= -S_1 F_{13} = -S_3 F_{31}
\end{aligned}
$$

and

$$
a_{23} = a_{32} = -S_2 F_{23} = -S_3 F_{32}
$$

The solution of Eqs. (10.53a) will yield the solution vector

$$
\mathbf{J} = \begin{bmatrix} J_1 \\ J_2 \\ J_3 \end{bmatrix}
$$

and, with this in hand, the total heat flow from the channel can be evaluated by summing the heat flow from nodes 1, 2, and 3 at potentials, J_1, J_2, and J_3:

$$q_T = K_u(J_1 - \sigma T_o^4) + K_v(J_2 - \sigma T_o^4) + K_v(J_3 - \sigma T_o^4)$$

or

$$q_T = K_u J_1 + K_v(J_2 + J_3) - (K_u + 2K_v)\sigma T_o^4 \tag{10.53 b}$$

To facilitate the evaluation of heat sink heat transfer that is due to radiation, it is helpful to reference the calculated radiated flux to the prime surface area. This can be done by defining an effective channel emittance, $\hat{E}$, as

$$\hat{E} = \frac{Q_T}{\sigma S_1(T_1^4 - T_o^4)} \tag{10.54}$$

where Q_T is the net emitted radiation given by Eq. (10.53b).

The resulting $\hat{E}$ values, as obtained by Bilitzky (1986), are shown in Figs. 10.9 through 10.13 for a representative range of channel geometries, and surface emissivities ranging from zero to unity. These figures are seen to display effective channel emittances that are dependent on the channel geometry and the surface emissivity. As expected, increasing emissivities lead to higher channel emittances and, hence, higher heat-rejection rates from each parallel-plate channel. Similar improvements result from increasing the channel depth, that is, increasing values of b/z. Somewhat surprisingly, as a consequence of the unconventional definition of the effective channel emittance [Eq. (10.54)], which includes radiation streaming out of the top (surface 6) and the bottom (surface 5), in addition to the open front end (surface 4), $\hat{E}$ is seen to attain values substantially greater than unity for short, relatively deep channels with L/b ratios approaching unity. This behavior is primarily due to the large open areas at the top and bottom, associated with such channels. The ratio of the front area to the top and bottom areas is directly proportional to L/b. For small values of this aspect ratio, the top and bottom edges of the channel cannot be neglected in the calculation of the emitted radiation.

Example 10.7: Determine the effective channel emittance of, as well as the net emitted radiation from, a parallel-plate channel 15 cm long ($L = 0.15$ m) and 3 cm high ($b = 0.03$ m) with a surface emissivity of 0.2. Assume that the surface is at 350 K, the surrounding ambient is at 300 K, and that the channel width, z, is 2 cm and 5 mm, respectively.

SOLUTION For the dimensions specified, L/b is equal to 5. Therefore refer to Figure 10.11 and observe that for an emissivity of 0.2, the effective channel emittance for the proper ratio b/z is approximately equal to

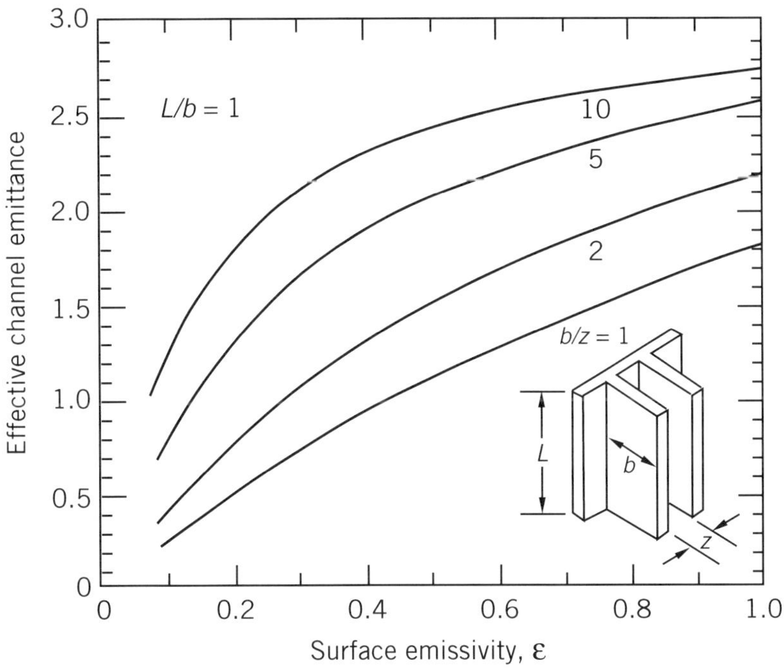

Figure 10.9 Channel radiative emmitance for $L/b = 1$.

$$\hat{E}(z = 2\,\text{cm}, b/z = 1.5) \quad = \quad 0.5$$

and

$$\hat{E}(z = 5\,\text{mm}, b/z = 6) \quad = \quad 0.9$$

The thermal radiation rejected by the heat sink channel for the conditions stated can be found via Eq. (10.54). With the base surface and the base temperature, respectively, designated as S_b and T_b,

$$\begin{aligned} Q_T &= \hat{E}\sigma S_b(T_b^4 - T_o^4) \\ &= \hat{E}(5.67 \times 10^{-8})(0.15z)(350^4 - 300^4) \\ &= 58.74\hat{E}z \end{aligned}$$

For the specified channel widths, the net radiated energy is found to be approximately

$$Q_T(z = 2\,\text{cm}, b/z = 1.5) = (58.7)(0.5)(0.02) \quad = \quad 0.587\,\text{W}$$

and

$$Q_T(z = 5\,\text{mm}, b/z = 6) = (58.7)(0.9)(0.005) \quad = \quad 0.264\,\text{W}$$

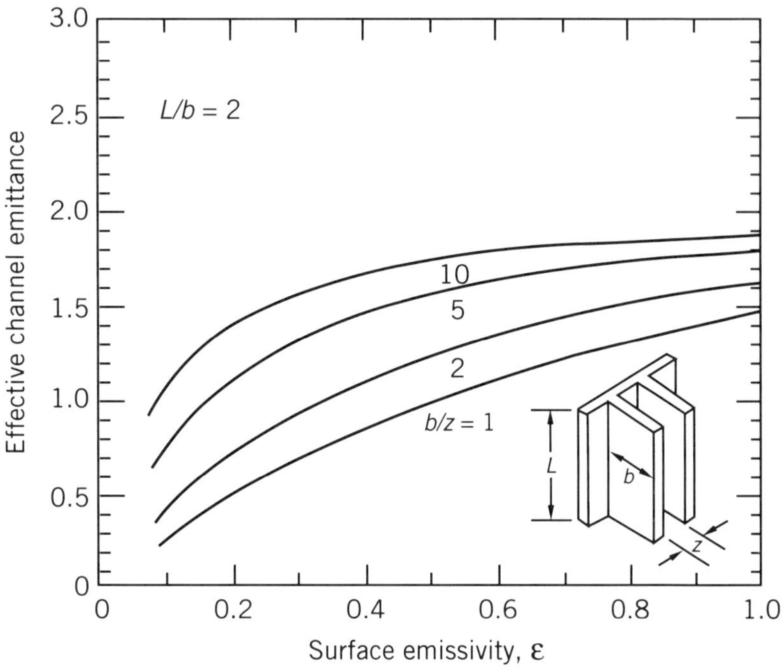

Figure 10.10 Channel radiative emmitance for $L/b = 2$.

It is interesting to compare the result in Ex. 10.7 to the natural convection values found in Ex. 10.6. For a heat sink channel of similar length and for an identical surface-to-air temperature difference, the natural convection heat-transfer coefficient was found to equal 6.05W/m^2-K for the 2-cm wide channel and 2.41 W/m^2-K for the 5-mm wide channel. Assuming that these heat sink plates are isothermal and applying these respective heat-transfer coefficients to the 3-cm high fin surface area and the exposed base area, the natural convection heat rejection rate is found to equal

$$\begin{aligned} Q_T &= h(S_b + S_f)(\Delta T) \\ Q_T(z = 2\,\text{cm}) &= 6.05\,[0.15(0.02) + 2(0.03)(0.15)]\,50 \\ &= 3.63\,\text{W} \end{aligned}$$

and

$$\begin{aligned} Q_T(z = 5\,\text{mm}) &= 2.41\,[0.15(0.005) + 2(0.03)(0.15)]\,50 \\ &= 1.17\,\text{W} \end{aligned}$$

Thus, it appears that for the specified channel geometry and a surface emissivity of 0.2, radiative heat rejection is approximately one-fifth of the buoyant convective heat transfer from the heat sink.

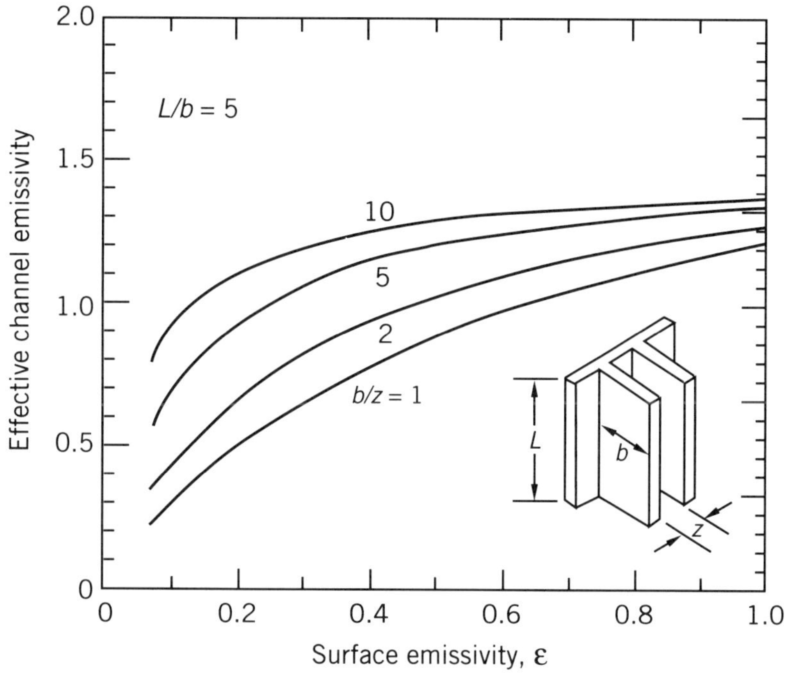

Figure 10.11 Channel radiative emmitance for $L/b = 5$.

Evaluation of the radiant heat transport from additional heat sink geometries reveals a similar pattern. When the finned surface is large relative to the base surface area, as typically associated with the use of many, thick, closely spaced fins, the radiative contribution to the overall thermal transport from the heat sink is small, and often even negligible, regardless of the surface emissivity. This is also the case for widely-spaced fins of low surface emissivity (polished aluminum with an emissivity of 0.05, for example). Alternatively, when the surface emissivity approaches unity, as attainable with anodized aluminum or painted surfaces, radiation from widely-spaced fins could rival heat transfer by natural convection.

Thus, although radiation is rarely the dominant heat-transfer mode in air-cooled heat sinks, precise analysis of heat sink performance does require evaluation of the radiant contribution. For small to modest radiative heat-transfer rates, the calculated Q_T can be divided by the fin surface area and wall-to-ambient temperature difference to yield a radiation heat-transfer coefficient, h_r. This radiation heat-transfer coefficient can then be added to the previously-discussed convective heat-transfer coefficient to form an effective heat sink heat-transfer coefficient for more complete, though necessarily approximate, calculations.

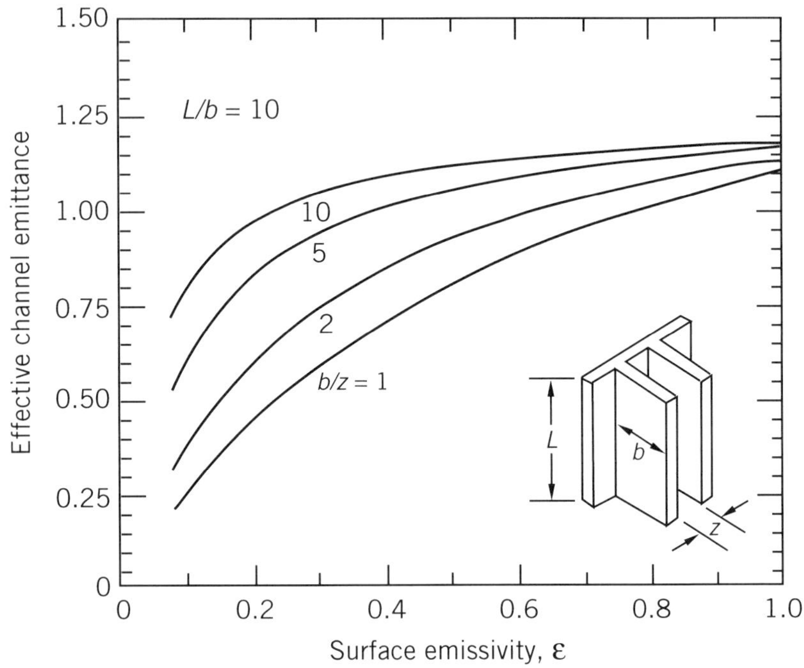

Figure 10.12 Channel radiative emmitance for $L/b = 10$.

10.5 Arrays of Longitudinal, Rectangular Fins in Natural Convection

10.5.1 Introduction

An understanding of natural convection heat transfer from isothermal, parallel-plate channels provides the theoretical underpinning for the conceptual design of natural convection cooled plate-fin heat sinks. However, detailed design and optimization of such fin structures requires an appreciation for the distinct characteristics of such phenomena as buoyancy-induced fluid flow in the interfin channels and conductive heat flow in the plate-fins.

The presence of the heat sink "base," or prime surface area, along one edge of the parallel-plate channel, contrasting with the open edge at the "tip" of the fins, introduces an inherent asymmetry in the flow field. The resulting three-dimensional flow pattern generally involves some inflow from (and possibly outflow through) the open edge. For relatively small fin spacings with long and low fins, this edge flow may result in a significant decrease in the air temperature between the fins and dramatically alter the performance of such heat sinks. For larger fin spacings with short, thick fins, the edge flow may well be negligible. Thus, Sparrow and Bahrami (1980), who studied 7.6-cm wide, vertically oriented, isothermal plates, found the edge flow to have a negligi-

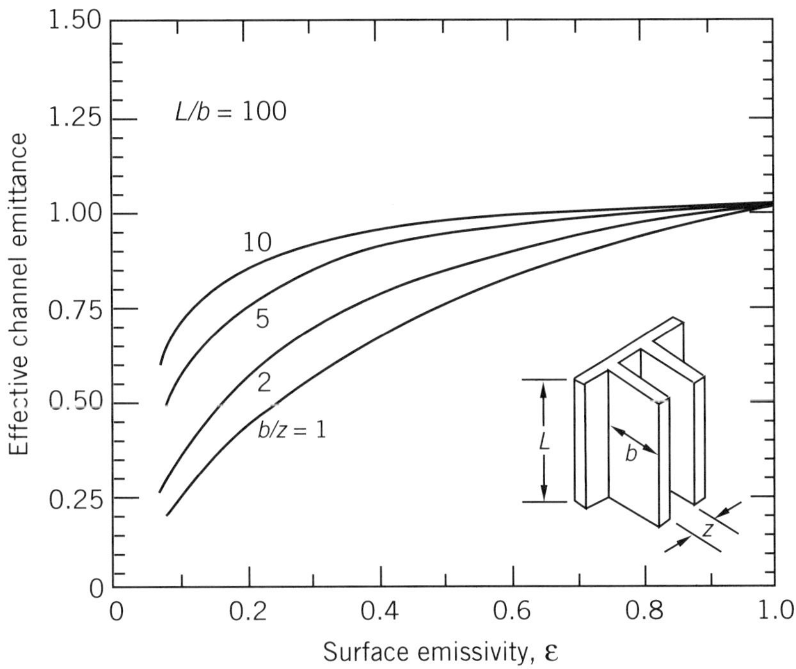

Figure 10.13 Channel radiative emmitance for $L/b = 100$.

ble effect on the heat-transfer coefficient for values of the Elenbaas number (channel Rayleigh numbers) greater than 10, but to produce deviations of up to 30 percent in the equivalent Nu_o when El was less than 4.

As discussed extensively in previous chapters, heat flow in extended surfaces must result in a temperature gradient at the fin base. When heat flow is from the base to the ambient, the temperature decreases along the fin, and the average fin surface temperature excess, $T_f - T_o$, is typically between 50 and 90 percent of the base temperature excess. As a consequence of the anisothermality of the fin surface, exact analytic determination of the heat sink capability requires a combined (or conjugate) solution of the fluid flow in the channel and heat flow in the fin. Due to the complexity of such a conjugate analysis, especially in the presence of three-dimensional flow effects, the thermal performance of heat sinks is frequently based on empirical results. In recent years, extensive use has also been made of detailed numerical solutions to quantify heat sink performance. Alternatively, a satisfactory estimate of heat sink capability can generally be obtained by decoupling the flow and temperature fields and using an average heat-transfer coefficient, along with an average fin surface temperature, to calculate the thermal transport from the fins to the ambient air.

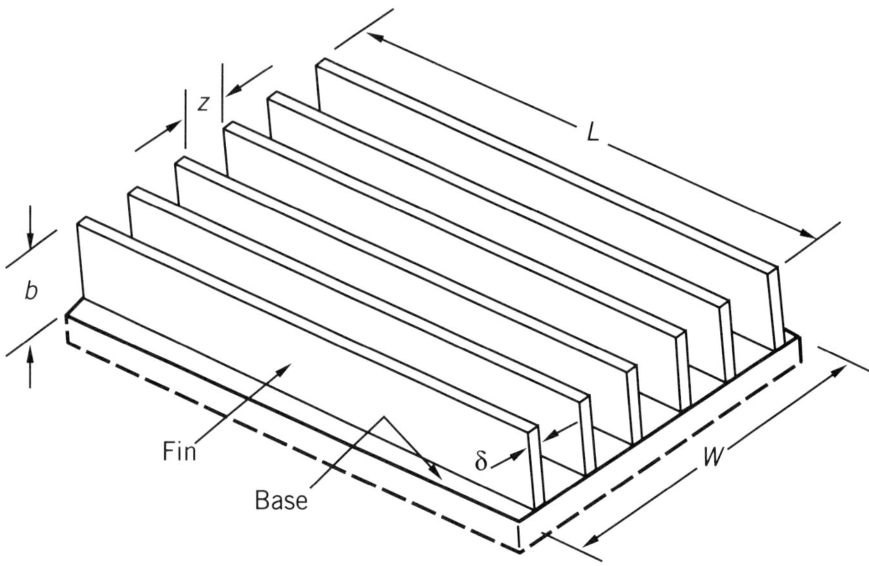

Figure 10.14 Geometric parameters for Starner and McManus (1963) fin arrays.

TABLE 10.2 Geometric Parameters for Starner and McManus (1963) Fin Arrays.

Array	N	L		W		b		z		δ	
		in	mm	in	mm	in	mm	in	mm	in	mm
1	17	10.0	254	5.0	127	0.5	12.7	0.25	6.35	0.04	1.016
2	17	10.0	254	5.0	127	1.5	38.1	0.25	6.35	0.04	1.016
3	17	10.0	254	5.0	127	0.25	6.35	0.25	6.35	0.04	1.016
4	14	10.0	254	5.0	127	1.0	25.4	0.313	7.95	0.04	1.016

10.5.2 Empirical Results

Starner and McManus (1963) were, perhaps, the first to investigate in detail the thermal performance of natural convection heat sinks, as a function of fin geometry (spacing and height) and angle of base plate orientation (vertical, horizontal, and 45°). Their configuration, with the present terminology, is shown in Fig. 10.14. For a base surface area of 254 mm by 127 mm, supporting 14 and 17, 1.02-mm fins (with the cases summarized in Table 10.2), Starner and McManus found that the measured heat-transfer coefficients for the vertical orientation were generally lower than the values expected for parallel-plate channels. The inclined orientation (45°) resulted in an additional 5 to 20 percent reduction in the heat-transfer coefficient. Results for the horizontal orientation showed a strong contribution from three-dimensional flow.

Welling and Wooldridge (1965) performed an extensive study of heat transfer from vertical arrays of 2.3-mm thick fins attached to an identical

203-mm by 66.3-mm base (the individual cases are summarized in Table 10.3). Their results revealed that, in the range of $0.6 < El_r < 100$ (El_r is the Elenbaas number based on the hydraulic radius of the channel), associated with 4.8-mm to 19-mm spacings and fin heights from 6.3 mm to 19 mm, the heat-transfer coefficients along the total wetted surface were lower than attained by an isolated, flat plate but generally above those associated with parallel-plate flow. This behavior was explained in terms of the competing effects of channel flow, serving to preheat the air, and inflow from the open edge, serving to mix the heated air with the cooler ambient fluid. In this study it was observed, for the first time, that for any given interfin spacing there is an optimum fin height, b, beyond which thermal performance, per unit surface area, deteriorates. A similar study of upward and downward facing fin arrays, on a horizontal base, was reported by Jones and Smith (1970). The Welling and Wooldridge (1965) data formed the basis for the Van de Pol and Tierney (1973) correlation discussed in a later section of this chapter.

In 1986, Bilitzky completed a comprehensive investigation of natural convection heat transfer from several heat sinks that differed, primarily, in fin height and spacing. The heat sinks were operated at different heat dissipations as well as different angles of inclination and orientation. Twelve distinct heat sinks and a flat plate were tested in a room within which extraneous convection had been suppressed. The range of angles is indicated in Fig. 10.15. The base was first kept vertical, while the fins were rotated through four different positions (90°, 60°, 30°, and 0°). Then, the base was tilted backwards towards the horizontal orientation through four different positions (90°, 60°, 30°, and 0°). Six of the heat sinks used 144-mm long by 115-mm wide bases to support plate fins, nominally 2 mm in thickness and 6 mm to 13.8 mm apart, ranging in height from 8.6 mm to 25.5 mm. Six additional heat sinks, with identical fin geometries, were supported on 280-mm long and 115-mm wide bases. The geometric parameters of the 12 heat sinks were selected to span the base and fin dimensions encountered in electronic cooling applications and are summarized in Table 10.4.

In all 12 heat sinks studied, the vertical-vertical orientation, that is, a vertically oriented base with vertical fins and channels, yielded the highest heat-transfer coefficients most often. However, in a relatively large number of situations, the thermal performance of the vertical-vertical array was indistinguishable from that attained by a vertical base plate with fins rotated 30° from the axis, a horizontal base plate, and a base plate inclined 60° from the horizontal with unrotated fins. On the other hand, the vertical-horizontal orientation, that is, the base plate vertical and the fins rotated 90° from the axis, led to the lowest heat-dissipation rates. For the unrotated fins, the lowest heat-transfer coefficients were almost always found to occur at a base plate angle of 30° from the horizontal. The use of smoke revealed a relatively complex three-dimensional flow pattern around the heat sinks, with very substantial inflow from the direction of the fin tips when the base plate was strongly inclined and when the heat sinks were in the vertical base-horizontal

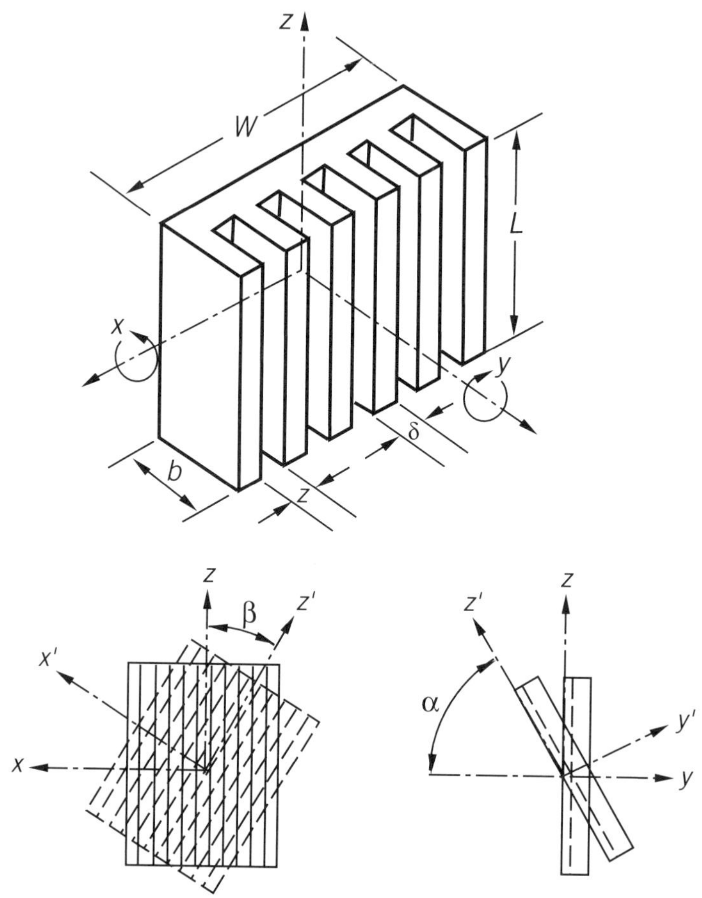

Figure 10.15 Geometry and orientation of Bilitzky (1986) fin arrays.

fins orientation.

The influence of the spacing, z, between the fins, for short and long base plates, was examined by comparing pairs of heat sinks that differed only in the geometric parameter, z (fin arrays 1 and 4, 2 and 5, 3 and 6, 7 and 10, 8 and 11, and 9 and 12 in Table 10.4). Bilitzky observed that, in nearly all of the configurations and operating conditions examined, the highest heat transfer coefficients were attained with the larger fin spacing. However, the improvement in the heat-transfer coefficient was not always sufficient to compensate for the loss of wetted fin surface area. Moreover, for the horizontal base plate configuration, as well as for the vertical base with horizontal fins, the total array dissipation appeared not to depend on this parameter.

A similar exploration of the fin height effect, by comparing the performance of four sets of three fins (arrays 1, 2, and 3, or 10, 11, and 12 in Ta-

TABLE 10.3 Geometric Parameters for Welling and Wooldridge (1965) Fin Arrays.

Array	N	L		W		b		z	S	δ	
		in	mm	in	mm	in	mm	in	mm	in	mm
1	4	8.0	203.2	2.61	66.3	0.75	19.05	0.75	19.05	0.09	2.29
2	6	8.0	203.2	2.61	66.3	0.75	19.05	0.414	10.52	0.09	2.29
3	8	8.0	203.2	2.61	66.3	0.75	19.05	0.27	6.86	0.09	2.29
4	10	8.0	203.2	2.61	66.3	0.75	19.05	0.19	4.83	0.09	2.29
5	4	8.0	203.2	2.61	66.3	0.5	12.7	0.75	19.05	0.09	2.29
6	6	8.0	203.2	2.61	66.3	0.5	12.7	0.414	10.52	0.09	2.29
7	8	8.0	203.2	2.61	66.3	0.5	12.7	0.27	6.86	0.09	2.29
8	10	8.0	203.2	2.61	66.3	0.5	12.7	0.19	4.83	0.09	2.29
9	4	8.0	203.2	2.61	66.3	0.25	6.35	0.75	19.05	0.09	2.29
10	6	8.0	203.2	2.61	66.3	0.25	6.35	0.414	10.52	0.09	2.29
11	8	8.0	203.2	2.61	66.3	0.25	6.35	0.27	6.86	0.09	2.29
12	10	8.0	203.2	2.61	66.3	0.25	6.35	0.19	4.83	0.09	2.29
13	0	8.0	203.2	2.61	66.3	0.00	0.00	2.61	66.3	0.00	0.00

TABLE 10.4 Geometric Parameters for Fin Arrays [Bilitzky (1986)].

Array	L		W		b		z		δ	
	in	mm	in	mm	in	mm	in	mm	in	mm
1	5.67	144	4.53	115	1.004	25.5	.236	6.0	.075	1.9
2	5.67	144	4.53	115	.677	17.2	.232	5.9	.079	2.0
3	5.67	144	4.53	115	.339	8.6	.228	5.8	.083	2.1
4	5.67	144	4.53	115	1.004	25.5	.547	13.9	.075	1.9
5	5.67	144	4.53	115	.677	17.2	.543	13.8	.079	2.0
6	5.67	144	4.53	115	.339	8.6	.539	13.7	.083	2.1
7	11.02	280	4.53	115	1.004	25.5	.236	6.0	.075	1.9
8	11.02	280	4.53	115	.669	17.0	.232	5.9	.079	2.0
9	11.02	280	4.53	115	.335	8.5	.228	5.8	.083	2.1
10	11.02	280	4.53	115	1.004	25.5	.547	13.9	.075	1.9
11	11.02	280	4.53	115	.669	17.0	.543	13.8	.079	2.0
12	11.02	280	4.53	115	.335	8.5	.539	13.7	.083	2.1

ble 10.4) revealed that, in most cases, the heat-transfer coefficient decreases with fin height. Nevertheless, for the shorter heat sinks (of length equal to 144 mm) with closely spaced fins, the heat-transfer coefficient attained a minimum value at the intermediate fin height (at 17 mm). An increase in the heat sink length was found to increase monotonically the total heat dissipation at identical base-to-air temperature differences and angles of orientation, but to decrease the actual convective heat-transfer coefficient along the wetted surfaces.

After reviewing his extensive database of empirical results, Bilitzky (1986) concluded that, for all of the geometries examined, the heat-transfer coefficients along the fin surfaces, for the most common vertical base–vertical fin orientation, lie between the values observed for flat plates and those asso-

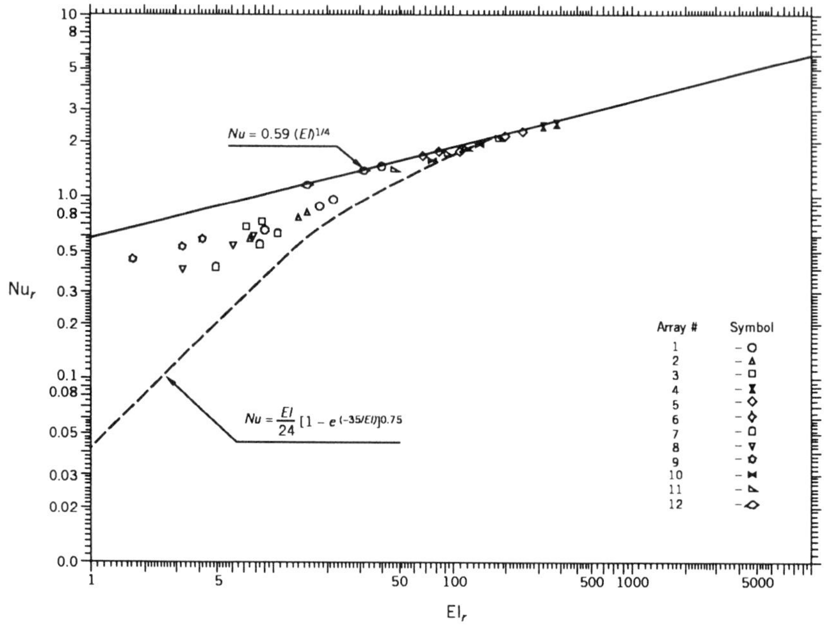

Figure 10.16 Nusselt numbers for Bilitzky (1986) vertical arrays.

ciated with flow in parallel-plate channels when the Nusselt and Elenbaas numbers are based on the hydraulic radius. However, as may be seen in Fig. 10.16, for this orientation, the experimental Nusselt numbers are substantially higher than the fully developed limit and those obtained from the Elenbaas correlation. This holds true, even for Elenbaas numbers below 10, which are normally associated with the fully developed flow limit. This behavior would appear to reflect the very strong impact of three-dimensional flow on heat transfer from the relatively low fins typical of the actual heat sink configurations examined in this study. A similar result can be seen in Fig. 10.17 where the relatively poor vertical base–horizontal fins orientation shows remarkably high Nu values, approaching 85 percent of the vertical flat-plate value at high Elenbaas numbers and still 20 percent of the vertical flat-plate value at Elenbaas numbers as low as 5. For the horizontal base plate orientations, shown in Fig. 10.18, the heat-transfer coefficients appear to straddle the horizontal flat-plate correlation, displaying 10 percent higher values at large El and 50 percent lower values at small El numbers.

Bilitsky (1986) also used his experimental results to examine the effectiveness with which the added mass was utilized in the various heat sink geometries. Defining a utility function, M, as

$$M \equiv \frac{Q_{\text{Heat Sink}}/Q_{\text{Base Plate}}}{\text{added mass of fins}} \tag{10.55}$$

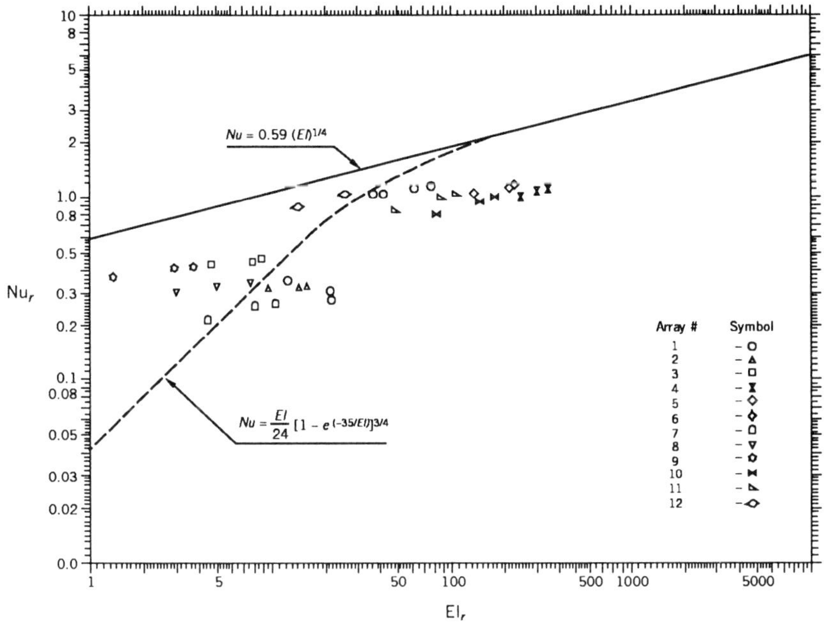

Figure 10.17 Nusselt numbers for Bilitzky (1986) vertical base–horizontal fin arrays.

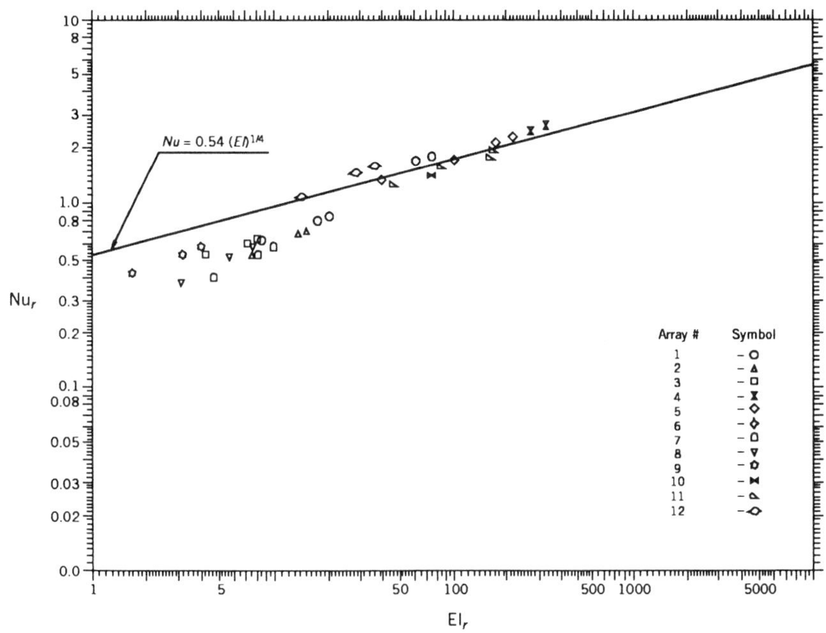

Figure 10.18 Nusselt numbers for Bilitzky (1986) horizontal fin arrays.

he found significant variations among the 12 heat sinks. The values of M were higher for the short heat sinks than for the long heat sinks (144 mm vs 280 mm) and peaked for the heat sink arrangement (number 6 in Table 10.4) with low, widely spaced fins.

10.5.3 Natural Convection Correlations

Yovanovich and Coworkers

The correlation of natural convection heat transfer from arbitrarily shaped bodies has been the subject of considerable research. Of particular importance are the studies by Yovanovich, (1987a), (1987b), and by Lee et al. (1991). The Yovanovich efforts have revealed that, in the domain of developing flow (that is, for relatively large Ra), the Nusselt Number for natural convection cooling of isothermal bodies can be obtained by the superposition of the diffusive and convective limits. When the characteristic length is taken as the square root of the wetted area, the Nusselt and Rayleigh numbers are defined by

$$\mathrm{Nu}_{\sqrt{A}} \equiv \frac{h\sqrt{A}}{k}$$

and

$$\mathrm{Ra}_{\sqrt{A}} \equiv \frac{g\beta \mathrm{Pr} \Delta T (\sqrt{A})^3}{\nu^2}$$

so that

$$\mathrm{Nu}_{\sqrt{A}} = \mathrm{Nu}_{\sqrt{A}}^{\infty} + f(\mathrm{Pr}) G_{\sqrt{A}} \mathrm{Ra}_{\sqrt{A}}^{0.25} \tag{10.56}$$

In this equation the "*universal*" Prandtl number function takes the form

$$f(\mathrm{Pr}) = \frac{0.67}{[1 + (0.5/\mathrm{Pr})^{9/16}]^{4/9}} \tag{10.57}$$

and $\mathrm{Nu}_{\sqrt{A}}^{\infty}$ is the diffusive limit, $\mathrm{Ra}_{\sqrt{A}}$ is the Rayleigh number, $G_{\sqrt{A}}$ is the body gravity function that is also based on the square root of the wetted area, A. Following Culham et al. (1994), Eq. (10.56) can be applied to a heat sink, with widely spaced fins by combining the diffusive limit, $Nu_{\sqrt{A}}^{\infty}$, for a cuboid with the body gravity function $G_{\sqrt{A}}$ for the wetted area. Thus, with the present nomenclature, for a heat sink structure consisting of a vertical base plate, the diffusive limit for a cuboid with dimensions, $L \times W \times H$ takes the form

$$\mathrm{Nu}_{\sqrt{A}}^{\infty} = \frac{3.192 + 1.868(H/L)^{0.76}}{[1 + 1.189(H/L)]^{1/2}} \tag{10.58}$$

Similarly, for an arbitrary array of rectangular profile fins, with base plate dimensions, $L \times W \times \delta_p$, and n fins, each with dimensions $b \times \delta \times L$, the body gravity function, $G_{\sqrt{A}}$, can be written as

$$G_{\sqrt{A}} = 1.0904 \left[\frac{L(bn + \delta_b + W)^2}{(\delta bn + \delta_b W + L(bn + \delta_b + W))^{3/2}} \right]^{1/4} \tag{10.59}$$

In this equation, the base plate and fin length, L, are parallel to the gravity vector. The thickness of the base plate and the fin are δ_b and δ, respectively, and it has been implicitly assumed that the wetted area of the heat sink surface is isothermal.

The results of applying this modeling approach to convective heat sinks with widely spaced fins can be seen in Fig. 10.19 which shows the Nu variation with Ra for four finned heat sinks (Karagiozis, 1991). All four heat sinks have an identical 220-mm wide, 150-mm high, and 9.5-mm thick base plate, but differ in the fin geometry; in the first and second heat sinks, 50-mm high, 9.5-mm thick fins were spaced 30 and 35 mm apart (Fig. 10.19a), while in the third and fourth heat sinks, 15-mm high by 15-mm thick fins were spaced 15 and 30 mm apart (Fig. 10.19b). Examination of these figures reveals that for relatively large Rayleigh numbers, where developing flow prevails, excellent agreement is found between the correlation and the data. However, as the Rayleigh number decreases, signifying an approach to fully-developed flow associated with smaller interfin spaces and lower base heat fluxes, the measured values fall below the correlation. It may also be noted that in keeping with the fundamental isothermal fin assumptions used in developing this correlation, all four heat sinks examined contain fins that are relatively thick and widely spaced. This geometry may be appropriate to cast structures but is unlike the optimum finned heat sink dimensions discussed in later sections of this chapter.

Example 10.8: Determine the effective natural convection heat-transfer coefficient and the total heat-transfer rate from a vertically-oriented, sand-cast aluminum heat sink on a 10-mm thick, 200-mm long, and 225-mm wide base. The 5-fins are 35 mm high, 10 mm thick, and spaced 35 mm apart. The heat sink can be assumed isothermal at a temperature of 350 K, and the ambient air is at 300 K.

SOLUTION This description suggests that the superposition relation, given by Eq. (10.56), can be used to determine the thermal performance of this isothermal heat sink with thick, widely spaced fins.

Using Eq. (10.58), the diffusive limit Nusselt number can be calculated. First determine the dimensions of the "cuboid." The base area, $L \times W$, is 200 mm by 225 mm, and the extension, H, is the base thickness plus the fin height or $H = 10\,\text{mm} + 35\,\text{mm} = 45\,\text{mm}$. Using these values in Eq. (10.58),

$$\begin{aligned}\text{Nu}^{\infty}_{\sqrt{A}} &= \frac{3.192 + 1.868(H/L)^{0.76}}{[1 + 1.189(H/L)]^{1/2}} \\ &= \frac{3.192 + 1.868(0.045/0.200)^{0.76}}{[1 + 1.189(0.045/0.200)]^{1/2}} \\ &= \frac{3.793}{1.126} = 3.369\end{aligned}$$

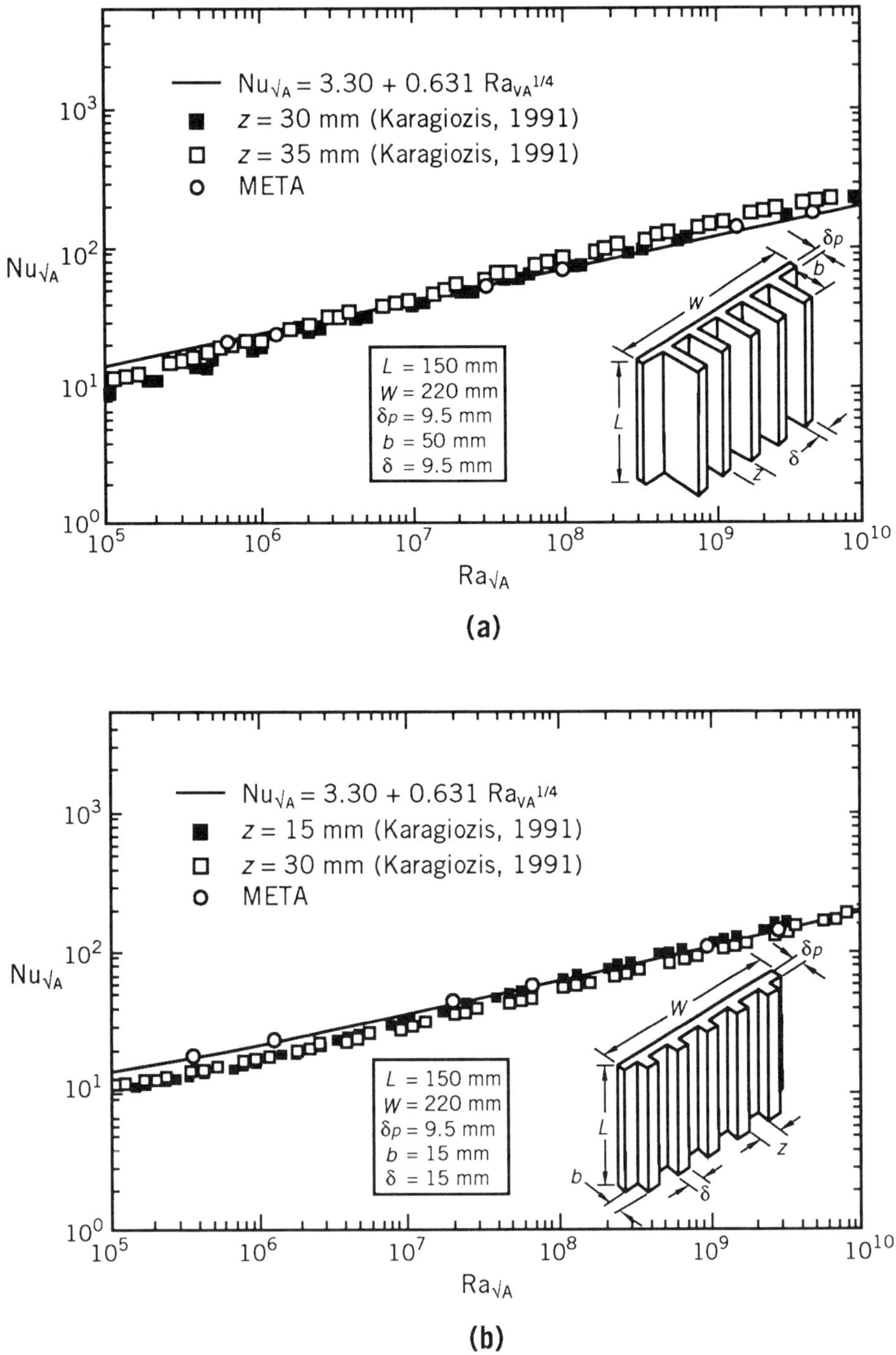

Figure 10.19 Rectangular fin assembly heat transfer for widely spaced fins (*a*) high fins and (*b*) low fins [Culham et al (1994)]. META is a computer code used at the University of Waterloo.

Next, find the body gravity function, $G_{\sqrt{A}}$, using Eq. (10.59) with the specified dimensions. With $\mathbf{\Lambda} = bn + \delta_b + W$ this may be written as

$$G_{\sqrt{A}} = 1.0904 \left[\frac{L\mathbf{\Lambda}^2}{[\delta bn + \delta_b W + L\mathbf{\Lambda}]^{3/2}} \right]^{1/4}$$

Here

$$\begin{aligned} \mathbf{\Lambda} &= bn + \delta_b + W \\ &= 0.035(5) + 0.010 + 0.225 \\ &= 0.410 \end{aligned}$$

so that

$$\begin{aligned} G_{\sqrt{A}} &= 1.0904 \left\{ \frac{0.2(0.41)^2}{[0.01(0.035)(5) + 0.01(0.225) + 0.2(0.41)]^{3/2}} \right\}^{1/4} \\ &= 1.0904 \left(\frac{0.03362}{0.02552} \right)^{1/4} \\ &= 1.1717 \cong 1.172 \end{aligned}$$

For an average fluid temperature of 325 K, the Prandtl number is 0.700 and the "universal" Prandtl number function, $f(\text{Pr})$, using Eq. (10.57), can be evaluated to yield $f(\text{Pr}) = 0.512$. Consequently, the superposition relation given by Eq. (10.56) takes the form

$$\text{Nu}_{\sqrt{A}} = 3.369 + 0.600\text{Ra}_{\sqrt{A}}^{1/4}$$

The Rayleigh number, $\text{Ra}_{\sqrt{A}}$, based on the wetted surface area A, is expressed as

$$\text{Ra}_{\sqrt{A}} = \frac{g\beta\text{Pr}\Delta T(\sqrt{A})^3}{\nu^2}$$

and the total wetted surface area of the heat sink is equal to the base surface area plus the product of the number of fins and the surface area of each fin:

$$\begin{aligned} S &= 0.200(0.225) + 5(2)(0.035)(0.200) \\ &= 0.045 + 0.070 \\ &= 0.115\,\text{m}^2 \end{aligned}$$

and

$$\sqrt{A} = 0.339\,\text{m}$$

Substitution of the fluid properties, evaluated at an average temperature of 325 K,

$$\begin{aligned}
\rho &= 1 \text{ kg/m}^3 \\
\beta &= 1/325 = 0.00308 \text{ 1/K} \\
c_p &= 1000 \text{ J/kg-K} \\
\mu &= 2 \text{ x } 10^{-5} \text{ N-s/m}^2 \\
k &= 0.028 \text{ W/m-K}
\end{aligned}$$

the calculated surface area, and a temperature difference of 50 K in the foregoing, yields the value of Ra as

$$\begin{aligned}
\text{Ra}_{\sqrt{A}} &= \frac{g\beta \text{Pr} \Delta T (\sqrt{A})^3}{\nu^2} \\
&= \frac{9.8(0.0031)(0.700)(50)(0.339)^3}{(18.4 \times 10^{-6})^2} \\
&= 1.226 \times 10^8
\end{aligned}$$

Returning to the basic relationship, the Nusselt number can now be found to equal

$$\begin{aligned}
\text{Nu} &= 3.369 + 0.600\text{Ra}^{1/4} \\
&= 3.369 + 0.600[1.226 \times 10^8]^{1/4} \\
&= 3.369 + (0.600)(105.22) \\
&= 66.49 \cong 66.5
\end{aligned}$$

so that

$$\begin{aligned}
h &= \text{Nu}\frac{k}{\sqrt{A}} \\
&= \frac{66.5(0.028)}{0.339} \\
&= 5.49 \text{ W/m}^2\text{-}K
\end{aligned}$$

Finally, using this heat-transfer coefficient, it is possible to determine the heat dissipation of the heat sink which is assumed to be isothermal.

$$Q = hS(T_b - T_{\text{air}}) = (5.49)(0.115)(50) = 31.6 \text{ W}$$

It is interesting to note that even this relatively "*coarse*" heat sink with, thick, widely spaced fins offers approximately 2.5 times the heat-transfer capability of the unfinned base area. As a first order correction for a non-isothermal heat sink, this value could be multiplied by a surface area-weight averaged fin efficiency.

Van De Pol and Tierney

A careful review of the Starner and McManus (1963) and the Welling and Wooldridge (1965) studies, as well as other data available in the literature at the time, led Van De Pol and Tierney in 1973 to suggest that successful application of the Elenbaas correlation to vertical heat sinks would require the insertion of channel aspect ratio corrections into Eq. (10.43):

$$\mathrm{Nu}_o = \frac{hz}{k} = \frac{\mathrm{El}}{24}\left(1 - e^{-35/\mathrm{El}}\right)^{3/4} \tag{10.43}$$

They proposed that the observed differences, between actual finned heat sinks and the relatively high plates studied by Elenbaas, be addressed by the introduction of a geometric parameter, ψ, into the Elenbaas equation, in the manner suggested by

$$\mathrm{Nu}_r = \frac{\mathrm{El}_r}{\psi}\left\{1 - e^{[-\psi(0.5/\mathrm{El}_r)^{3/4}]}\right\} \tag{10.60}$$

Here, the Nusselt and Elenbaas numbers are based on a hydraulic radius, $r = 2A/P$, (twice the cross-sectional area divided by the wetted perimeter) and the Elenbaas number has replaced the channel Rayleigh number originally used in this correlation. The geometric parameter, ψ, takes the form

$$\psi = \frac{24\left(1 - 0.483e^{-0.17/a}\right)}{\left\{\left[1 + \frac{a}{2}\right]\left[1 + \left(1 - e^{-0.83a}\right)\left(9.14a^{1/2}e^{-VS} - 0.61\right)\right]\right\}^3} \tag{10.61}$$

where a is the channel aspect ratio, z/b, and VS is a dimensional constant equal to 0.4646 mm^{-1}.

The Van de Pol and Tierney (1973) relationship was shown to fit the data of Welling and Wooldridge (1965) with a standard error of just 0.039 in Nusselt number (see Figure 10.20). Furthermore, the form of this correlation ensures that at very small aspect ratios, Eq. (10.61) approaches the original Elenbaas correlation, and that at very large values it merges smoothly into the isolated plate limit.

It should be noted that, in keeping with the data of Welling and Wooldridge (1965), all fluid properties in Eq. (10.60) are to be evaluated at the wall temperature, with the exception of the volumetric expansion coefficient that is taken at the fluid temperature.

Bilitzky

Analysis of data for thermal transport from his vertically oriented heat sinks led Bilitzky (1986) to recognize that closely spaced low fins, typical of actual heat sinks in the electronic industry, display substantially higher heat-transfer coefficients than predicted by the Elenbaas correlation. In particular, as displayed in Fig. 10.16 and echoing earlier results of Sparrow and Bahrami (1980), the empirical data for the Nusselt number do not appear to approach

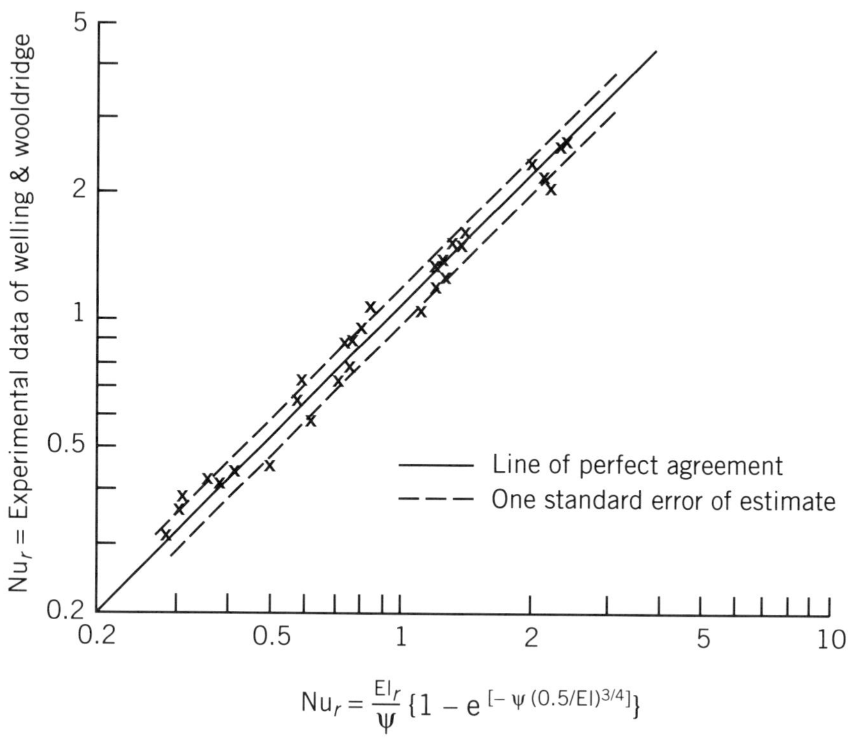

Figure 10.20 Comparison of predicted and measured fin array Nusselt numbers (Van de Pol and Tierney, 1973).

the anticipated fully developed limit at small Elenbaas numbers and appear, instead, to approach an asymptote nearly parallel to, though lower than, the isolated plate limit. Flow visualization studies of these heat sinks revealed the presence of a significant inflow from the open front edge of the interfin channels. This inflow increased in importance as the fin height and fin spacing decreased.

Based on these observations, Bilitzky (1986) modified the Van de Pol and Tierney (1973) geometric factor, ψ, given by Eq. (10.61) by replacing the VS product with a purely geometric factor, B, reflecting the contribution of fin tip inflow to the overall channel flow. This parameter is equivalent to 1.25 times the ratio of the hydraulic radius associated with the fin tip opening to that of the vertical channel (with attention paid to the actual wetted perimeter):

$$B = 1.25\frac{r_{\text{open}}}{r_{\text{vertical}}} = 1.25\left[1 + \frac{z}{2b}\right] \tag{10.62}$$

Insertion of this correction into the Van de Pol and Tierney (1973) geometric parameter of Eq. (10.61) yields

$$\psi_B = \frac{24\mathbf{\Lambda}_1}{\left[\left(1+\frac{a}{2}\right)(1+\mathbf{\Lambda}_2\mathbf{\Lambda}_3)\right]^2} \tag{10.63}$$

where, with $a = z/b$,

$$\begin{aligned}\mathbf{\Lambda}_1 &= 1 - 0.483e^{-0.17/a} \\ \mathbf{\Lambda}_2 &= \left(1 - e^{-0.83a}\right)\end{aligned}$$

and

$$\mathbf{\Lambda}_3 = 9.14a^{1/2}e^{-B} - 0.61$$

Use of this correction in Eq. (10.60) was reported by Bilitzky (1966) to yield agreement within 5 percent for his data, approximately 10 percent for the Welling and Wooldridge (1965) data and similar agreement with the results of Starner and McManus (1963).

Example 10.9: Determine the effective natural convection heat-transfer coefficient and the total heat-transfer rate from a vertically oriented, extruded aluminum heat sink on a 10-mm thick, 225-mm wide (W), and 200-mm long base (L). The 32-fins are 15 mm high (b), 2 mm thick (δ), and spaced 5 mm apart (z). The heat sink can be assumed to be isothermal at a temperature of 350 K, and the ambient air is at 300 K.

SOLUTION It appears that the Bilitzky-modified version of the Van de Pol and Tierney (1973) Nusselt number correlation [Eq. (10.60)], with ψ evaluated in accordance with Eq. (10.63), is appropriate for this calculation.

To follow this procedure, it is necessary to evaluate several parameters. First,

$$\begin{aligned}B &= 1.25\left[1 + \frac{z}{2b}\right] \\ &= 1.25\left[1 + \frac{0.005}{2(0.015)}\right] \\ &= 1.25(1.167) \\ &= 1.458\end{aligned}$$

next

$$\begin{aligned}a &= \frac{z}{b} \\ &= \frac{0.005}{0.015}\end{aligned}$$

$$= \quad 0.333$$

then, noting that the wetted perimeter of the channel is $2b + z$

$$\begin{aligned} r &= \frac{2bz}{2b + z} \\ &= \frac{2(0.015)(0.005)}{2(0.015) + 0.005} \\ &\cong 0.00429\,\mathrm{m} \end{aligned}$$

and finally

$$\mathrm{El}_r = \frac{\rho^2 g \beta c_p (T_w - T_o) r^4}{\mu k L}$$

For the average temperature of 325 K, the fluid properties are found to equal:

$$\begin{aligned} \rho &= 1\ \mathrm{kg/m^3} \\ \beta &= 1/325 = 0.00308\ \mathrm{K^{-1}} \\ c_p &= 1000\ \mathrm{J/kg\ K} \\ \mu &= 2 \times 10^{-5}\ \mathrm{N\text{-}s/m^2} \\ k &= 0.028\ \mathrm{W/m\text{-}K} \end{aligned}$$

The Elenbaas number is evaluated at 325 K, where the temperature difference is $\Delta T = 50$ K, the hydraulic radius, r, is 4.29 mm, and the channel length, L is 0.2 m,

$$\begin{aligned} \mathrm{El}_r &= 5.385 \times 10^7 \Delta T r^4 / L \\ &= 5.385 \times 10^7 (50)(0.00429)^4 / 0.2 \\ &= 4.56 \end{aligned}$$

Insertion of all these values into Eq. (10.63) yields the modified ψ parameter (ψ_B):

$$\begin{aligned} \boldsymbol{\Lambda}_1 &= 1 - 0.483 e^{-0.17/a} \\ &= 1 - 0.483 e^{-0.17/0.333} \\ &= 1 \quad 0.483(0.600) \\ &= 0.710 \\ \boldsymbol{\Lambda}_2 &= \left(1 - e^{-0.83a}\right) \\ &= \left(1 - e^{-0.83(0.333)}\right) \\ &= 1 - 0.758 \\ &= 0.242 \end{aligned}$$

and

$$\begin{aligned}\mathbf{\Lambda}_3 &= 9.14a^{1/2}e^{-B} - 0.61 \\ &= 9.14(0.333)^{1/2}e^{-1.458} - 0.61 \\ &= 1.228 - 0.61 = 0.618\end{aligned}$$

Then

$$\begin{aligned}\psi_B &= \frac{24\mathbf{\Lambda}_1}{\left[\left(1+\frac{a}{2}\right)(1+\mathbf{\Lambda}_2\mathbf{\Lambda}_3)\right]^3} \\ &= \frac{24(0.710)}{\left\{\left(1+\frac{0.333}{2}\right)[1+0.242(0.618)]\right\}^3} \\ &= \frac{17.040}{[1.167(1.150)]^3} \\ &= \frac{17.040}{2.411} = 7.066\end{aligned}$$

Returning now to Eq. (10.60), with $\psi = \psi_B = 7.066$

$$\begin{aligned}\mathrm{Nu}_r &= \frac{\mathrm{El}_r}{\psi}\left\{1 - e^{[-\psi(0.5/\mathrm{El}_r)^{3/4}]}\right\} \\ &= \frac{4.56}{7.066}\left[1 - e^{-7.066(0.5/4.56)^{3/4}}\right] \\ &= 0.648\left[1 - e^{-1.346}\right] \\ &= 0.648(0.740) = 0.479\end{aligned}$$

It is now possible to determine the heat-transfer coefficient for the finned area of the heat sink from this Nusselt number as

$$h = \frac{\mathrm{Nu}_r k}{r} = \frac{0.479(0.028)}{0.00429} = 3.13\,\mathrm{W/m^2\text{-}K}$$

The total heat dissipation from all of the wetted surfaces in the heat sink is based on the assumption that a single value of h can be employed. With $h = 3.13\,\mathrm{W/m^2\text{-}K}$, and, assuming a fin efficiency near unity

$$\begin{aligned}Q_T &= hA(T_w - T_{\mathrm{air}}) \\ &= (3.13)[0.225(0.200) + 32(2)(0.015)(0.200](50) \\ &= (3.13)(0.045 + 0.192)(50) \\ &= 37.1\,\mathrm{W}\end{aligned}$$

It is interesting to note that this heat sink of relatively thin, narrow, and closely spaced fins is thus capable of dissipating approximately 17 percent

TABLE 10.5 Nusselt Number Correlations for Bilitzky (1986) Heat Sinks with Selected Orientations (for geometric parameters, see Table 10.4).

Array	Vertical Base		Horizontal Base	
	$\mathrm{Nu}_r = C(\mathrm{El}_r)^m$		$\mathrm{Nu}_r = C(\mathrm{El}_r)^m$	
	C	m	C	m
1	0.259	0.420	0.311	0.327
2	0.231	0.447	0.243	0.385
3	0.373	0.295	0.371	0.256
4	0.629	0.227	0.304	0.368
5	0.457	0.290	0.218	0.440
6	0.530	0.271	0.272	0.437
7	0.166	0.556	0.191	0.484
8	0.216	0.496	0.228	0.474
9	0.373	0.273	0.356	0.365
10	0.423	0.303	0.193	0.459
11	0.451	0.295	0.256	0.413
12	0.558	0.257	0.359	0.415

more heat than the heat sink with the few relatively massive fins examined in the previous example (31.4 W in Ex. 10.8). The careful reader will note that the geometry used in this example is similar to array-8 studied by Bilitzky (Table 10.4). The values of the Nusselt number for this array were correlated directly by Bilitzky (see Table 10.4) as

$$\mathrm{Nu}_r = 0.216(\mathrm{El}_r)^{0.496}$$

Substituition of $\mathrm{El}_r = 4.6$ into this correlation yields a value of Nu_r equal to 0.46 or nearly the same value as calculated in the foregoing example.

Bilitzky (1986) also provided distinct Nusselt number correlations for all 12 heat sinks covered in his study, in the vertical base–vertical fins, and horizontal-base orientations. These correlations are shown in Table 10.5 where all of the properties are taken at the wall temperature, except for the volumetric expansion coefficient, β, which is evaluated at the fluid temperature.

10.6 NATURAL CONVECTION IN PIN-FIN ARRAYS

Although the preceding sections have focussed on natural convection heat transfer from arrays of plate-fins, it is appropriate to note the growing commercial use of cylindrical or *pin*-fin arrays. For identical conditions, such pin-fin heat sinks can often transfer considerably more heat than plate-fins (Zografos and Sunderland, 1990a) per unit of prime or base surface area and generally attain a superior heat-transfer rate per unit of fin mass.

Natural convection Nusselt number correlations for an isolated cylinder can serve as a starting point for pin-fin array analysis. But, due to the complex flow pattern that can develop in these arrays in both horizontal and vertical orientations, such relations cannot be used to predict the array heat transfer accurately. The thermal performance of cylindrical fin heat sinks is influenced by the pin diameter and length, the lateral and vertical pin spacing, the "staggering" of the pins, and the base plate inclination. However, it appears that the ratio of the fin diameter to the lateral fin spacing is the single most important geometric parameter.

Thus, experimental and numerical results suggest that, for relatively large arrays that are 203 × 203 mm square, a ratio of 0.333 yields the optimum combination of individual heat-transfer coefficients and array surface area. In this optimum configuration, the total heat dissipation was found to be 2 to 3 times higher than denser fin arrays (Zografos and Sunderland, 1990a, b). For a far smaller array, approximately 76 × 76 mm square, the optimum value of the ratio of fin diameter to lateral fin spacing was found to be closer to 0.50 (Sparrow and Vemuri, 1986). It is interesting to note that the orientation of the pin-fin array has been found to exert only a minor influence on array heat transfer. Zografos and Sunderland (1990a) noted that the thermal transport from arrays inclined by 15° to 30° from the vertical was identical to that of the vertical array. Sparrow and Vemuri (1986) observed that heat transfer from an optimal array was 15 percent higher than for the vertical when the base plate was horizontal with the fins pointing upward and only 5 percent lower than the vertical when the base plate was horizontal with the fins pointing downward. These researchers also found radiation to be a most significant contribution to array heat transfer.

Much of the thermofluid behavior of a vertical-base, pin-fin array can be related to the downstream velocity and temperature fields. In a dense array, the detrimental heating effect of the warm wake on the downstream fins overshadows the beneficial effect of greater buoyancy-driven mass flow through the array. On the other hand, in a more sparse array, it is the beneficial bouyancy effect that dominates. Similar behavior, first observed in single, vertical rows of discrete heat sources (Lieberman and Gebhart, 1969) and cylinders (Sparrow and Chrysler, 1981), also occurs in natural convection flow through parallel-plate channels. In the isolated plate limit of channel flow, it is the bouyancy effects that dominate, whereas in the fully developed limit, the downstream heating constrains the rate of heat transfer.

Analytical relations that decribe these two competing effects are incorporated into the Elenbaas correlation [Eq. (10.43)] as well as subsequent correlations for natural convection flow through channels. These previously discussed expressions (Sec. 10.3) can thus serve to undergird the analysis of natural convection through the somewhat amorphous (or "leaky") channels present in pin-fin arrays.

In the most detailed study of pin-fin arrays to date, Aihara et al (1990) did, in fact, succeed in developing just such a correlation for the average convective heat-transfer coefficient. Using a form similar to that of Eq. (10.43) for parallel-plate channels with an effective array length of L and a lateral fin spacing of s_h all of their measured values were correlated to within 10 percent by[1]

$$\left(\frac{\pi d}{2s_v}\right)\mathrm{Nu}_{p,s} = \frac{(\eta \mathrm{El}_s^*)^{3/4}}{20}\left[1 - e^{-120/\eta \mathrm{El}_s^*}\right]^{1/2} + \frac{(\eta \mathrm{El}_s^*)^{1/4}}{200}$$

for the range $\eta\mathrm{El}_s^* = 0.3$ to 600, $d/s_v = 0.28$ to 0.60, and $b/L = 0.16$ to 1.37. Here, El_s^* is the *pin-fin channel Elenbaas number,*

$$\mathrm{El}_s^* \equiv \frac{\mathrm{Pr}_b g\beta(T_b - T_{fl})s_h^4}{L\nu_b^2}$$

where Pr_b is the Prandtl number with fluid properties taken at the base temperature. In addition, s_v and s_h are the center-to-center horizontal and vertical fin spacings, $\mathrm{Nu}_{p,s}$ is the array Nusselt number defined as

$$\mathrm{Nu}_{p,s} \equiv \frac{h_p s_h}{k_b}$$

where h_p is the fin heat transfer coefficient, L is the length of the array, b is the pin height, and η is the pin fin efficiency. Here S_h is the center-to-center spacing between adjacent rows, while S_v is the center-to-center spacing between alternate rows.

10.7 Optimum (Least Material) Heat Sinks

10.7.1 Introduction

The opening sections of this chapter dealt with the fundamentals of thermal transport from parallel-plate channels, the geometrical "*building block*" for the most common of heat sink configurations. Attention then focussed on the thermofluid analysis and correlation of empirical results for such heat sinks, those composed of arrays of longitudinal fins of rectangular profile. The foregoing thus provides the tools needed for the first-order design of such heat sinks. But in weight-sensitive applications, further design and optimization is required to minimize the effect of added mass. This has grown in importance in all cost-driven products and spans nearly the entire range of industrial activity. The design of such "optimum" heat sinks is especially important for avionics and for electronic packages that are surface mounted on printed circuit boards.

The optimization of single fins, discussed in Chap. 9, provides the starting point for the heat sink or array optimization. For a specified convective heat-transfer coefficient, it is possible to determine the optimum dimensions

[1]With the present nomenclature.

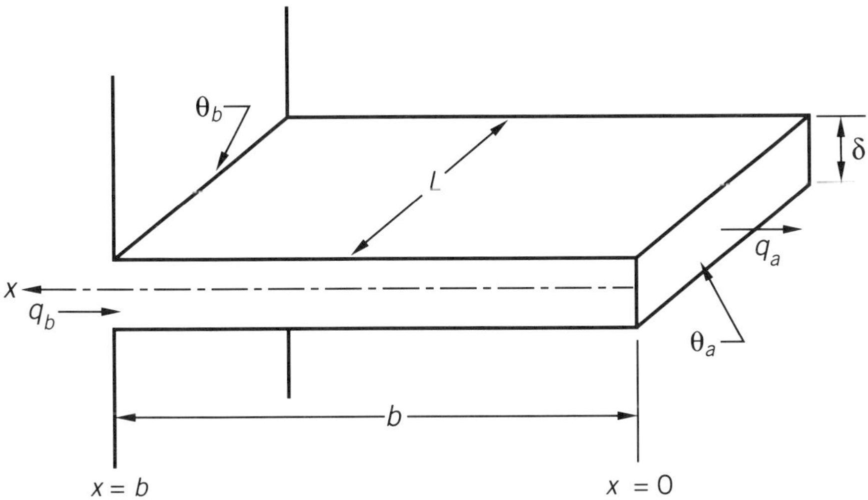

Figure 10.21 Longitudinal fin of rectangular profile.

of the rectangular, longitudinal fin, as well as several other classes of fin geometries. However, because the convective heat-transfer coefficient varies with the spacing between the fins, heat sink optimization involves an understanding of convection in the interfin channels, as well as conduction in the fins.

10.7.2 Theoretical Background

Following the method outlined by Kern and Kraus (1972) and employing the Murray-Gardner assumptions (Murray, 1938, and Gardner, 1945), an energy balance on the differential fin element shown in Fig. 10.21 results in the familiar governing differential equation for temperature excess, $\theta = T - T_s$, given by Eq. (2.1) and repeated here as Eq. (10.64):

$$\frac{d^2\theta}{dx^2} - m^2\theta = 0 \tag{10.64}$$

Observe in Fig. 10.21 that, for this longitudinal fin of rectangular profile, the height coordinate, x, is directed from fin tip to fin base,

$$m = \left(\frac{2h}{k\delta}\right)^{1/2}$$

and that θ is the excess temperature or temperature difference between the fin surface and ambient fluid. As is common in fin analyses, this equation neglects heat transfer from the prime surface area exposed directly to the fluid and is valid when one-dimensional conduction (in the x-direction), as well as a uniform wall temperature condition, is known to prevail.

Application of the uniform and constant wall temperature boundary condition, at the fin base,

$$\theta(x = b) = \theta_b$$

together with an assumed adiabatic tip,

$$\left.\frac{d\theta}{dx}\right|_{x=0} = 0$$

yields the familiar fin temperature excess profile

$$\theta = \theta_b \frac{\cosh mx}{\cosh mb} \tag{10.65}$$

Using the Fourier conduction equation at the fin base, (at $x = b$), with the wall temperature gradient obtained via differentiation of Eq. (10.65), the heat flow per unit length from the fin (for $\theta_b > 0$) is found to equal

$$\frac{q}{L} = k\delta\theta_b m \tanh mb \tag{10.66}$$

For convenience in later calculations, this equation can be re-expressed in terms of the fin profile area, $A_p \equiv \delta b$, (shown in Fig. 10.21), as

$$\frac{q}{L} = \left(\frac{2hkA_p}{b}\right)^{1/2} \theta_b \tanh\left[b^{3/2}\left(\frac{2h}{kA_p}\right)^{1/2}\right] \tag{10.67}$$

Examination of Eq. (10.67) suggests that for a fixed fin profile area, the fin heat flow will depend on the fin height, b, approaching zero for both very low and very high fins and peaking at an intermediate value of b. This "optimum" value of b can be found by differentiating Eq. (10.67) with respect to b and determining where this derivative vanishes. The result of this operation with

$$\beta \equiv mb = b^{3/2}(2h/kA_p)^{1/2}$$

is the transcendental equation, Eq. (9.3a), repeated here, with some refined nomenclature, as Eq. (10.68)

$$(\tanh \beta - 3\beta \operatorname{sech}^2 \beta)_{s,\text{opt}} = 0 \tag{10.68}$$

where the subscript, s, opt, is intended to mean single fin optimum.

As shown in Sec. 9.2.1, the solution of Eq. (10.68) can be expected to yield the rectangular fin geometry that will maximize the heat-transfer rate for a given profile area, volume, or mass. Alternatively, this fin geometry can be expected to yield the least material fin for a specified heat-transfer rate. The single fin, optimum value of $\beta_{s,\text{opt}}$, is found in Chap. 9 to equal 1.4192.

With $\beta_{s,\text{opt}} = 1.4192$, the aspect ratio, δ/b, of the single, optimum fin can be found to equal

$$\left(\frac{\delta}{b}\right)_{s,\text{opt}} = \frac{2}{(1.4192)^2}\frac{hb}{k} = 0.993\frac{hb}{k} \tag{10.69}$$

and the optimum single fin heat-transfer rate to be given by

$$\left(\frac{q}{L}\right)_{s,\text{opt}} = 1.258\theta_b(h^2 A_p k)^{1/3} \tag{10.70}$$

10.7.3 Optimum Multiple Fin Array

Individually Optimum Fins

The results of Sec. 10.7.2 can be applied to the optimization of an array of fins by assuming that the array is to be composed of n individually optimum fins. These fins are to possess an aspect ratio determined by Eq. (10.69) and to be capable of transferring heat according to Eq. (10.70). In such an array, the required number of fins, n, is approximately equal to the desired or design point total heat dissipation divided by the heat-transfer rate of each fin, Q/q. In examining Eq. (10.70), it may be observed that the optimum fin profile area, A_p, is proportional to q^3/h^2. Thus, the profile area of the entire array of fins, nA_p, is proportional to Qq^2/h^2.

As a consequence of this proportionality, minimization of the total array profile area and mass would appear to require either an increase in the heat-transfer coefficient (an obvious approach) or a decrease in the heat dissipation of each fin by reducing the fin dimensions. Either approach, carried to its logical conclusion, would yield a thermally optimum array of a near-infinite number of vanishingly thin fins, which would, together, have a near-zero profile area and a near-zero mass!

Unfortunately, manufacturing constraints and available pumping power (or the buoyant driving force) generally limit the minimum fin thickness and minimum interfin spacing and thus make the realization of a near-zero added mass, optimum array impossible. Instead, the thermal optimization must seek to find the minimum viable array profile area (or mass) consistent with the appropriate system constraints. When the fin material, the prime mover, and the working fluid have been determined, the requisite analysis can often be cast in the form of minimizing the total fin profile area (or mass) for a specified number of fins and a given heat-transfer coefficient. A least material array optimization, consistent with this definition, is presented in succeeding paragraphs.

In optimizing a longitudinal fin array, as may be done in the design of a heat sink, it must be recognized that the base temperature, the local fluid temperature, and the surface heat-transfer coefficient may all vary in the flow

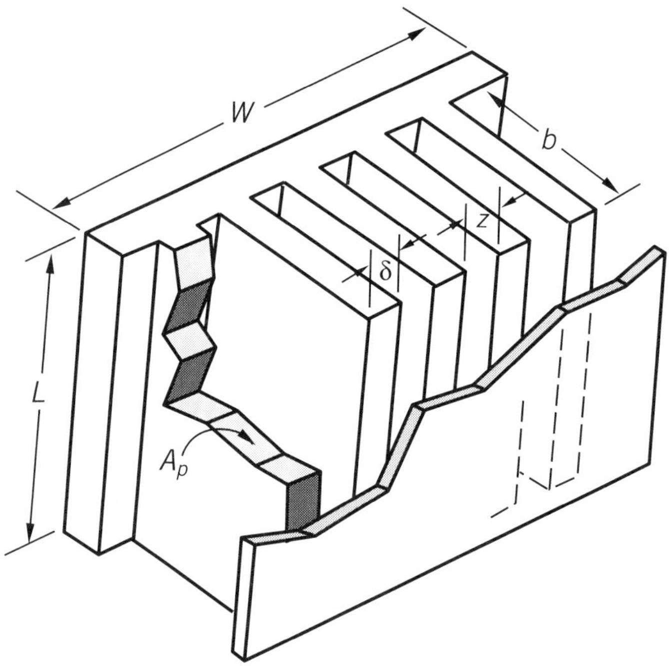

Figure 10.22 Geometric definition of a longitudinal fin array.

direction. Consequently, in applying Eqs. (10.69) and (10.70) to a single fin in an array, the designer can elect to use average values for θ_b and h or to perform his or her calculations in discrete steps in the flow direction. Such discretization can, of course, follow naturally from the geometry of the array when an interrupted fin structure is used.

It must be noted that while the fin heat-transfer rate is proportional to both θ_b and $h^{2/3}$, the optimum single fin aspect ratio, as indicated by Eq. (10.69), is independent of θ_b and, in fact, becomes invariant in the flow direction as the fully developed condition (h = constant) is attained. As a consequence, the use of average values can be expected to introduce only a minor error in the determination of the optimum aspect ratio for such a fin.

The Optimum Array

The rate of heat transfer from a shrouded array of longitudinal, rectangular, fins, as shown in Fig. 10.22, can be found by summing the thermal contribution of the fins and the interfin area, according to

$$\frac{Q}{L} = n\left[\left(\frac{2hkA_p}{b}\right)^{1/2}\theta_b \tanh\beta + h_w\theta_b\left(\frac{W}{n} - \frac{A_p}{b}\right)\right] \tag{10.71}$$

where A_p, as before, is the profile area of the single fin and equal to δb, W is the array width, and h_w the heat-transfer coefficient along the interfin surfaces (the prime or *wall* surfaces). While provision has been made in Eq. (10.71) for differences between the heat-transfer coefficient prevailing along the fin surfaces and in the interfin spaces, it has been assumed that the primary surface area (including both the fin base and interfin surface area) is isothermal. The validity of this common assumption can be expected to improve as the heat-transfer coefficient and the fin spacing decreases and the thermal conductivity, as well as the thickness of the fin support plate, increases.

To find the maximum array dissipation at a fixed total profile area, nA_p, or (with the number of fins predetermined by system considerations) for a fixed A_p, it is possible to differentiate $Q/Ln\theta_b$ with respect to b and to set the derivative equal to zero. On so doing, it is found that the array optimum is governed by

$$\left[(2hkA_p)^{1/2}\left(-\frac{1}{2}b^{-3/2}\tanh\beta + b^{-1/2}\frac{3\beta}{2}\,\text{sech}^2\beta\right) + h_w A_p b^{-2}\right]_{a,\text{opt}} = 0 \tag{10.72}$$

Or, after multiplying through by $2b^3/kA_p$ and cancelling common terms,

$$(\beta\tanh\beta - 3\beta^2\,\text{sech}^2\beta)_{a,\text{opt}} = \frac{2h_w b}{k} \tag{10.73}$$

Comparison of Eqs. (10.73) and (10.68) reveals the relation for the optimum array value of β to differ from the single fin optimization by a term, $2h_w b/k$, which pertains to the heat transfer in the interfin spaces. Indeed, for h_w equal to zero, Eq. (10.73) reduces to Eq. (10.68) and $\beta_{a,\text{opt}} = \beta_{s,\text{opt}} = 1.4192$. The iteratively obtained solution of Eq. (10.73) for nonzero values of h_w is shown in Fig. 10.23 and may be approximated by the linear relation

$$\beta_{a,\text{opt}} = 1.4192 + 1.125\frac{h_w b}{k} \tag{10.74}$$

in the range $1.4192 < \beta_{a,\text{opt}} < 1.6$.

Example 10.10: Determine the array-optimum fin parameter, $\beta_{a,\text{opt}}$, for typical heat sink configurations and compare its value to the single fin optimum value. The cases to be considered are:

a. A 3-cm high (b) aluminum fin cooled by forced convection in air

b. A 20-cm high (b) cast-iron fin cooled by natural convection in air

c. A 3-cm high (b) aluminum fin cooled by natural convection in air

SOLUTION

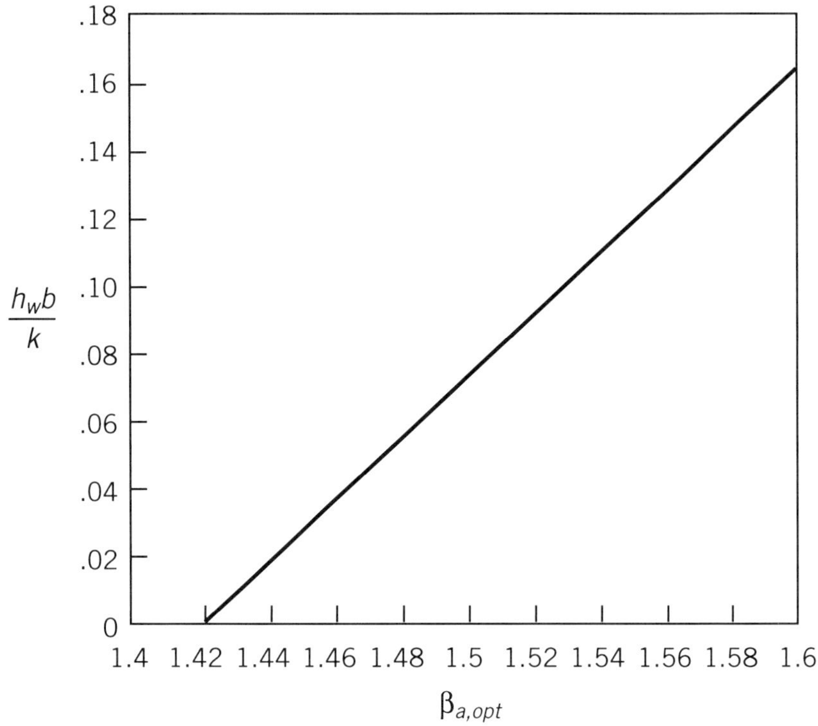

Figure 10.23 Optimizing value of β for a fin array.

a. For the 3-cm high aluminum fin cooled by forced convection in air, suppose that $k = 175$ W/m-K and $h = 100$ W/m^2-K. Then

$$\begin{aligned}\beta &= 1.4192 + 1.125\left(\frac{hb}{k}\right)\\ &= 1.4192 + 1.125\left[\frac{100(0.03)}{175}\right]\\ &= 1.438\end{aligned}$$

and the ratio of the array-optimum, β, to the single fin optimum, is $1.438/1.4192 = 1.0136$. The array-optimum value is 1.4 percent higher.

b. For the 20-cm high cast-iron fin cooled by natural convection in air, suppose that $k = 30$ W/m-K and $h = 6$ W/m^2-K. Then

$$\begin{aligned}\beta &= 1.4192 + 1.125\left(\frac{hb}{k}\right)\\ &= 1.4192 + 1.125\left[\frac{6(0.20)}{30}\right]\end{aligned}$$

$$= \quad 1.464$$

and the ratio of the array-optimum β to the single fin optimum is 1.464 / 1.4192 = 1.0317. The array-optimum value is 3.2 percent higher.

c. For the 3-cm high aluminum fin cooled by natural convection in air, suppose that $k = 175$ W / m-K and $h = 6$ W / m^2-K. Then

$$\begin{aligned} \beta &= 1.4192 + 1.125\left(\frac{hb}{k}\right) \\ &= 1.4192 + 1.125\left[\frac{6(0.03)}{175}\right] \\ &= 1.420 \end{aligned}$$

and the ratio of the array-optimum β to the single fin optimum is 1.420 / 1.4192 = 1.0008. The array-optimum value is essentially the same as the individual fin optimum.

Because the fin aspect ratio, δ/b, is, as shown in Eq. (10.68), inversely proportional to β, the marginally higher values for the optimum array imply that the array-optimum fins can be expected to be somewhat thinner than those based on the individual fin optimization. However, judging by the values shown in Ex. 10.10, this difference may often be sufficiently small to justify the use, in commercial practice, of the individual optimum fin values. The implications of this conclusion are explored in the next two sections, dealing with natural and forced convection optimum fin arrays in air, respectively.

As noted earlier, the validity of the foregoing analysis is limited to configurations in which the primary surface area is (or can be assumed) isothermal, as is often the case in the thermal control of electronic components. Moreover, this analysis is limited to situations in which the heat-transfer coefficient along the finned surface can be assumed constant.

10.7.4 Optimum Natural Convection Heat Sinks

Introduction

The design of natural convection fin arrays or heat sinks is complicated by the coupling between the flow and temperature fields. This is a characteristic of all natural convection heat transfer. Thermal optimization of such an array of fins requires selection of both the optimum fin spacing and optimum fin thickness, each determined to maximize total heat dissipation for a given added mass or profile area.

In natural convection, a balance between viscous dissipation and buoyant potential, rather than pump or blower characteristics, establishes the desired interfin channel width. It may be anticipated that as the channel width or plate spacing is reduced, the heat-transfer coefficient would decrease below the value associated with an isolated plate in an infinite medium. Based on his experimental results, and the correlation given in Eq. (10.43), Elenbaas (1942) determined that the optimum spacing between parallel, isothermal, thin, vertical plates in air could be obtained by setting the channel Rayleigh number (based on plate spacing and channel aspect ratio and here referred to as the Elenbaas number) equal to 50. At this spacing, the product of the total plate surface area and the heat-transfer coefficient is maximized, yielding the peak volumetric heat dissipation rate for the entire array. The optimum Nusselt number was found by Elenbaas (1942) to equal 1.25, thus providing an optimum average heat-transfer coefficient along the plate surfaces of

$$h_{\text{opt}} = 1.25 \frac{k_g}{z_{\text{opt}}} \tag{10.75}$$

In Bar-Cohen (1979), the Elenbaas' optimum spacing was extended to optimum fin arrays by inserting a fin efficiency correction, to yield

$$z_{\text{opt}} = \left(\frac{50 L \nu^2}{\eta g \beta \theta_b \text{Pr}} \right)^{1/4} = 2.99P \tag{10.76}$$

where β is the volumetric expansion coefficient and P is an array parameter

$$P \equiv \left(\frac{L \nu^2}{g \beta \theta_b \text{Pr}} \right)^{1/4}$$

and where η has been set equal to 0.626, the fin efficiency of the single optimum fin. The extension of such an optimization approach to fins of nonnegligible thickness is facilitated by the use of the composite Nussselt number equation [Eq. (10.47)] and is discussed by Bar-Cohen and Rohsenow (1983).

It is interesting to recall that the Elenbaas results were obtained with relatively high plates where inflow from the open edges had a negligible impact on the heat-transfer rate. However, studies on actual heat sinks, most recently by Bilitzky (1986), as well as of low vertical plates by Sparrow and Bahrami (1980), have shown that the Elenbaas correlation, Eq. (10.43), underpredicts the heat-transfer coefficients for such configurations at low Elenbaas numbers. As displayed in Fig. 10.16 for closely spaced low fins, the Nusselt numbers lie between the isolated plate limit and the Elenbaas correlation, thereby reflecting the contribution of inflow from the open edge. As the Elenbaas number increases, the Elenbaas correlation asymptotically approaches the isolated plate limit and in the range of El = 50, corresponding to the optimum fin spacing, the isolated plate relation yields Nusselt numbers that are just 25 percent higher than Eq. (10.43). In this same intermediate range of Elenbaas numbers, the empirical heat-transfer coefficients, for the heat sinks studied by Bilitzky (1986), are approximately 15 percent above the values predicted by Eq. (10.43). Moreover, examination of Eq. (10.69) reveals that the "least-material" optimization of a longitudinal fin of rectangular profile yields relatively high fins (for example, a b of 10 cm for a 1-mm thick fin with a thermal conductivity of 50 W/m-K).

Consideration of these twin factors, intermediate Elenbaas numbers, and relatively high fins suggests that the optimization of a multiple fin array can proceed on the basis of the Elenbaas correlation for the natural convection heat-transfer coefficient. Furthermore, because the previous section has shown the optimizing parameter for the fin aspect ratio, b/δ, to be nearly independent of the fin array characteristics, it would now appear possible to use Eqs. (10.70), (10.75), and (10.76) to determine the geometric parameters of an optimum heat sink using vertical, natural convection air-cooled fins.

Fin Array Dissipation

To obtain the heat dissipation of an array of optimally spaced, least material fins, Eq. (10.70) with Eq. (10.75) can be used to quantify the heat flow through each fin, while Eq. (10.76) sets the interfin spacing. Moreover, assuming, that h_{opt} can also be used to characterize the heat-transfer coefficient in the area between the fins and that the number of fins is equal to the number of fin spaces (as is nearly the case for a large number of fins), the total heat dissipation of the array is expressible as

$$\frac{Q}{L} = \frac{(q_{\text{opt}}/L + h_{\text{opt}} z_{\text{opt}} \theta_b)W}{z_{\text{opt}} + \delta_{\text{opt}}} \tag{10.77}$$

Following substitution for q_{opt}/L, h_{opt}, and z_{opt}, and using the fin efficiency of 0.626 as is appropriate for optimum rectangular fins, Eq. (10.77) can be modified to yield

$$\frac{Q}{LW\theta_b} = \frac{0.813(k_g k \delta_{\text{opt}}/P^{1/2}) + 1.25k_g}{2.99P + \delta_{\text{opt}}} \tag{10.78}$$

A graphical presentation of Eq. (10.78) for aluminum, brass, and steel arrays of least material fins, dissipating heat to ambient air, is shown in Figs. 10.24 through 10.26, respectively, for a typical parametric range. Array dissipation is seen to vary greatly with both the array parameter, P, and the thickness of the optimum fin, δ_{opt}. Furthermore, a maximizing value of δ_{opt} appears to exist for each value of P.

The locus of the array dissipation maxima, shown graphically in Figs. 10.24 through 10.26, can be obtained by differentiation of Eq. (10.78) or, more simply, by differentiation of an approximate form of this relation, in which heat transfer from the area between the fins is neglected. With this additional assumption, clearly justified by the large surface area of the optimum natural convection fin, Eq. (10.77) yields

$$\frac{Q}{LW\theta_b} \cong 1.258 \frac{(hk\delta_{\text{opt}})^{1/2}}{z_{\text{opt}} + \delta_{\text{opt}}} \tag{10.79}$$

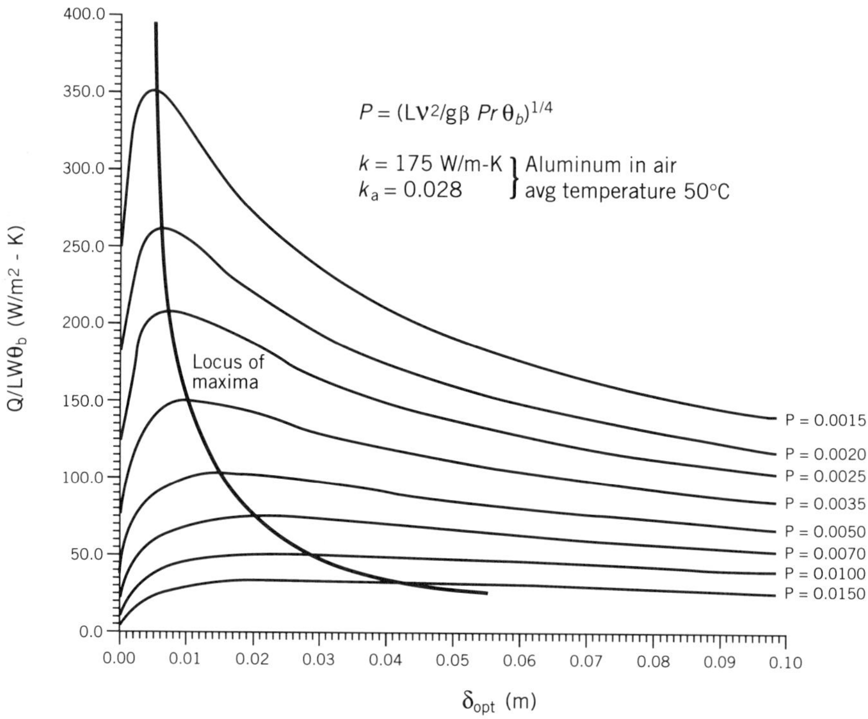

Figure 10.24 Natural convection heat dissipation of an array of optimally spaced, vertical, rectangular profile longitudinal fins of least material aluminum fins.

Following Bar-Cohen (1979), differentiating Eq. (10.79) with respect to δ_{opt} and setting the derivative to zero, the maximizing value of δ_{opt} is found to equal z_{opt}.

This particular value of fin thickness defines an array geometry that maximizes heat dissipation from a prime area of given length, L, width, W, and specified temperature excess, θ_b, while using least-material fins of rectangular profile. Substitution of $\delta_{\text{opt}} = z_{\text{opt}}$ into Eq. (10.78) yields the heat-transfer rate for the maximum dissipation, optimum array as

$$\left(\frac{Q}{LW\theta_b}\right)_{\text{opt}} = 0.236\left(k\frac{k_g}{P^2}\right)^{1/2} \tag{10.80}$$

where k is the thermal conductivity of the fin and k_g the thermal conductivity of the fluid.

Design Considerations

Examination of Figs. 10.24 through 10.26 and Eq. (10.78) shows $Q/LW\theta_b$ (constituting an effective array heat-transfer coefficient referenced to the primary

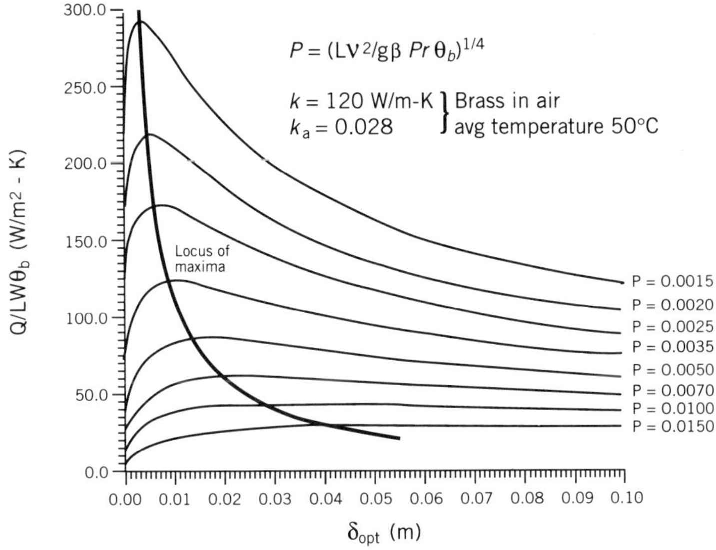

Figure 10.25 Natural convection heat dissipation of an array of optimally spaced, vertical, rectangular profile longitudinal fins of least material brass fins.

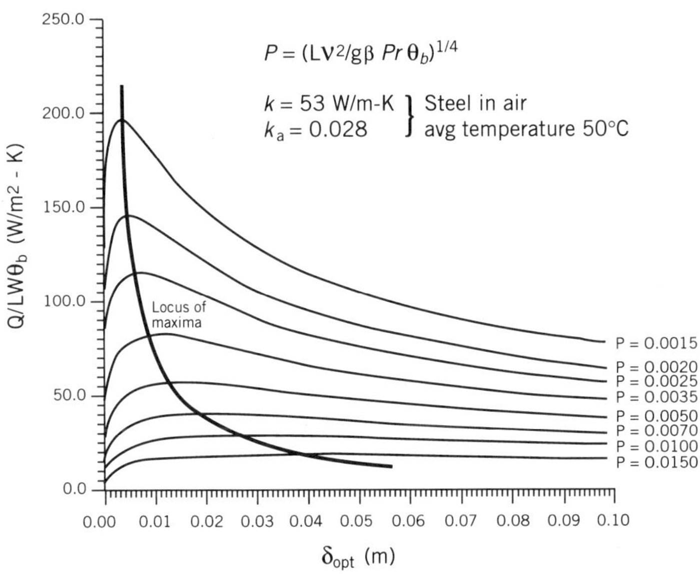

Figure 10.26 Natural convection heat dissipation of an array of optimally spaced, vertical, rectangular profile longitudinal fins of least material steel fins.

area) to be governed by two key parameters: P, embodying the influence of geometric, thermal, and air property variations, and δ_{opt}, the individual optimum fin thickness. At fixed values of P, a maximizing value of δ_{opt} exists, although at larger P values, the array dissipation is only slightly dependent on δ_{opt}, and a maximizing "*plateau*" is encountered.

Examining the array configuration at which the peak heat dissipation rates are encountered, it is apparent that by proper design it is possible to attain effective heat sink heat-transfer coefficients in natural convection, $Q/LW\theta_b$, that are even greater than the values normally associated with forced convection transfer (in air) from the primary surface alone. In particular, for aluminum fins (k= 175 W/m-K), in air, the maximum effective heat-transfer coefficient, as presented in Fig. 10.24, is seen to range from 100 to 350 W/m^2-K in the parametric range of interest, providing a 15- to 45-fold improvement over natural convection heat transfer from an unfinned surface. The transfer rates attainable with an optimum array of brass (k = 120 W/m-K) and steel fins (k = 53 W/m-K) are plotted in Figs. 10.25 and 10.26, respectively. While the improvement over the unfinned surface is seen to decrease, the margin nevertheless remains most significant even for the low conductivity steel array. These transfer rates are attained when the fins are optimally spaced (based on the modified Elenbaas criteria), the thickness of each fin closely approximates the optimum spacing, and the height of the fins is determined by the least-material criterion. Array design that compromises any one of these three requirements cannot be expected to yield the foregoing maximum heat-transfer rates.

A summary of a maximum dissipation design of a heat sink using longitudinal, rectangular aluminum fins, transferring heat by natural convection to ambient air, is given in Table 10.6. This array of fins is located on a base area 0.21 m wide and 0.08 m long, constituting the upper half of the rear panel of an electronic module. The maximum allowable fin base temperature has been set at 70° C, the ambient temperature is 30°C, and it is desirable to transfer as much heat as possible. For the condition stated, the fin array parameter, P, is found to equal 2.50×10^{-3}. At this value, the optimum fin spacing is found, via Eq. (10.76), to equal 7.5 mm, and the peak dissipation of an array of least-material fins on the base area is found, via Eq. (10.80), to equal 140 W (exclusive of radiative transfer and subject to the assumptions previously discussed). While this heat transfer rate, associated with an effective heat-transfer coefficient, $Q/LW\theta_b$, of 210 W/m^2- K, is rather remarkable for a natural convection array, the 53-cm fin height and 12-kg array mass may be too large to be accommodated.

The heat-transfer rate and the values of other parameters associated with optimum fin heat sinks of varying fin thickness are also shown in Table 10.6. It may be noted that while the use of relatively thin fins ($\delta < z_{opt}$) does lower the array heat dissipation, it dramatically reduces the mass of the array. Consequently, as discussed in Sec. 10.7.3, the design of practical natural convection fin arrays can be expected to revolve around relatively thin fins.

TABLE 10.6 Natural Convection Arrays of Longitudinal, Rectangular, Least Material Fins in Air[b].

Parameter	Maximum dissipation optimum array	Thin fin optimum arrays		Thick fin optimum arrays	
Thickness, δ (mm)	7.5	1	2	10	14
Height, b (mm)	531	194	274	614	726
Number of fins	14	25	22	12	10
Mass of fins (kg)	12	1.0	2.6	15.9	21.4
Spacing, z (mm)	7.5	7.5	7.5	7.5	7.5
Effective heat transfer coefficient, h (W/m^{2}K)	210	138	173	207	199
Total convective dissipation[a] (W)	141	93	116	139	134
Dissipation/mass (W/kg)	12	90	44	8.8	6.3

[a] Based on Eq. 10.78 not corrected for fin number round off.
[b] Array geometry: width W = 0.21 m, height L = 0.08m; material properties: conductivity k = 175 W/m-K, density ρ = 2700 kg/m^3; operating parameters: excess temperature θ_b = 40K average temperature T = 50°C.

Non-Optimum Fins

The preceding development deals exclusively with arrays of least-material or optimum fins. However, this same methodology can be used to provide an upper bound on the heat dissipation from an array of infinitely high fins. This upper-bound expression can be obtained by taking the asymptotic limit of q/L given by Eq. (10.67) as b increases.

Returning to Eq. (10.79), the upper-limit heat-transfer rate of an array with infinitely high fins is found by replacing the 1.258 coefficient by $\sqrt{2}$ as appropriate to a fin of height, $b = \infty$ [see Eq. (10.67)]. Maximization of $Q/LW\theta_b$ for this array again results in $\delta = z_{\text{opt}}$ and yields

$$\left(\frac{Q}{LW\theta_b}\right)_{\max} \leq 0.265 \left(k \frac{k_g}{P^2}\right)^{1/2} \tag{10.81}$$

A comparison of Eqs. (10.80) and (10.81) shows that an improvement of not more than 12.3 percent can be obtained by the use of infinitely high fins.

10.7.5 Optimum Forced Convection Arrays

Introduction

As previously noted, the added mass of a forced convection–cooled heat sink is inversely related to the square of the heat-transfer coefficient. Because the heat-transfer coefficient, in all but fully developed laminar flow, is itself dependent on the fluid velocity, the added mass is constrained by the available pumping power. As a consequence, it is not possible to define a single optimum forced convection array geometry, and an optimization calculation must be performed for each combination of parameters. The effort needed to find the optimum array can be reduced considerably by using a computerized search procedure, for example, FINSYS developed by Jelinek (1981). Given the fin material, the air flow rate, available pressure head, and the desired fin thickness, FINSYS will attempt to determine the second fin dimension, the required number of fins, and the added mass of the fin array needed to dissipate a specified heat rate. If such an array can be found, at or below the given pressure head, the optimization will be complete, and the printout will include the necessary pressure head. If the pressure head is insufficient, the printout will state that there is no solution and include an estimate of the pressure head required to achieve the stated thermal performance. In using FINSYS, it is also possible to specify the desired number of fins or the desired added mass and obtain the appropriate characterization, if possible, of the optimum fin array.

As can be anticipated from Eqs. (10.69) and (10.74), use of FINSYS provides fin δ/b ratios that are less than those obtained from an optimization based on $\beta_{s,\mathrm{opt}}$. This difference increases for higher heat-transfer coefficients, larger fin heights, and/or lower thermal conductivity fins. Two examples of forced convection array optimizations, which serve to illustrate the approach and quantify the trends, are presented in the following sections.

The Air-Cooled Cold Plate

Convectively cooled cold plates, of the type shown in Fig. 10.27, are frequently used to provide thermal control for electronic components. Due to the relatively modest dissipation rates of individual components and their careful distribution along the cold plate surface, isothermal conditions often prevail at the base of such an array of fins. In a particular application, where the base of the fins was, on average, 50°C above the air temperature, fin height was limited by space considerations to 40 mm. The high-velocity flow of air through the 15-mm wide channels between the fins provided an average surface heat-transfer coefficient of 80 W/m^2-K. The cold plate was 196 mm wide by 150 mm long and made of aluminum with a thermal conductivity of 100 W/m-K. The fin thickness, heat dissipation rate, and added mass for an array of individually optimum fins and a truly optimum array for this application are shown in Table 10.7.

Inspection of Table 10.7 reveals that, while the optimum array appears to

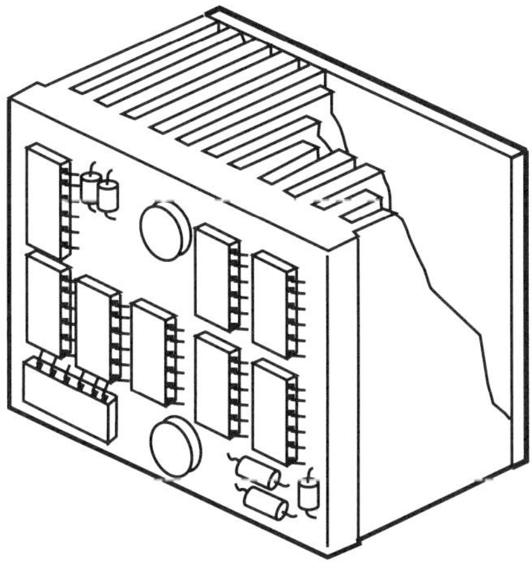

Figure 10.27 Artist's conception of a finned cold-plate used for thermal control of electronic components.

TABLE 10.7 Optimum Forced Convection, Cold Plate Fin Array[a]

Parameter	Individual Fin optimum	Array optimum
β	1.419	1.455
δ	1.27 mm	1.21 mm
η	0.626	0.617
Number of Fins	12	12
Added Mass	246.9 g	234.8 g
Array dissipation		
Fins	360.8 W	354.8 W
Unfinned Surface	108.5 W	108.9 W
Total	469.3 W	463.7 W
Added Mass/additional dissipation	0.702 g/W	0.679 g/W

[a]Heat transfer coefficient, h = 80 W/m^{2}K; fin thermal conductivity, k = 100 W/m−k; fin density, ρ = 2700kg/m^3; excess temperature, $\theta_b = 50°C$; fin thickness, b = 40 mm; array length, L = 150 mm; array width, W = 196 mm.

TABLE 10.8 Optimum Array for 25 kW Air-Cooled Power Tube[a]

Parameter	London (1954)	Fin height similitude	Fin thickness similitude	Flow condition similitude
Thickness, δ (mm)	1.04	1.09	1.04	0.75
Height, b (mm)	47.4	47.8	45.0	35.0
Number of fins	144	118	116	125
Mass of fins (kg)	6030	5228	4599	2799
Inlet volumetric flow Rate (m^3/s)	0.80	0.84	0.86	0.80
Array pressure drop (N/m^2)	1100	594	681	1079
Pumping Power (W)	880	499	586	863

[a] Fin thermal conductivity, $k = 386$ W/mK; fin density, $\rho = 8950$ kg/m^3; wall temperature = 258°C; inlet air temperature = 49°C.

require thinner fins and necessitate less added mass, on both an absolute basis as well as when referenced to the heat dissipation, the differences are less than 5 percent in absolute terms and approximately 3 percent on a per-watt basis. Consequently, for the stated conditions, only a modest improvement can be obtained by optimizing the array using Eq. (10.74) instead of relying on the use of individually optimum fins via Eq. (10.68).

10.7.6 A 25 kW Power Tube

In a classic paper on electronic cooling, London (1954) presented the design of a compact fin array for the air cooling of a 25-kW power tube. The fin array, shown schematically in Fig. 10.28, consisted of rectangular longitudinal copper fins integral to the cylindrical anode section of the power tube (95 mm long and 102 mm in diameter). Ambient air at 49°C was available for removing the 25-kW dissipation, and the anode wall temperature was to be maintained at or below 258°C. The recommended design, detailed in the first column of Table 10.8, included 144 fins, each 1.04 mm thick and 47.4 mm high, for a total added mass of 6.035-kg and required a flow rate of 0.82 kg/s air. In flowing through the array, this air was found to experience an 1100 N/m^2 pressure drop, exclusive of entrance and exit losses.

The foregoing analysis can be used to design an optimum fin array for this application, subject to the assumption of a uniform wall temperature and an average heat-transfer coefficient. The results obtained via FINSYS are shown in Table 10.8, where the second column presents the least-material array for

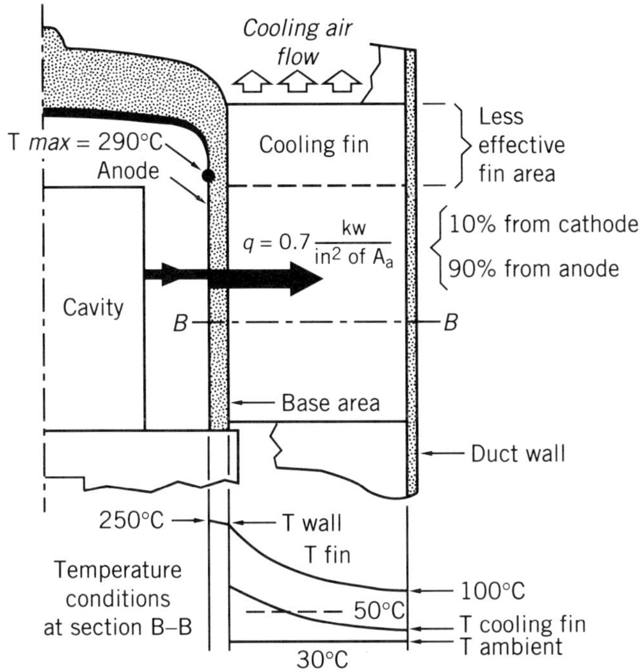

Figure 10.28 Geometric definition of a 25 kW vacuum tube (London, 1954).

nearly the fin height selected by London, the third column characterizes an optimum array of fins equal in thickness to London's value, and the fourth column shows an optimum array operating at the flow rate and slightly below the pressure drop of the original design.

As can be seen clearly in Table 10.8, an array optimization, which is based on both fin geometry and pumping power considerations, can yield significant reductions in the total mass required to dissipate the prescribed heat load. While some of the mass difference between the London (1954) array and the optimized arrays can be related to the uniform wall temperature assumption and the use of an average heat-transfer coefficient along the fin surfaces, it may be noted that for the coolant flow conditions of interest, the optimum fins are significantly thinner and narrower, resulting in a substantially lower mass fin array than selected by London (1954).

10.8 CLOSURE

The plate-fin heat sink, perhaps the most commonly recognized fin structure, has found wide application in the electronic industry for the thermal management of microprocessors, logic and memory devices, power transistors, telecommunication modules, and laboratory measurement equipment,

The design of such heat sinks requires an understanding of both natural and forced convection flow through parallel-plate channels, thermal conduction in extended surfaces, and radiative heat transfer within and from such channels. The interplay of these heat-transfer modes also creates a fertile ground for thermofluid heat sink optimization when mass, volume, and/or pumping power must be conserved. Due to the importance of parallel-plate heat sinks, this chapter has explored these topics in considerable detail.

The design and optimization of passively cooled (natural convection) parallel-plate heat sinks have received extensive attention in this chapter. In recognition of the modeling difficulty presented by the close coupling between thermal transport and bouyant flow, as well as the presence of three-dimensional flow patternes, emphasis has been placed on the available empirical results. Specific experimental data, correlations for the effective natural convection heat-transfer coefficient and for the heat rejection capability of typical heat sink configurations, have been provided and discussed.

Mastery of this chapter will permit the reader to understand, analyze, design, and optimize parallel-plate heat sinks for a wide variety of applications and environments.

REFERENCES

Aihara, T., S. Maruyama, and S. Kobayakawa (1990). Free Convective/Radiative Heat Transfer from Pin-Fin Arrays with a Vertical Base Plate (General Representation of Heat Transfer Performance). *Int. J. Heat and Mass Trans.*, 33 (6), 1223–1232.

Antonetti, V. W., and R. E. Simons (1985). Bibliography of Heat Transfer in Electronic Equipment. *Trans. IEEE CHMT, CHMT-8* (2), 289–295.

Ashiwake, N., W. Nakayama, T. Daikoku, and F. Kobayashi (1983). Forced Convective Heat Transfer from LSI Packages in Air Cooled Wiring Card Array. *Heat Transfer in Electronic Equipment - 1983. ASME Publication HTD-52*, 57–66.

Aung, W. (1972). Fully Developed Laminar Free Convection Between Vertical Plates Heated Asymmetrically. *Int. J. Heat Mass Trans.*, 15, 1577–1580.

Aung, W., L. S. Fletcher, and V. Sernas (1972). Developing Laminar Free Convection Between Vertical Flat Plates with Asymmetric Heating. *Int. J. Heat Mass Trans.*, 15, 2293–2308.

Aziz, A. (1992). Optimum Dimensions of Extended Surfaces Operating in a Convective Environment. *ASME Appl. Mech. Rev.*, 45 (5), 155–173.

Bar-Cohen, A. (1979). Fin Thickness for an Optimized Natural Convection Array of Rectangular Fins. *J. Heat Trans.*, 101, 564–566.

Bar-Cohen, A. and W. M. Rohsenow (1983). Thermally Optimum Arrays of Cards and Fins in Natural Convection. *Trans. IEEE CHMT, CHMT-6*, 154–158.

Bar-Cohen, A. and W. M. Rohsenow (1984). Thermally Optimum Spacing of Vertical, Natural Convection Cooled, Parallel Plates. *J. Heat Trans.*, 106, 116–122.

Bilitsky, A. (1986). The Effect of Geometry on Heat Transfer by Free Convection from a Fin Array. MS Thesis, Department of Mechanical Engineering, Ben-Gurion University of the Negev, Beer Sheva, Israel.

Bodoia, J. R. and J. F. Osterle (1964). The Development of Free Convection Between Heated Vertical Plates. *J. Heat Trans.*, 84, 40–44.

Brown, A. (1965). Optimum Dimensions of Uniform Annular Fins. *Int. J. Heat Mass Trans*, 8, 655–661.

Chapman, A. J. (1987). *Fundamentals of Heat Transfer, 4th Ed.* Macmillan, New York, N Y,

Chen, S. Y., and G. L. Zyskowski (1963). Steady State Heat Conduction in a Straight Fin with Variable Heat Transfer Coefficient. ASME Paper 63-HT-1, Sixth ASME-AIChE Heat Transfer Conference, Boston, MA.

Colburn, A. P. (1933). A Method of Correlating Forced Convection Heat Transfer Data and a Comparison with Fluid Friction. *Trans. AIChE*, 29, 174–210.

Culham, J. R., S. Lee, and M. M. Yovanovich (1993a). Conjugate Heat Transfer from a Raised Isothermal Heat Source Attached to a Vertical Board. Proc. 9th Annual IEEE Semi-Therm Symp., 151–159.

Culham, J. R., M. M. Yovanovich, and S. Lee (1993b). Thermal Modeling of Isothermal Cuboids and Rectangular Heat Sinks Cooled by Natural Convection. IEEE Intersociety Conf. on Thermal Performance, 73–82.

Cumo, M., S. Lopez, and G. C. Pinchera (1965). Numerical Calculation of Extended Surface Efficiency. *Chem. Eng. Prog. Symposium Series*, 61 (59), 225.

Eckert, E. R. G., and R. M. Drake (1972). *Analysis of Heat and Mass Transfer*, McGraw-Hill, New York, N Y.

Elenbaas, W. (1942). Heat Distribution of Parallel Plates by Free Convection. *Physica*, 9 (1), 665–671.

Ellison, G. (1993). Methodologies for Thermal Analysis of Electronic Components and Systems. *Advances in Thermal Modeling of Electronic Compnents and Systems, Vol III*, A. Bar-Cohen and A. D. Kraus, Eds, ASME Press, New York, NY, 153–256.

Fortescue, P. (1957). Some Characteristics of Reactor Gas Cooling Systems. *Nuclear Power*, 2 (13), 188.

Gardner, K. A. (1945). Efficiency of Extended Surface. *Trans. ASME*, 67, 621.

Gardner, K. A. (1951). Discussion on paper by M. L. Ghai, General Discussion on Heat Transfer, Institute of Mechanical Engineers, London, 214.

Gardner, K. A. (1964). Rational Design Temperature Limits for Interference-Fit Finned Tubing. ASME Symp. Volume on Air Cooled Heat Exchangers, ASME, New York, N Y.

Gardner, K. A., and T. C. Carnavos (1960). Thermal Contact Resistance in Finned Tubing. *J. Heat Trans.*, 82, 279.

Ghai, M. L. (1951). Heat Transfer in Straight Fins. General Discussion of Heat Transfer, Institute of Mechanical Engineers, London, 180.

Godsell, S., B. Richard, J. Dischler, and S. M. Westbrook (1990). Implementing a Packaging Strategy for High Performance Computers. *High Performance Systems*, Jan, 28–31.

Han, L. S. and S. G. Lefkowitz (1960). Constant Cross Section Fin Efficiencies for Non-uniform Surface Heat Transfer Coefficients. ASME Paper 60-WA-41.

Hausen, H. (1943). Darstelling des Warmeübergangs in Rohren Durch Verallgemeinerte Potenzbeziehungen. *V. ZDI. Beih. Verfahrenstechnik*, 4, 91.

Hildebrand, F. B. (1976). *Advanced Calculus for Applications*. Prentice-Hall, Englewood Cliffs, NJ

Hughes, D. F. and M. R. Slack (1958). A Correlation of Heat Transfer for Finned Fuel Elements for Carbon Dioxide Reactors. Proc. 2nd United Nations Int. Conf. on the Peaceful Uses of Atomic Energy, 7, 717.

Jelinek, M. (1981). Optimum Array of Cooling Fins, MS Thesis, Department of Mechanical Engineering, Ben-Gurion University of the Negev, Beer Sheva, Israel.

Jones, C. D., and L. F. Smith (1970). Optimum Arrangement of Rectangular Fins on Horizontal Surfaces for Free Convection Heat Transfer. *J. Heat Trans.*, 92, 6.

Karagiozis, A. N. (1991). An Investigation of Laminar Free Convection Heat Transfer from Isothermal Finned Surfaces, PhD Thesis, University of Waterloo, Waterloo, Ontario.

Kays, W. M. (1960). Basic Heat Transfer and Flow Friction Characteristics of Six High Performance Heat Transfer Surfaces. *J. Eng. Power*, 82, 27.

Kays, W. M., and A. L. London (1984). Compact Heat Exchangers 3rd Ed., Mcgraw-Hill, New York, N Y.

Kern, D. Q., and A. D. Kraus (1972). *Extended Surface Heat Transfer*, McGraw-Hill, New York, NY.

Kraus, A. D., and S. Specht (1974). The Effect of Bond Resistance on the Overall Efficiency of a Heat Exchanger Core, Proc. 5th Int. Heat Transfer Conference, Tokyo, (V), 165.

Kraus, A. D., A. D. Snider, and L. F. Doty (1978). An Efficient Algorithm for Evaluating Arrays of Extended Surface. *J. Heat Trans.*, 100, 288.

Kraus, A. D., and A. D. Snider (1980). New Parameterizations for Heat Transfer in Fins and Spines, *J. Heat Trans.*, 102, 415.

Kraus, A. D. (1982). *Analysis and Evaluation of Extended Surface Thermal Systems*, Hemisphere Publishing Corp., New York, NY.

Kraus, A. D., A. D. Snider, and F Landis (1982). The Reciprocity of Extended Surface and the Node Analysis of Finned Arrays, Proc. 7th Int. Heat Transfer Conf., Munich, FRG., 6, 223.

Kraus, A. D., and A. Bar-Cohen (1983). *Thermal Analysis and Control of Electronic Equipment*, Hemisphere Publishing Corp., New York, NY.

Kraus, A. D., and A. D. Snider (1985). The Choking and Optimization of Extended Surface Arrays. *J. Heat Trans.*, 107, 746.

Kraus, A. D. (1988). Sixty Five Years of Extended Surface Technology. *ASME Appl. Mech. Rev.*, 41 (9), 321.

Kraus, A. D. (1990). Analysis of Extended Surface Arrays for Air Cooled Electronic Equipment. *Proc. Int Cent. for Heat and Mass Trans*, A. E. Bergles, Ed, Hemisphere Publishing Corp., Washington DC, 85–112.

Lee, S., M. M. Yovanovich, and K. Jafarpur (1991). Effects of Geometry and Orientation on Laminar Natural Convection from Isothermal Bodies. *AIAA J. of Thermophysics and Heat Trans.*, 5, 208-216.

Lieberman, J. and B. Gebhart (1969). Interactions in Natural Convection from an Array of Heated Elements - Experimental. *Int. J. Heat and Mass Trans.*, 12, 1385-1396.

London, A. L. (1954). Air-Coolers for High Power Vacuum Tubes. *Trans. IRE, ED-1*, 9–26.

Lundberg, R. E., W. C. Reynolds, and W. M. Kays (1963). Heat Transfer with Laminar Flow in Concentric Annuli with Constant and Variable Wall Temperature with Heat Flux. Technical report, NASA TN-D1972, National Technical Information Service, Springfield, Va.

McAdams, W. H. (1954). *Heat Transmission*, McGraw-Hill, New York, NY.

Melese, G. B. (1957). Efficacité d'Une Ailette Longitudinale avec Variation du Coefficient d'Échange de Chaleur le Long de l'Ailette. *J. Nuclear Energy*, 5 (3-4), 285.

Mikk, I. (1980). Convective Fin of Minimum Mass. *Int. J. Heat Mass Trans.*, 23, 707.

Miyatake, O., and T. Fujii (1972). Free Convective Heat Transfer Between Vertical Parallel Plates - One Plate Isothermally Heated and the Other Thermally Insulated. *Heat Trans. Jpn. Res.*, 3, 30–38.

Miyatake, O., T. Fujii, and H. Tanaka, H (1973). Natural Convective Heat Transfer Between Vertical Parallel Plates - One Plate with a Uniform Heat Flux and the Other Thermally Insulated. *Heat Trans. Jpn. Res.*, 4, 25–33.

Murray, W. M. (1938). Heat Transfer Through an Annular Disk or Fin of Uniform Thickness. *Trans. ASME, J. Applied Mech.*, 60, A78.

Negus, K.J., Yovanovich, M.M., and Beck, J.V., 1989, On the Non-Dimensionalization of Constriction Resistance for Semi-Infinite Heat Flux Tubes, ASME JHT, 3,804–807.

Netrakanti, M. N. and C. L. D. Huang (1985). Optimization of Annular Fins with Variable Thermal Parameters by Invariant Imbedding. *J. Heat Trans.*,

Rabenstein, A. L. (1966). *Introduction to Ordinary Differential Equations*, Academic Press, New York NY.

Razelos, P. and K. Imre (1980). The Optimum Dimensions of Circular Fins with Variable Thermal Parameters. *J. Heat Trans.*, 102, 420.

Rohsenow, W. M. (1952). A Method of Correlating Heat Transfer Data for Surface Boiling of Liquids. *Trans ASME*, 74. 969–976

Rohsenow, W. M., and H. Choi (1961). *Heat, Mass and Momentum Transfer*, Prentice-Hall, Englewood Cliffs, NJ.

Sheffield, J. W. (1982). Effects of Double-Connected Finned Surfaces on Heat Transfer During Solidification, AIAA Paper, AIAA 82-0847.

Sheffield, J. W., B. D. Stafford, and H. J. Sauer, Jr (1985) Finned-Tube Contact Conductance: Investigation pf Contacting Surfaces. *Trans ASHRAE*, 91, 442

Sieder, E. N. and G. E. Tate (1936). Heat Transfer and Pressure Drop of Liquids in Tubes. *Ind. Eng. Chem.*, 28, 1429–1435.

Siegel, R., and J. R. Howell (1992). *Thermal Radiation Heat Transfer* 3rd *Ed*, Hemisphere Publishing Corp., Washington, DC.

Snider, A. D., and A. D. Kraus (1981). A General Extended Surface Analysis Method. *J. Heat Trans.*, 103, 699.

Snider, A. D., and A. D. Kraus (1982). Correction for the Variability of the Heat Transfer Coefficient in Extended Surface Analysis, Proc. 7th Int. Heat Transfer Conf., Munich, FRG., 6, 239.

Snider, A. D., and A. D. Kraus (1983). Recent Developments in the Analysis and Design of Extended Surface. *J. Heat Trans.*, 105, 302.

Sobel, N., F. Landis, and W. K. Mueller (1966). Natural Convection Heat Transfer in Short Vertical Channels Including the Effect of Stagger, Proc. 3rd Int. Heat Transfer Conf., Chicago, 2, 121–125.

Sonn, A., and A. Bar-Cohen (1980). Optimum Cylindrical Pin Fin. *J. Heat Trans.*, 103, 814.

Sparrow, E. M., B. R. Baliga, and S. V. Patankar (1978). Forced Convection Heat Transfer from a Shrouded Fin Array with and without Tip Clearance. *J. Heat Trans.*, 100, 572.

Sparrow, E. M., and P. A. Bahrami (1980). Experiments on Natural Convection from Vertical Parallel Plates with Either Open or Closed Edges. *J. Heat Trans.*, 102, 221–227.

Sparrow, E. M., and G. M. Chrysler (1981). Natural Convection Heat Transfer Coefficients for a Short Horizontal Cylinder Attached to a Vertical Plate. *J. Heat Trans.*, 103, 630–637.

Sparrow, E. M., and G. M. Chrysler (1981). Natural Convection Heat Transfer Coefficients for a Short Horizontal Cylinder Attached to a Vertical Plate. *J. Heat Trans.*, 103, 630–637.

Sparrow, E. M., J. E. Niethhammer, and A. Chaboki (1982). Heat Transfer and Pressure Drop Characteristics of Arrays of Rectangular Modules Encountered in Electronic Equipment. *Int. J. Heat and Mass Trans.*, 25 (7), 961-973.

Sparrow, E. M., and S. B. Vemuri (1986). Orientation Effects on Natural Convection/Radiation Heat TRansfer from Pin-Fin Arrays. *Int. J. of Heat and Mass Trans.*, 29 (3), 359–368.

Starner, K. E., and H. N. McManus (1963). An Experimental Investigation of Free Convection Heat Transfer from Rectangular Fin Arrays. *J Heat Trans.*, 85, 273–278.

Ullman, A., and H. Kalman, H (1989). Efficiency and Optimized Dimensions of Annular Fins of Different Cross Sections. *Int. J. Heat and Mass Trans.*, 32, 1105.

Van de Pol, D. W., and J. K. Tierney (1973). Free Convection Nusselt Number for Vertical U-Shaped Channels. *J. Heat Trans.*, 95, 542–543.

Welling, J. R., and C. B. Wooldridge (1965). Free Convection Heat Transfer Coefficients From Rectangular Vertical Fins. *J. Heat Trans.*, 87, 439–444.

Wirtz, R. A., and P. Dykshoorn (1984). Heat Transfer from Arrays of Flatpacks in Channel Flow, Proc. 4th Int. Electronic Packaging Soc. Conf., New York, 318-326.

Yovanovich, M. M. and V. W. Antonetti (1988). Application of Thermal Contact resistance Theory to Electronic Packages, *Advances in Thermal Modeling of Electronic Components and Systems*, A. Bar-Cohen and A. D. Kraus Eds., Hemisphere Publishing Corp., New York, NY, 79-128.

Yovanovich, M. M. (1987a). Natural Convection from Isothermal Spheroids in the Conductive to Laminar Flow Regimes, AIAA Paper AIAA-87-1587.

Yovanovich, M. M. (1987b). On the Effect of Shape, Aspect Ratio and Orientation Upon Natural Convection from Isothermal Bodies," ASME HTD, vol 82.

Zografos, A. I., and J. E. Sunderland (1990a). Natural Convection from Pin Fins. *Experimental Thermal and Fluid Sci.*, 3, 440–449.

Zografos, A. I., and J. E. Sunderland (1990b). Numerical Simulation of Natural Convection from Pin-Fin Arrays, ASME HTD-157, 55–66.

Appendix

A

MATRICES AND DETERMINANTS

A.1 INTRODUCTION

Before the advent of the digital computer, the hand-held calculator, and the desk-top computer, one could talk about the solution of large-scale systems of linear algebraic equations. But few had ever attempted a solution of such a system, and those who had made the attempt were confronted with a most laborious and detailed computational procedure. Now, as a result of the technological revolution that began in the late 1940s and early 1950s, large systems of n algebraic equations in n unknowns (with n in the hundreds and even the thousands) can be formed expeditiously and solved efficiently using the modern, high-speed digital computer.

The formulation and solution of such systems of equations relies on an ordered methodology that is matrix oriented. Moreover, a knowledge of the fundamentals of the theory of matrices is important in the use of this book. This appendix seeks to provide a capsule review of this theory.

A.2 BASIC CONCEPTS

A system of n linear, algebraic equations in n unknowns $x_1, x_2, x_3, \cdots, x_n$ such as

$$\begin{array}{cccccccc} a_{11}x_1 & +a_{12}x_2 & +a_{13}x_3 & +\cdots & +a_{1n}x_n & = & y_1 \\ a_{21}x_1 & +a_{22}x_2 & +a_{23}x_3 & +\cdots & +a_{2n}x_n & = & y_2 \\ a_{31}x_1 & +a_{32}x_2 & +a_{33}x_3 & +\cdots & +a_{3n}x_n & = & y_3 \\ & & \cdots & \cdots & & & \\ a_{n1}x_1 & +a_{n2}x_2 & +a_{n3}x_3 & +\cdots & +a_{nn}x_n & = & y_n \end{array}$$

can conveniently be represented by the matrix equation

$$\begin{bmatrix} a_{11} & a_{12} & a_{13} & \cdots & a_{1n} \\ a_{21} & a_{22} & a_{23} & \cdots & a_{2n} \\ a_{31} & a_{32} & a_{33} & \cdots & a_{3n} \\ & \cdots & \cdots & \cdots & \\ a_{n1} & a_{n2} & a_{n3} & \cdots & a_{nn} \end{bmatrix} \begin{bmatrix} x_1 \\ x_2 \\ x_3 \\ \cdots \\ x_n \end{bmatrix} = \begin{bmatrix} y_1 \\ y_2 \\ y_3 \\ \cdots \\ y_n \end{bmatrix}$$

or more simply by

$$\mathbf{AX} = \mathbf{Y}$$

where $\mathbf{A}$ is a rectangular matrix (in this case square) having elements a_{ij} and where $\mathbf{X}$ and $\mathbf{Y}$ are column vectors with elements x_i and y_i, respectively. The foregoing representations imply that

$$\sum_{j=1}^{n} a_{ij}x_i = y_i \qquad i = 1, 2, 3, \cdots, n$$

The matrix $\mathbf{A}$ is called the *coefficient matrix*. If it is desired to associate the elements of $\mathbf{Y}$ with the coefficient matrix, $\mathbf{A}$, one may augment $\mathbf{A}$ and define an augmented matrix

$$\mathbf{A}_a = \begin{bmatrix} a_{11} & a_{12} & a_{13} & \cdots & a_{1n} & y_1 \\ a_{21} & a_{22} & a_{23} & \cdots & a_{2n} & y_2 \\ a_{31} & a_{32} & a_{33} & \cdots & a_{3n} & y_3 \\ & \cdots & \cdots & \cdots & \cdots & \\ a_{n1} & a_{n2} & a_{n3} & \cdots & a_{nn} & y_n \end{bmatrix}$$

which has n rows and $n+1$ columns. This matrix may be written more simply as the augmented matrix

$$\mathbf{A}_a = [\mathbf{A}|\mathbf{Y}]$$

where the subscript means *augmented* and where the idea of a partitioned matrix is apparent. For example, in the system of linear algebraic equations,

$$\begin{aligned} 8x_1 \quad -3x_2 \quad +2x_3 &= 14 \\ -3x_1 \quad +7x_2 \quad -x_3 &= -3 \\ x_1 \quad -6x_2 \quad +11x_3 &= 17 \end{aligned}$$

the matrix

$$\begin{bmatrix} 8 & -3 & 2 \\ -3 & 7 & -1 \\ 1 & -6 & 11 \end{bmatrix}$$

is called the *coefficient matrix,* $\mathbf{A}$, of the system

$$\mathbf{AX} = \mathbf{B}$$

and the matrix

$$\begin{bmatrix} 8 & -3 & 2 & 14 \\ -3 & 7 & -1 & -3 \\ 1 & -6 & 11 & 17 \end{bmatrix}$$

which contains the constant terms, in addition to the elements of $\mathbf{A}$, is called the *augmented matrix* of the system. Moreover, the unknowns and the constant terms form two column vectors $\mathbf{X}$ and $\mathbf{B}$.

In the representation

$$\mathbf{AX} = \mathbf{B}$$

$\mathbf{A}$ is said to premultiply $\mathbf{X}$ ($\mathbf{A}$ is a premultiplier) and $\mathbf{X}$ is said to postmultiply $\mathbf{A}$ ($\mathbf{X}$ is a postmultiplier).

A.3 MATRIX AND VECTOR TERMINOLOGY

A matrix of order $m \times n$,

$$\begin{bmatrix} a_{11} & a_{12} & a_{13} & \cdots & a_{1n} \\ a_{21} & a_{22} & a_{23} & \cdots & a_{2n} \\ a_{31} & a_{32} & a_{33} & \cdots & a_{3n} \\ & \cdots & \cdots & \cdots & \\ a_{m1} & a_{m2} & a_{m3} & \cdots & a_{mn} \end{bmatrix}$$

is a rectangular ordered array of a total of mn entries arranged in m rows and n columns. The order of this matrix is $m \times n$ which is often written as (m, n).

If $m = n$, the matrix is square of order $n \times n$ (or of n or of nth order):

$$\begin{bmatrix} a_{11} & a_{12} & a_{13} & \cdots & a_{1n} \\ a_{21} & a_{22} & a_{23} & \cdots & a_{2n} \\ a_{31} & a_{32} & a_{33} & \cdots & a_{3n} \\ & \cdots & \cdots & \cdots & \\ a_{n1} & a_{n2} & a_{n3} & \cdots & a_{nn} \end{bmatrix}$$

In both the rectangular and square matrices, a_{ij} is called the ijth element of $\mathbf{A}$. If the matrix is square and $i = j$, the element is said to define and be located on the *principal diagonal*. The elements designated by $a_{n1}, a_{(n-1),2}, a_{(n-2),3}, \cdots, a_{1n}$ are located on and constitute the *secondary diagonal*.

All elements where $i \neq j$ are considered to be off-diagonal, subdiagonal if $i > j$, and superdiagonal if $i < j$. The sum of the elements on the principal diagonal of $\mathbf{A}$ is called the *trace* of $\mathbf{A}$

$$\text{tr}(\mathbf{A}) = \sum_{k=1}^{n} a_{kk}$$

For example: The matrix

$$\mathbf{A} = \begin{bmatrix} 4 & 2 & 0 & 1 \\ -3 & 4 & 8 & 0 \\ -1 & -1 & 6 & 2 \\ -2 & 7 & 1 & 9 \end{bmatrix}$$

is square and is of fourth order (4×4). The elements 4, 4, 6, and 9 constitute the principal diagonal, and the elements $-2, -1, 8$, and 1 constitute the secondary diagonal. The element 8 is the a_{23} element, which lies at the intersection of the second row and third column. The trace of $\mathbf{A}$ is

$$\text{tr}(\mathbf{A}) = 4 + 4 + 6 + 9 = 23$$

A vector is a matrix containing a single row or a single column. If it is a $1 \times n$ matrix (a matrix of order $1 \times n$), it is a row vector:

$$\mathbf{V} = [v_1 \ v_2 \ v_3 \ \cdots \ v_n]$$

If the vector is an $m \times 1$ vector (order $m \times 1$), it is a column vector:

$$\mathbf{V} = \begin{bmatrix} v_1 \\ v_2 \\ v_3 \\ \cdots \\ v_m \end{bmatrix}$$

This concept and the usual one regarding a vector have certain similarities. These similarities are the reason why the elements of a vector are frequently called *components.* However, caution is necessary because the usual three-dimensional space does not imply that m or n (for column or row vectors, respectively) are limited to an upper bound of 3.

A.4 SOME SPECIAL MATRICES

An $m \times n$ matrix, such as the one displayed in the previous section, is called a *null* matrix if every element in the matrix is identically equal to zero. For example, the 2×3 matrix

$$\begin{bmatrix} 0 & 0 & 0 \\ 0 & 0 & 0 \end{bmatrix}$$

is null.

The *transpose* of an $m \times n$ matrix is an $n \times m$ matrix with the rows and columns of the original matrix interchanged. For the 3×4 matrix

$$\mathbf{A} = \begin{bmatrix} 3 & 4 & -2 & 1 \\ 2 & 2 & 1 & 0 \\ 1 & -4 & -3 & 1 \end{bmatrix}$$

the transpose is 4×3

$$\mathbf{A}^T = \begin{bmatrix} 3 & 2 & 1 \\ 4 & 2 & -4 \\ -2 & 1 & -3 \\ 1 & 0 & 1 \end{bmatrix}$$

Note the use of the superscript T to indicate the transpose, and recognize that the transpose of the transpose is the original matrix

$$[\mathbf{A}^T]^T = \mathbf{A}$$

The nth order square matrix

$$\begin{bmatrix} a_{11} & a_{12} & a_{13} & \cdots & a_{1n} \\ a_{21} & a_{22} & a_{23} & \cdots & a_{2n} \\ a_{31} & a_{32} & a_{33} & \cdots & a_{3n} \\ & \cdots & \cdots & \cdots & \\ a_{n1} & a_{n2} & a_{n3} & \cdots & a_{nn} \end{bmatrix}$$

is said to be *diagonal* or a *diagonal matrix* if $a_{ij} = 0$ for all $i \neq j$:

$$\mathbf{A} = \begin{bmatrix} a_{11} & 0 & 0 & \cdots & 0 \\ 0 & a_{22} & 0 & \cdots & 0 \\ 0 & 0 & a_{33} & \cdots & 0 \\ & \cdots & \cdots & \cdots & \\ 0 & 0 & 0 & \cdots & a_{nn} \end{bmatrix}$$

If all a_{ij} are equal for all $i = j$ (that is, $a_{ij} = \alpha; i = j$) and $a_{ij} = 0; i \neq j$, then the resulting matrix is said to be a *scalar matrix*, which is a diagonal matrix with all principal diagonal elements equal:

$$\mathbf{A} = \begin{bmatrix} \alpha & 0 & 0 & \cdots & 0 \\ 0 & \alpha & 0 & \cdots & 0 \\ 0 & 0 & \alpha & \cdots & 0 \\ & \cdots & \cdots & \cdots & \\ 0 & 0 & 0 & \cdots & \alpha \end{bmatrix}$$

If all α in the scaler matrix are equal to unity ($\alpha = 1$), the scalar matrix becomes the *identity matrix*:

$$\mathbf{I} = \begin{bmatrix} 1 & 0 & 0 & \cdots & 0 \\ 0 & 1 & 0 & \cdots & 0 \\ 0 & 0 & 1 & \cdots & 0 \\ & \cdots & \cdots & \cdots & \\ 0 & 0 & 0 & \cdots & 1 \end{bmatrix}$$

A.5 MATRIX EQUALITY

A matrix $\mathbf{A} = [a_{ij}]_{m\times n}$ will be equal to a matrix $\mathbf{B} = [b_{ij}]_{m\times n}$ if and only if $a_{ij} = b_{ij}$ for all i and j. This essentially states that two matrices will be equal if and only if they are of the same order and corresponding elements are equal.

A.6 MATRIX ADDITION AND SUBTRACTION

A matrix $\mathbf{A} = [a_{ij}]_{m\times n}$ may be added to a matrix $\mathbf{B} = [b_{ij}]_{m\times n}$ to form a matrix $\mathbf{C} = [c_{ij}]_{m\times n} = [a_{ij} + b_{ij}]_{m\times n}$. This points out that in order to form the sum of two matrices, the matrices must be of the same order and that the elements of the sum are determined by adding the corresponding elements of the matrices forming the sum.

Example A.1: If

$$\mathbf{A} = \begin{bmatrix} 1 & 2 & 3 \\ 4 & 5 & 6 \end{bmatrix}$$

$$\mathbf{B} = \begin{bmatrix} -1 & 0 & 4 \\ 3 & -2 & -4 \end{bmatrix}$$

and

$$\mathbf{C} = \begin{bmatrix} 2 & 1 \\ 3 & -1 \end{bmatrix}$$

find $\mathbf{A} + \mathbf{B}$ and $\mathbf{A} + \mathbf{C}$.

SOLUTION

$$\mathbf{A}+\mathbf{B}=\begin{bmatrix} 1 & 2 & 3 \\ 4 & 5 & 6 \end{bmatrix}+\begin{bmatrix} -1 & 0 & 4 \\ 3 & -2 & -4 \end{bmatrix}$$

$$=\begin{bmatrix} (1-1) & (2+0) & (3+4) \\ (4+3) & (5-2) & (6-4) \end{bmatrix}=\begin{bmatrix} 0 & 2 & 7 \\ 7 & 3 & 2 \end{bmatrix}$$

The sum $\mathbf{A}+\mathbf{C}$ does not exist because the order of $\mathbf{C}$ does not equal the order of $\mathbf{A}$.

Matrix addition is both commutative and associative

$$\mathbf{A}+\mathbf{B}=\mathbf{B}+\mathbf{A}$$

$$\mathbf{A}+(\mathbf{B}+\mathbf{C})=(\mathbf{A}+\mathbf{B})+\mathbf{C}$$

In addition, the sum $\mathbf{A}+\mathbf{C}$ is equal to the sum $\mathbf{B}+\mathbf{C}$ if and only if $\mathbf{A}=\mathbf{B}$. This is the cancellation law for addition.

The matrix $\mathbf{B}=[b_{ij}]_{m\times n}$ may be subtracted from the matrix $\mathbf{A}=[a_{ij}]_{m\times n}$ to form the matrix $\mathbf{D}=[d_{ij}]_{m\times n}$. This indicates that two matrices of the same order may be subtracted by forming the difference between the corresponding elements of the minuend and the subtrahend. Moreover, with

$$\mathbf{A}+\mathbf{B}=\mathbf{C}$$

then

$$\mathbf{A}=\mathbf{C}-\mathbf{B}$$

Finally, it may be observed that a square matrix possesses a unique decomposition into a sum of a subdiagonal, a diagonal, and a superdiagonal matrix. For example

$$\mathbf{A}=\begin{bmatrix} 1 & 2 & 3 \\ 5 & 6 & 7 \\ 8 & 5 & 6 \end{bmatrix}=\begin{bmatrix} 0 & 0 & 0 \\ 5 & 0 & 0 \\ 8 & 5 & 0 \end{bmatrix}+\begin{bmatrix} 1 & 0 & 0 \\ 0 & 6 & 0 \\ 0 & 0 & 6 \end{bmatrix}+\begin{bmatrix} 0 & 2 & 3 \\ 0 & 0 & 7 \\ 0 & 0 & 0 \end{bmatrix}$$

A.7 MATRIX MULTIPLICATION

A matrix may be multiplied by a scalar or by another matrix. If $\mathbf{A}=[a_{ij}]$ and α is a scalar, then

$$\alpha\mathbf{A}=[\alpha a_{ij}]$$

This shows that multiplication by a scalar is commutative and that multiplication by a scalar involves the multiplication of each and every element of the matrix by the scalar. In addition,

$$(\alpha+\beta)\mathbf{A}=\alpha\mathbf{A}+\beta\mathbf{A}$$

$$\alpha(\mathbf{A}+\mathbf{B})=\alpha\mathbf{A}+\alpha\mathbf{B}$$

and

$$\alpha(\beta\mathbf{A}) = (\alpha\beta)\mathbf{A}$$

Observe that a scalar matrix is equal to the product of the scalar and the identity matrix. For example

$$\begin{bmatrix} 3 & 0 & 0 \\ 0 & 3 & 0 \\ 0 & 0 & 3 \end{bmatrix} = 3 \begin{bmatrix} 1 & 0 & 0 \\ 0 & 1 & 0 \\ 0 & 0 & 1 \end{bmatrix}$$

A modest effort must be expended to use the terminology *multiplication by a scalar* in order to avoid confusion with the process known as scalar multiplication.

The product of a row matrix of order $1 \times n$ and a column matrix of order $n \times 1$ forms a 1×1 matrix that has no important property that is not possessed by a scalar. This product is therefore called the *scalar* or *dot* product (some sources also use the terminology *inner product*). It is called for through the use of a dot placed between the two matrices in the product; that is, if $\mathbf{A}$ and $\mathbf{B}$ are column vectors,

$$\mathbf{A} \cdot \mathbf{B} = [a_{ij}]_{1\times n} \cdot [b_{ij}]_{1\times n} = \mathbf{AB}^T = \mathbf{BA}^T = \gamma$$

and where γ is a scalar obtained from

$$\gamma = \sum_{k=1}^{n} a_k b_k$$

If the scalar product of two vectors is uniquely equal to zero, then the vectors are said to be *orthogonal*.

Example A.2: If

$$\mathbf{A} = \begin{bmatrix} 1 \\ 2 \\ 3 \\ 4 \\ 2 \end{bmatrix}$$

and

$$\mathbf{B} \begin{bmatrix} 4 \\ -5 \\ -3 \\ -2 \\ 7 \end{bmatrix}$$

what is the dot product $\mathbf{A} \cdot \mathbf{B}$?

SOLUTION

$$\mathbf{A} \cdot \mathbf{B} = 1(4) + 2(-5) + 3(-3) + 4(-2) + 2(7)$$
$$= 4 - 10 - 9 - 8 + 14 = -9$$

In Sec. A.2, a set of linear simultaneous algebraic equations was shown to be represented by the notation

$$\mathbf{AX} = \mathbf{Y}$$

where $\mathbf{A}$ was the $n \times n$ coefficient matrix and $\mathbf{X}$ and $\mathbf{Y}$ were $n \times 1$ column vectors. In order to obtain the original set of equations from a set where $n = 3$,

$$\begin{bmatrix} a_{11} & a_{12} & a_{13} \\ a_{21} & a_{22} & a_{23} \\ a_{31} & a_{32} & a_{33} \end{bmatrix} \begin{bmatrix} x_1 \\ x_2 \\ x_3 \end{bmatrix} = \begin{bmatrix} y_1 \\ y_2 \\ y_3 \end{bmatrix}$$

a row by column element product and sum operation are clearly evident,

$$\begin{bmatrix} a_{11}x_1 & +a_{12}x_2 & +a_{13}x_3 & = & y_1 \\ a_{21}x_1 & +a_{22}x_2 & +a_{23}x_3 & = & y_2 \\ a_{31}x_1 & +a_{32}x_2 & +a_{33}x_3 & = & y_3 \end{bmatrix}$$

and it is observed that each element of y is obtained by multiplying the corresponding elements of $\mathbf{A}$ by the elements of $\mathbf{X}$ and adding the results. Notice that the foregoing procedure will not be possible if the number of columns of $\mathbf{A}$ does not equal the number of rows of $\mathbf{X}$. In this event, there will not always be corresponding elements to multiply. Moreover it should be noted that $\mathbf{Y}$ contains the same number of rows as both $\mathbf{A}$ and $\mathbf{X}$.

This suggests a general definition for the multiplication of two matrices. If $\mathbf{A}$ is $m \times n$ and $\mathbf{B}$ is $p \times q$, $\mathbf{AB} = \mathbf{C}$ will exist if $n = p$, in which case the matrix $\mathbf{C}$ will be $m \times q$ with elements given by

$$[c_{ij}]_{m \times q} = \sum_{k=1}^{n=p} a_{ik} b_{kj} \qquad \begin{cases} i = 1, 2, 3, \cdots, m \\ j = 1, 2, 3, \cdots, q \end{cases}$$

When $n = p$, the matrices $\mathbf{A}$ and $\mathbf{B}$ are said to be conformable for multiplication.

Example A.3: If

$$\mathbf{A} = \begin{bmatrix} 1 & 2 & 3 & 4 \\ -1 & 3 & -2 & 0 \end{bmatrix}$$

$$\mathbf{B} = \mathbf{A}^T = \begin{bmatrix} 1 & -1 \\ 2 & 3 \\ 3 & -2 \\ 4 & 0 \end{bmatrix}$$

and

$$\mathbf{C} = \begin{bmatrix} 1 & 2 \\ 3 & 4 \end{bmatrix}$$

find $\mathbf{AB}$, $\mathbf{BA}$, and $\mathbf{AC}$.

SOLUTION The product $\mathbf{AB}$ exists because $\mathbf{A}$ is 2×4 and $\mathbf{B}$ is 4×2. The result, $\mathbf{P}$, will be 2×2:

$$\mathbf{P} = \mathbf{AB} = \begin{bmatrix} 1 & 2 & 3 & 4 \\ -1 & 3 & -2 & 0 \end{bmatrix} \begin{bmatrix} 1 & -1 \\ 2 & 3 \\ 3 & -2 \\ 4 & 0 \end{bmatrix}$$

$$= \begin{bmatrix} (1+4+9+16) & (-1+6-6+0) \\ (-1+6-6+0) & (1+9+4+0) \end{bmatrix} = \begin{bmatrix} 30 & -1 \\ -1 & 14 \end{bmatrix}$$

The product $\mathbf{BA}$ also exists:

$$\mathbf{BA} = \begin{bmatrix} 1 & -1 \\ 2 & 3 \\ 3 & -2 \\ 4 & 0 \end{bmatrix} \begin{bmatrix} 1 & 2 & 3 & 4 \\ -1 & 3 & -2 & 0 \end{bmatrix}$$

$$\begin{bmatrix} (1+1) & (2-3) & (3+2) & (4+0) \\ (2-3) & (4+9) & (6-6) & (8+0) \\ (3+2) & (6-6) & (9+4) & (12+0) \\ (4+0) & (8+0) & (12+0) & (16+0) \end{bmatrix} = \begin{bmatrix} 2 & -1 & 5 & 4 \\ -1 & 13 & 0 & 8 \\ 5 & 0 & 13 & 12 \\ 4 & 8 & 12 & 16 \end{bmatrix}$$

Notice that $\mathbf{AB} \neq \mathbf{BA}$, which shows that matrix multiplication, in general, is not commutative. Notice also that the product $\mathbf{AC}$ will not exist because $\mathbf{A}$ and $\mathbf{C}$ are not conformable for multiplication ($\mathbf{A}$ is 2×4 and $\mathbf{C}$ is 2×2).

While the commutative law does not hold, the multiplication of matrices is associative,

$$(\mathbf{AB})\mathbf{C} = \mathbf{A}(\mathbf{BC})$$

and matrix multiplication is distributive with respect to addition,

$$\mathbf{A}(\mathbf{B} + \mathbf{C}) = \mathbf{AB} + \mathbf{AC}$$

assuming that conformability exists for both addition and multiplication.

If the product $\mathbf{AB}$ is null, that is, $\mathbf{AB} = \mathbf{0}$, it cannot be concluded that either $\mathbf{A}$ or $\mathbf{B}$ is null. Furthermore, if $\mathbf{AB} = \mathbf{AC}$ or $\mathbf{CA} = \mathbf{BA}$, it cannot be concluded that $\mathbf{B} = \mathbf{C}$. This means that, in general, cancellation of matrices is not permissible.

The transpose of a product of matrices is equal to the product of the individual transposes taken in reverse order:

$$(\mathbf{AB})^T = \mathbf{B}^T\mathbf{A}^T$$

A.8 MATRIX DIVISION AND MATRIX INVERSION

Matrix division is not defined. Instead, use is made of a process called *matrix inversion,* which relies on the existence of the identity matrix that is related to a square matrix $\mathbf{A}$ by

$$\mathbf{AI} = \mathbf{IA} = \mathbf{A}$$

Consider the identity for addition, 0, so that it has the property for all scalars α,

$$\alpha + 0 = 0 + \alpha = \alpha$$

and an identity element for multiplication, 1, so that

$$\alpha 1 = 1\alpha = \alpha$$

The scalar most certainly possesses a reciprocal or multiplicative inverse, $1/\alpha$, which when multiplied by α yields the identity element for scalar multiplication:

$$(1/\alpha)\alpha = (\alpha^{-1}\alpha) = 1$$

This reasoning may be extended to the $n \times n$ matrix $\mathbf{A}$ and the pair of identity matrices; the $n \times n$ identity matrix for multiplication $\mathbf{I}$ and the $n \times n$ identity matrix for addition $\mathbf{0}$ (a null matrix). Thus, as already noted,

$$\mathbf{AI} = \mathbf{IA} = \mathbf{A}$$

and

$$\mathbf{A} + \mathbf{0} = \mathbf{0} + \mathbf{A} = \mathbf{A}$$

If there is an $n \times n$ matrix $\mathbf{A}$ that pre- and postmultiplies $\mathbf{A}$ such that

$$\mathbf{A}^{-1}\mathbf{A} = \mathbf{AA}^{-1} = \mathbf{I}$$

then $\mathbf{A}^{-1}$ is an inverse of $\mathbf{A}$ with respect to matrix multiplication. The matrix $\mathbf{A}$ is said to be invertible or *nonsingular* if $\mathbf{A}^{-1}$ exists and singular if $\mathbf{A}^{-1}$ does not exist.

For example: The 3×3 matrix

$$\mathbf{A} = \begin{bmatrix} 3 & -2 & 0 \\ -2 & 4 & -1 \\ 0 & -1 & 6 \end{bmatrix}$$

can be shown to possess the inverse,

$$\mathbf{A} = \begin{bmatrix} 23/45 & 4/15 & 2/45 \\ 4/15 & 2/5 & 1/15 \\ 2/45 & 1/15 & 8/45 \end{bmatrix}$$

A simple multiplication will produce the identity matrix

$$\mathbf{AA}^{-1} = \begin{bmatrix} 3 & -2 & 0 \\ -2 & 4 & -1 \\ 0 & -1 & 6 \end{bmatrix} \begin{bmatrix} 23/45 & 4/15 & 2/45 \\ 4/15 & 2/5 & 1/15 \\ 2/45 & 1/15 & 8/45 \end{bmatrix} = \begin{bmatrix} 1 & 0 & 0 \\ 0 & 1 & 0 \\ 0 & 0 & 1 \end{bmatrix}$$

The reader may wish to verify that the identity is also produced if the product $\mathbf{A}^{-1}\mathbf{A}$ is taken.

The inverse of a product of matrices is equal to the product of the individual inverses taken in reverse order:

$$(\mathbf{AB})^{-1} = \mathbf{B}^{-1}\mathbf{A}^{-1}$$

A.9 DETERMINANTS

A.9.1 Definitions and Terminolgy

A square matrix of order n (an $n \times n$ matrix),

$$\mathbf{A} = \begin{bmatrix} a_{11} & a_{12} & a_{13} & \cdots & a_{1n} \\ a_{21} & a_{22} & a_{23} & \cdots & a_{2n} \\ a_{31} & a_{32} & a_{33} & \cdots & a_{3n} \\ & \cdots & \cdots & \cdots & \\ a_{n1} & a_{n2} & a_{n3} & \cdots & a_{nn} \end{bmatrix}$$

possesses a uniquely defined scalar (a single number), which is designated as the determinant of $\mathbf{A}$ or merely the determinant

$$\det \mathbf{A} = |A|$$

where the order of the determinant is the same as the order of the matrix from which it derives. Observe that only square matrices possess determinants the use of vertical lines and not brackets to designate determinants, and that the elements of the determinant are identical to the elements of the matrix:

$$\det \mathbf{A} = \begin{vmatrix} a_{11} & a_{12} & a_{13} & \cdots & a_{1n} \\ a_{21} & a_{22} & a_{23} & \cdots & a_{2n} \\ a_{31} & a_{32} & a_{33} & \cdots & a_{3n} \\ & \cdots & \cdots & \cdots & \\ a_{n1} & a_{n2} & a_{n3} & \cdots & a_{nn} \end{vmatrix}$$

A determinant of the first order consists of a single element a and has, therefore, the value $\det \mathbf{A} = a$. A determinant of the second order contains four elements in a 2×2 square array:

$$\det \mathbf{A} = |A| = \begin{vmatrix} a_{11} & a_{12} \\ a_{21} & a_{22} \end{vmatrix}$$

A determinant of the third order is described in similar fashion. It is a 3×3 square array containing nine elements:

$$\det \mathbf{A} = |A| = \begin{vmatrix} a_{11} & a_{12} & a_{13} \\ a_{21} & a_{22} & a_{23} \\ a_{31} & a_{32} & a_{33} \end{vmatrix}$$

One may deduce that a determinant of nth order consists of a square array of $n \times n$ elements, a_{ij}, and that the total number of elements in an nth order determinant is n^2. Although this representation of the determinant looks to be purely abstract, the determinant can be proved to be a very rational function that can be evaluated in a number of ways. Moreover, the value of the use of determinants in the taking of matrix inverses and in the solution of simultaneous, linear algebraic equations, although somewhat computationally inefficient, cannot and should not be underemphasized.

A.9.2 Determinant Evaluation

Consider, for example, a pair of simultaneous linear algebraic equations that are presumed to be linearly independent,

$$a_{11}x_1 + a_{12}x_2 = b_1 \tag{A.1a}$$

$$a_{21}x_1 + a_{22}x_2 = b_2 \tag{A.1b}$$

and observe that they may also be written in the matrix form $\mathbf{AX} = \mathbf{B}$

$$\begin{bmatrix} a_{11} & a_{12} \\ a_{21} & a_{22} \end{bmatrix} \begin{bmatrix} x_1 \\ x_2 \end{bmatrix} = \begin{bmatrix} b_1 \\ b_2 \end{bmatrix}$$

In Eqs. (A.1), the x's are the unknowns and the a's form the coefficient matrix $\mathbf{A}$. If $\det \mathbf{A} \neq 0$, the equations are said to be *linearly independent* and one method of solving this second-order system is to multiply Eq. (A.1a) by a_{22} and Eq. (A.1b) by a_{12}:

$$\begin{array}{llll} a_{22}a_{11}x_1 & +a_{22}a_{12}x_2 & = & = a_{22}b_1 \\ a_{12}a_{21}x_1 & +a_{12}a_{22}x_2 & = & = a_{12}b_2 \end{array}$$

A subtraction then yields

$$(a_{22}a_{11} - a_{12}a_{21})x_1 = a_{22}b_1 - a_{12}b_2$$

and then x_1 is obtained

$$x_1 = \frac{a_{22}b_1 - a_{12}b_2}{a_{22}a_{11} - a_{12}a_{21}} \tag{A.2a}$$

A similar procedure yields x_2

$$x_2 = \frac{a_{11}b_2 - a_{21}b_1}{a_{22}a_{11} - a_{12}a_{21}} \tag{A.2b}$$

Observe that the denominators of the equations that yield x_1 and x_2 can be represented by the determinant

$$\begin{vmatrix} a_{11} & a_{12} \\ a_{21} & a_{22} \end{vmatrix} = a_{22}a_{11} - a_{12}a_{21}$$

and that the numerators of these equations can be represented by

$$\begin{vmatrix} b_1 & a_{12} \\ b_2 & a_{22} \end{vmatrix} = b_1a_{22} - b_2a_{21}$$

and

$$\begin{vmatrix} a_{11} & b_1 \\ a_{21} & b_2 \end{vmatrix} = a_{11}b_2 - a_{21}b_1$$

This must always hold unless the determinant in the denominators is equal to zero, which is ruled out because the discussion began originally with the statement that the two equations to be solved were linearly independent.

Thus, one may write the solutions for x_1 and x_2 in Eqs. (A.1) as

$$x_1 = \frac{\begin{vmatrix} b_1 & a_{12} \\ b_2 & a_{22} \end{vmatrix}}{\begin{vmatrix} a_{11} & a_{12} \\ a_{21} & a_{22} \end{vmatrix}} \tag{A.3a}$$

and

$$x_2 = \frac{\begin{vmatrix} a_{11} & b_1 \\ a_{21} & b_2 \end{vmatrix}}{\begin{vmatrix} a_{11} & a_{12} \\ a_{21} & a_{22} \end{vmatrix}} \tag{A.3b}$$

and this is a demonstration of a method of solution of simultaneous, linear algebraic equations known as *Cramer's rule*.

The foregoing reasoning applies equally well to a set of n simultaneous algebraic equations. For a set of three equations in three unknowns,

$$\begin{aligned} a_{11}x_1 \quad +a_{12}x_2 \quad +a_{13}x_3 &= y_1 \\ a_{21}x_1 \quad +a_{22}x_2 \quad +a_{23}x_3 &= y_2 \\ a_{31}x_1 \quad +a_{32}x_2 \quad +a_{33}x_3 &= y_3 \end{aligned}$$

which are assumed to be linearly independent and which may be written in matrix form as

$$\begin{bmatrix} a_{11} & a_{12} & a_{13} \\ a_{21} & a_{22} & a_{23} \\ a_{31} & a_{32} & a_{33} \end{bmatrix} \begin{bmatrix} x_1 \\ x_2 \\ x_3 \end{bmatrix} = \begin{bmatrix} b_1 \\ b_2 \\ b_3 \end{bmatrix}$$

it can be shown that x_1 can be evaluated from

$$x_1 = \frac{\begin{array}{l} b_1 a_{22} a_{33} + b_3 a_{12} a_{23} + b_2 a_{13} a_{32} \\ \qquad - b_3 a_{22} a_{13} - b_1 a_{32} a_{23} - b_2 a_{12} a_{33} \end{array}}{\begin{array}{l} a_{11} a_{22} a_{33} + a_{12} a_{23} a_{31} + a_{13} a_{21} a_{32} \\ \qquad - a_{31} a_{22} a_{13} - a_{32} a_{23} a_{11} - a_{33} a_{21} a_{12} \end{array}}$$

Both the numerator and denominator can be rearranged by employing a little algebra,

$$x_1 = \frac{\begin{array}{l} b_1(a_{22}a_{33} - a_{32}a_{23}) - b_2(a_{12}a_{33} - a_{32}a_{13}) \\ \qquad + b_3(a_{12}a_{23} - a_{22}a_{13}) \end{array}}{\begin{array}{l} a_{11}(a_{22}a_{33} - a_{32}a_{23}) - a_{21}(a_{12}a_{33} - a_{32}a_{13}) \\ \qquad + a_{31}(a_{12}a_{23} - a_{22}a_{13}) \end{array}} \tag{A.4}$$

and an inspection of the terms within the parentheses shows that the solution for x_1 cannot only be written (Cramer's rule) as the quotient of two determinants but also each of the determinants can be represented in terms of three second-order determinants:

$$x_1 = \frac{\begin{vmatrix} b_1 & a_{12} & a_{13} \\ b_2 & a_{22} & a_{23} \\ b_3 & a_{32} & a_{33} \end{vmatrix}}{\begin{vmatrix} a_{11} & a_{12} & a_{13} \\ a_{21} & a_{22} & a_{23} \\ a_{31} & a_{32} & a_{33} \end{vmatrix}}$$

$$x_1 = \frac{b_1 \begin{vmatrix} a_{22} & a_{23} \\ a_{32} & a_{33} \end{vmatrix} - b_2 \begin{vmatrix} a_{12} & a_{13} \\ a_{32} & a_{33} \end{vmatrix} + b_3 \begin{vmatrix} a_{12} & a_{13} \\ a_{22} & a_{23} \end{vmatrix}}{a_{11} \begin{vmatrix} a_{22} & a_{23} \\ a_{32} & a_{33} \end{vmatrix} - a_{21} \begin{vmatrix} a_{12} & a_{13} \\ a_{32} & a_{33} \end{vmatrix} + a_{31} \begin{vmatrix} a_{12} & a_{13} \\ a_{22} & a_{23} \end{vmatrix}} \tag{A.5}$$

The expansion in Eq. (A.5) is known as the *Laplace expansion* or the *Laplace development*.

The method of evaluating second-order determinants is suggested in Eqs. (A.2) and (A.3). The second-order determinant is evaluated as the remainder of the product resulting from the multiplication of the upper-left and lower-right elements (the principal diagonal elements) minus the product of the lower-left and the upper-right elements (the secondary diagonal elements). This procedure is demonstrated in Fig A.1a.

The third-order determinant may be evaluated by taking the products and then the sums and differences of the elements shown in Fig A.1b. This procedure may be assisted by rewriting the first two columns of the determinant and then proceeding as indicated in Fig A.1c. It is important to note that, for this purpose, the diagonals of the third-order determinant are continuous; that is, the last column is followed by the first column.

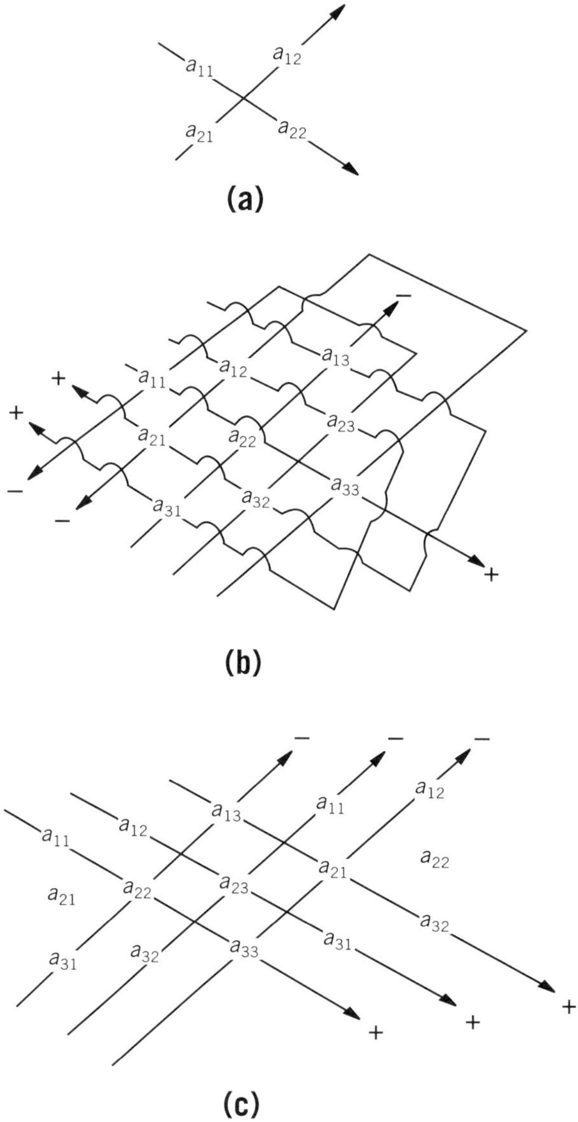

Figure A.1 (*a*) Procedure for evaluating a second-order determinant; (*b*) and (*c*) equivalent procedures for evaluating a third-order determinant.

Caution is necessary: fourth- and higher-order determinants may not be evaluated by following the procedures displayed in Fig A.1. The Laplace expansion or *pivotal condensation* must be employed in these cases.

Example A.4: Evaluate the determinants

$$|A| = \begin{vmatrix} 2 & 3 \\ 1 & 4 \end{vmatrix}$$

and

$$|B| = \begin{vmatrix} 3 & 2 & 1 \\ 1 & 4 & 2 \\ 1 & 3 & 1 \end{vmatrix}$$

SOLUTION The second-order determinant is evaluated by the procedure indicated in Fig A.1a:

$$|A| = \begin{vmatrix} 2 & 3 \\ 1 & 4 \end{vmatrix} = 2(4) - 3(1) = 8 - 3 = 5$$

The third-order determinant is evaluated in accordance with Figs A.1b or A.1c as

$$|B| = \begin{vmatrix} 3 & 2 & 1 \\ 1 & 4 & 2 \\ 1 & 3 & 1 \end{vmatrix}$$

$$= 3(4)(1) + 2(2)(1) + 1(1)(3) - 1(4)(1) - 3(2)(3) - 1(1)(2)$$

or

$$B = 12 + 4 + 3 - 4 - 18 - 2 = -5$$

A.9.3 Pivotal Condensation

The evaluation of a determinant by the Laplace expansion can be a long, tedious, and laborious procedure. Assuming that third order determinants can be evaluated quickly, a fifth-order determinant containing no zero elements requires the evaluation of $5 \times 4 = 20$ third-order determinants. For a sixth-order determinant, this number becomes $6 \times 5 \times 4 = 120$. In general, if $n > 6$, the evaluation of an nth order determinant can require the evaluation of $(n-1)!$ third-order determinants.

Pivotal condensation is a much more efficient process. Take the determinant

$$\det \mathbf{A} = \begin{vmatrix} a_{11} & a_{12} & a_{13} & \cdots & a_{1n} \\ a_{21} & a_{22} & a_{23} & \cdots & a_{2n} \\ a_{31} & a_{32} & a_{33} & \cdots & a_{3n} \\ & \cdots & \cdots & \cdots & \\ a_{n1} & a_{n2} & a_{n3} & \cdots & a_{nn} \end{vmatrix}$$

The element a_{11} is selected as the element in the *pivotal position*. It is called the *pivotal element* or merely the *pivot* in the following development. The objective is to find a determinant $|B|$ that is one order less than $|A|$ by operating on $|A|$ in such a manner as to produce a column of zeroes in the column containing the pivot. If $a_{11} = 0$, a row or column interchange can be performed to put a nonzero element in the pivotal position.

The condensation process that brings an nth order determinant down to an $(n-1)^{\text{th}}$ order determinant is continued until the order is reduced to 3 or 2. Then the evaluation can be accomplished by the methods provided in the previous subsection. The entire condensation procedure can be handled by the computationally efficient matrix relationship

$$\det \mathbf{A} = \frac{1}{a_{11}^{n-2}} \det[a_{11}\mathbf{P} - \mathbf{Q}] \tag{A.6}$$

where the $n-1$ matrices $\mathbf{P}$ and $\mathbf{Q}$ are given by

$$\mathbf{P} = \begin{bmatrix} a_{22} & a_{23} & \cdots & a_{2n} \\ a_{32} & a_{33} & \cdots & a_{3n} \\ & \cdots & \cdots & \\ a_{n2} & a_{n3} & \cdots & a_{nn} \end{bmatrix}$$

and

$$\mathbf{Q} = \begin{bmatrix} a_{21} \\ a_{31} \\ \cdots \\ a_{n1} \end{bmatrix} \begin{bmatrix} a_{12} & a_{13} & \cdots & a_{1n} \end{bmatrix}$$

Example A.5: Use pivotal condensation to evaluate the determinant

$$|A| = \begin{vmatrix} 2 & -1 & 1 & 1 & 2 \\ 0 & 2 & 3 & 2 & 1 \\ 0 & 1 & 2 & 1 & 2 \\ 0 & 1 & -1 & -1 & 3 \\ 0 & 2 & 1 & 1 & -2 \end{vmatrix}$$

SOLUTION By pivotal condensation,

$$|A| = \frac{1}{2^3} \det \left[2 \begin{bmatrix} 2 & 3 & 2 & 1 \\ 1 & 2 & 1 & 2 \\ 1 & -1 & -1 & 3 \\ 2 & 1 & 1 & -2 \end{bmatrix} - \begin{bmatrix} 0 \\ 0 \\ 0 \\ 0 \end{bmatrix} \begin{bmatrix} -1 & 1 & 1 & 2 \end{bmatrix} \right]$$

or

$$|A| = \frac{1}{8} \det \begin{bmatrix} 4 & 6 & 4 & 2 \\ 2 & 4 & 2 & 4 \\ 2 & -2 & -2 & 6 \\ 4 & 2 & 2 & -4 \end{bmatrix}$$

Then

$$|A| = \frac{1}{8(4)^2} \det \left[4 \begin{bmatrix} 4 & 2 & 4 \\ -2 & -2 & 6 \\ 2 & 2 & -4 \end{bmatrix} - \begin{bmatrix} 2 \\ 2 \\ 4 \end{bmatrix} \begin{bmatrix} 6 & 4 & 2 \end{bmatrix} \right]$$

or

$$|A| = \frac{1}{128} \det \left[\begin{bmatrix} 16 & 8 & 16 \\ -8 & -8 & 24 \\ 8 & 8 & -16 \end{bmatrix} - \begin{bmatrix} 12 & 8 & 4 \\ 12 & 8 & 4 \\ 24 & 16 & 8 \end{bmatrix} \right]$$

$$= \frac{1}{128} \det \begin{bmatrix} 4 & 0 & 12 \\ -20 & -16 & 20 \\ -16 & -8 & -24 \end{bmatrix}$$

The third-order determinant is easily evaluated:

$$|A| = \frac{1}{128}(1536 + 0 + 1920 - 3072 + 640 - 0) = \frac{1024}{128} = 8$$

A.9.4 Additional Properties

Several rules pertaining to the simplification and manipulation of determinants will be presented in this subsection without formal proof.

- Interchanging any row (or column) of a determinant with its immediately adjacent row (or column) alters the sign of the determinant.
- The multiplication of any single row (column) of a determinant by a scalar constant is equivalent to the multiplication of the entire determinant by the scalar. Observe that this differs from the multiplication of a matrix by a scalar; the multiplication of a matrix by a scalar results in the multiplication of each and every element of the matrix by the scalar.
- If every element in an nth order determinant is multiplied by the same scalar, α, the value of the determinant is multiplied by α^n.
- If any two rows (columns) of a determinant are identical, the value of the determinant is zero, and the matrix from which the determinant derives is said to be *singular*.
- If any row (or column) of a determinant contains nothing but zeroes, the value of the determinant is zero.
- If any two rows (columns) of a determinant are proportional, the determinant is equal to zero. In this case, the two rows (columns) are said to be *linearly dependent*.

- If the elements of any row (column) of a determinant are added to or subtracted from the corresponding elements of another row (column), the value of the determinant is unchanged.

- If the elements of any row (column) of a determinant are multiplied by a constant and then added to or subtracted from the corresponding elements of another row (column), the value of the determinant is unchanged.

- The value of the determinant of a diagonal matrix is equal to the product of the terms on the diagonal.

- The value of the determinant of a matrix is equal to the value of the determinant of the transpose of the matrix.

- The determinant of the product of two matrices is equal to the product of the determinants of the two matrices.

- If the determinant of the product of two square matrices is zero, then at least one of the matrices is singular, and the value of its determinant is equal to zero.

A.10 MINORS AND COFACTORS

Consider the nth-order determinant

$$\det \mathbf{A} = \begin{vmatrix} a_{11} & a_{12} & a_{13} & \cdots & a_{1n} \\ a_{21} & a_{22} & a_{23} & \cdots & a_{2n} \\ a_{31} & a_{32} & a_{33} & \cdots & a_{3n} \\ & \cdots & \cdots & \cdots & \\ a_{n1} & a_{n2} & a_{n3} & \cdots & a_{nn} \end{vmatrix} \tag{A.7}$$

which will be used for proposing two useful quantities.

The $(n-1)$th-order *minor* of an nth-order determinant $|A|$ is the determinant formed by deleting one row and one column from $|A|$. The minor, designated by $|M|_{ij}$, is the determinant formed by deleting the ith row and the jth column from $|A|$

The *cofactor*, designated as A_{ij}, without vertical rules and with a double subscript is the *signed* $(n-1)$th-order minor formed from the nth-order determinant. If the minor has been formed by deleting the ith row and the jth column from $|A|$, then

$$A_{ij} = (-1)^{i+j} M_{ij} \tag{A.8}$$

The sign of the cofactor can be determined from Eq. (A.8) or from the *checkerboard rule* summarized in Fig A.2.

$$\begin{array}{ccccccc} + & - & + & - & + & \cdots & + \\ - & + & - & + & - & \cdots & - \\ + & - & + & - & + & \cdots & + \\ - & + & - & + & - & \cdots & - \\ & & & \cdots & & & \\ + & - & + & - & + & \cdots & + \end{array} \qquad \begin{array}{ccccccc} + & - & + & - & + & \cdots & - \\ - & + & - & + & - & \cdots & + \\ + & - & + & - & + & \cdots & - \\ - & + & - & + & - & \cdots & + \\ & & & \cdots & & & \\ + & - & + & - & + & \cdots & + \end{array}$$

(a) **(b)**

Figure A.2 The checkerboard rule for finding the sign of a cofactor of an nth-order determinant, (a) for n odd and (b) for n even.

Example A.6: Consider the fourth-order determinant

$$|A| = \det \begin{bmatrix} 1 & 3 & -1 & 2 \\ 4 & 1 & 1 & 3 \\ 3 & 1 & -2 & 1 \\ 1 & 3 & 2 & 5 \end{bmatrix}$$

What is the minor and cofactor formed by deleting the third row and fourth column?

SOLUTION

$$|M|_{34} = \det \begin{bmatrix} 1 & 3 & -1 \\ 4 & 1 & 1 \\ 1 & 3 & 2 \end{bmatrix} = 2 + 3 - 12 + 1 - 3 - 24 = -33$$

The cofactor is the *signed* minor. By the checkerboard rule of Fig A.2 or by Eq. (A.8),

$$A_{34} = (-1)^{3+4}(-33) = -(-33) = 33$$

A.11 THE COFACTOR MATRIX

A square nth-order matrix

$$\mathbf{A} = \begin{bmatrix} a_{11} & a_{12} & a_{13} & \cdots & a_{1n} \\ a_{21} & a_{22} & a_{23} & \cdots & a_{2n} \\ a_{31} & a_{32} & a_{33} & \cdots & a_{3n} \\ & \cdots & \cdots & \cdots & \\ a_{n1} & a_{n2} & a_{n3} & \cdots & a_{nn} \end{bmatrix}$$

possesses a cofactor matrix with elements indicated by capital letters with double subscripts:

$$\mathbf{A}_c = \begin{bmatrix} A_{11} & A_{12} & A_{13} & \cdots & A_{1n} \\ A_{21} & A_{22} & A_{23} & \cdots & A_{2n} \\ A_{31} & A_{32} & A_{33} & \cdots & A_{3n} \\ & \cdots & \cdots & \cdots & \\ A_{n1} & A_{n2} & A_{n3} & + \cdots & A_{nn} \end{bmatrix}$$

Example A.7: Show the cofactor matrix for the third-order symmetrical matrix matrix:

$$\begin{bmatrix} 3 & -2 & 0 \\ -2 & 4 & -1 \\ 0 & -1 & 6 \end{bmatrix}$$

SOLUTION The nine cofactors with signs determined by Eq. (A.8) or from the checkerboard rule in Fig. A.2 are formed from the nine possible second-order minors:

$$A_{11} = +|M|_{11} = \begin{vmatrix} 4 & -1 \\ -1 & 6 \end{vmatrix} = +(24 - 1) = 23$$

$$A_{12} = -|M|_{12} = \begin{vmatrix} -2 & -1 \\ 0 & 6 \end{vmatrix} = -(-12) = 12$$

$$A_{13} = +|M|_{13} = \begin{vmatrix} -2 & 4 \\ 0 & -1 \end{vmatrix} = 2$$

$$A_{21} = -|M|_{21} = \begin{vmatrix} -2 & 0 \\ -1 & 6 \end{vmatrix} = -(-12) = 12$$

$$A_{22} = +|M|_{22} = \begin{vmatrix} 3 & 0 \\ 0 & 6 \end{vmatrix} = 18$$

$$A_{23} = -|M|_{23} = \begin{vmatrix} 3 & -2 \\ 0 & -1 \end{vmatrix} = -(-3) = 3$$

$$A_{31} = +|M|_{31} = \begin{vmatrix} -2 & 0 \\ 4 & -1 \end{vmatrix} = 2$$

$$A_{32} = -|M|_{32} = \begin{vmatrix} 3 & 0 \\ -2 & -1 \end{vmatrix} = -(-3) = 3$$

$$A_{33} = +|M|_{33} = \begin{vmatrix} 3 & -2 \\ -2 & 4 \end{vmatrix} = +(12 - 4) = 8$$

Thus

$$\mathbf{A}_c = \begin{bmatrix} 23 & 12 & 2 \\ 12 & 18 & 3 \\ 2 & 3 & 8 \end{bmatrix}$$

and this confirms that symmetrical matrices possess symmetrical cofactor matrices.

A.12 THE LAPLACE EXPANSION

In the denominator of Eq. (A.5), the third-order determinant **A** was shown to be equal to some function of three second-order determinants:

$$\begin{vmatrix} a_{11} & a_{12} & a_{13} \\ a_{21} & a_{22} & a_{23} \\ a_{31} & a_{32} & a_{33} \end{vmatrix}$$

$$= a_{11}\begin{vmatrix} a_{22} & a_{23} \\ a_{32} & a_{33} \end{vmatrix} - a_{21}\begin{vmatrix} a_{12} & a_{13} \\ a_{32} & a_{33} \end{vmatrix} + a_{31}\begin{vmatrix} a_{12} & a_{13} \\ a_{22} & a_{23} \end{vmatrix} \qquad \text{(A.9)}$$

Notice that each of the second-order determinants is a second-order minor of **A**. This means that three cofactors exist and hence, Eq. (A.9) gives a rule for the evaluation of a third-order determinant that can be extended to an nth order determinant. For the ith row,

$$|A| = \sum_{j=1}^{j=n} (-1)^{i+j} a_{ij} |M|_{ij}$$

or

$$|A| = \sum_{j=1}^{j=n} a_{ij} A_{ij} \qquad (A.10a)$$

and for the jth column,

$$|A| = \sum_{i=1}^{i=n} (-1)^{i+j} a_{ij} |M|_{ij}$$

or

$$|A| = \sum_{i=1}^{i=n} a_{ij} A_{ij} \qquad (A.10b)$$

Equations (A.10) describe a procedure known as the *Laplace development* or the *Laplace expansion*.

Example A.8: Evaluate the determinant

$$|A| = \begin{vmatrix} 6 & 3 & 0 & 1 \\ 2 & 5 & 1 & -2 \\ 1 & 0 & -1 & 1 \\ 1 & 0 & 3 & 4 \end{vmatrix}$$

SOLUTION Expand the second column to reduce the labor (two zeroes occur in this column),

$$|A| = a_{21}|A|_{21} + a_{22}|A|_{22}$$

The cofactors derive from the appropriate minors with their sign determined from Eq. (A.8) or from the checkerboard rule illustrated in Fig. A.2:

$$\begin{aligned} A_{21} &= -|M|_{21} = \begin{vmatrix} 2 & 1 & -2 \\ 1 & -1 & 1 \\ 1 & 3 & 4 \end{vmatrix} \\ &= -(-8+1-6-2-6-4) = -(1-26) = 25 \end{aligned}$$

and

$$\begin{aligned} A_{22} &= = +|M|_{22} = \begin{vmatrix} 6 & 0 & 1 \\ 1 & -1 & 1 \\ 1 & 3 & 4 \end{vmatrix} \\ &= (-24+0+3+1-18-0) = 4-42 = -38 \end{aligned}$$

The value of the determinant is

$$|A| = a_{21}A_{21} + a_{22}A_{22} = 3(25) + 5(-38) = 75 - 190 = -115$$

It should be noted that if the elements of a row or column of a determinant are multiplied by cofactors of the corresponding elements of a different row or column, the resulting sum of these products is zero:

$$\sum_{i=1}^{i=n} (-1)^{k+j} a_{kj}|M|_{kj} = 0 \qquad (i \neq k) \qquad (A.11a)$$

and

$$\sum_{j=1}^{j=n} (-1)^{i+k} a_{ik}|M|_{ik} = 0 \qquad (j \neq k) \qquad (A.11b)$$

A.13 MATRIX INVERSION

An nth-order set of simultaneous, linear algebraic equations in the n unknowns, $x_1, x_2, x_3, \cdots, x_n$, can conveniently be represented by the matrix equation

$$\mathbf{AX} = \mathbf{Y} \qquad (A.12)$$

where $\mathbf{A}$, as indicated in Sec. A.2, is a square matrix of coefficients having elements a_{ij}, and where $\mathbf{X}$ and $\mathbf{Y}$ are $n \times 1$ column vectors with elements x_i and y_i, respectively. Because division of matrices is not permitted, one method for the solution of matrix equations such as the one shown in Eq. (A.12) is called *matrix inversion.*

If Eq. (A.12) is premultiplied by an $n \times n$ square matrix $\mathbf{B}$ so that

$$\mathbf{BAX} = \mathbf{BY}$$

a solution for the unknowns, $\mathbf{X}$, will evolve if the product $\mathbf{BA}$ is equal to the identity matrix $\mathbf{I}$:

$$\mathbf{BAX} = \mathbf{IX} = \mathbf{BY}$$

or

$$\mathbf{X} = \mathbf{BY} \tag{A.13}$$

If

$$\mathbf{BA} = \mathbf{AB} = \mathbf{I}$$

the matrix $\mathbf{B}$ is said to be the inverse of $\mathbf{A}$

$$\mathbf{B} = \mathbf{A}^{-1} \tag{A.14a}$$

and, of course, the inverse of the inverse is the matrix itself:

$$\mathbf{A} = \mathbf{B}^{-1} \tag{A.14b}$$

or

$$(\mathbf{A})^{-1} = \mathbf{A}$$

It may be recalled that, in general, matrix multiplication is not commutative. The multiplication of a matrix by its inverse is one specific case where matrix multiplication is commutative:

$$\mathbf{AA}^{-1} = \mathbf{A}^{-1}\mathbf{A} = \mathbf{I}$$

A.13.1 Properties of the Inverse

The inverse of a product of two matrices is the product of the inverses taken in reverse order. This is easily proved. Consider the product $\mathbf{AB}$ and postmultiply by $\mathbf{B}^{-1}\mathbf{A}^{-1}$. Because matrix multiplication is associative, this product can be taken with a rearrangement of the parentheses and then by a straightforward application of the definition of the matrix inverse:

$$\mathbf{AB}(\mathbf{B}^{-1}\mathbf{A}^{-1}) = \mathbf{A}(\mathbf{BB}^{-1})\mathbf{A}^{-1} = \mathbf{AIA}^{-1} = \mathbf{AA}^{-1} = \mathbf{I}$$

Thus

$$(\mathbf{AB})^{-1} = \mathbf{B}^{-1}\mathbf{A}^{-1}$$

In addition, the inverse of the transpose of a matrix is equal to the transpose of its inverse:

$$(\mathbf{A}^{T})^{-1} = (\mathbf{A}^{-1})^{T}$$

Negative powers of a matrix are related to its inverse

$$\mathbf{A}^{-n} = (\mathbf{A}^{-1})^n$$

and the determinant of the product of a matrix and its inverse must be equal to unity:

$$\det(\mathbf{A}\mathbf{A}^{-1}) = \det \mathbf{I} = 1$$

If a matrix does not possess an inverse, it is said to be *singular*, but if a matrix does possess an inverse, the inverse is unique.

A.13.2 The Adjoint Matrix

The *adjoint matrix*, which is sometimes called the *adjugate matrix* and which, in this development will be referred to merely as the *adjoint*, applies only to a square matrix and is the transpose of the cofactor matrix,

$$\text{adj } \mathbf{A} = \mathcal{A} = (\mathbf{A}_c)^T \tag{A.15}$$

and because symmetrical matrices possess symmetrical cofactor matrices, the adjoint of a symmetrical matrix is the cofactor matrix itself.

The matrix that is of nth order,

$$\mathbf{A} = \begin{bmatrix} a_{11} & a_{12} & a_{13} & \cdots & a_{1n} \\ a_{21} & a_{22} & a_{23} & \cdots & a_{2n} \\ a_{31} & a_{32} & a_{33} & \cdots & a_{3n} \\ & \cdots & \cdots & \cdots & \\ a_{n1} & a_{n2} & a_{n3} & \cdots & a_{nn} \end{bmatrix}$$

has been observed to possess a cofactor matrix,

$$\mathbf{A}_c = \begin{bmatrix} A_{11} & A_{12} & A_{13} & \cdots & A_{1n} \\ A_{21} & A_{22} & A_{23} & \cdots & A_{2n} \\ A_{31} & A_{32} & A_{33} & \cdots & A_{3n} \\ & \cdots & \cdots & \cdots & \\ A_{n1} & A_{n2} & A_{n3} & + \cdots & A_{nn} \end{bmatrix}$$

and this cofactor matrix has an adjoint

$$\text{adj } \mathbf{A} = (\mathbf{A}_c)^T = \begin{bmatrix} A_{11} & A_{21} & A_{31} & \cdots & A_{n1} \\ A_{12} & A_{22} & A_{32} & \cdots & A_{n2} \\ A_{13} & A_{23} & A_{33} & \cdots & A_{n3} \\ & \cdots & \cdots & \cdots & \\ A_{1n} & A_{2n} & A_{3n} & \cdots & A_{nn} \end{bmatrix}$$

A.13.3 One Method for the Determination of the Inverse

Suppose an $n \times n$ matrix, $\mathbf{A}$, is postmultiplied by its adjoint and that the product is designated as $\mathbf{P}$:

$$\mathbf{A}(\text{adj }\mathbf{A}) = \begin{bmatrix} a_{11} & a_{12} & a_{13} & \cdots & a_{1n} \\ a_{21} & a_{22} & a_{23} & \cdots & a_{2n} \\ a_{31} & a_{32} & a_{33} & \cdots & a_{3n} \\ \cdots & \cdots & \cdots & & \\ a_{n1} & a_{n2} & a_{n3} & \cdots & a_{nn} \end{bmatrix} \begin{bmatrix} A_{11} & A_{21} & A_{31} & \cdots & A_{n1} \\ A_{12} & A_{22} & A_{32} & \cdots & A_{n2} \\ A_{13} & A_{23} & A_{33} & \cdots & A_{n3} \\ & \cdots & \cdots & \cdots & \\ A_{1n} & A_{2n} & A_{3n} & \cdots & A_{nn} \end{bmatrix} = \mathbf{P}$$

The elements of $\mathbf{P}$ may be divided into two categories: those that lie on its principal diagonal (of which p_{22} is typical) and those that do not. For the principal diagonal element p_{22},

$$p_{22} = a_{21}A_{21} + a_{22}A_{22} + a_{23}A_{23} + \cdots + a_{2n}A_{2n}$$

and by Eq. (A.10a), it is seen that

$$p_{22} = |A|$$

For the off-diagonal element of which p_{13} is typical,

$$p_{13} = a_{11}A_{31} + a_{12}A_{32} + a_{13}A_{33} + \cdots + a_{1n}A_{3n}$$

and by Eq. (A.11a), it is seen that

$$p_{13} = 0$$

Thus the product of A and its adjoint is

$$\mathbf{A}(\text{adj }\mathbf{A}) = \begin{bmatrix} |A| & 0 & 0 & +\cdots & 0 \\ 0 & |A| & 0 & +\cdots & 0 \\ 0 & 0 & |A| & +\cdots & 0 \\ & \cdots & \cdots & \cdots & \\ 0 & 0 & 0 & +\cdots & |A| \end{bmatrix} = |A|\mathbf{I}$$

If this is put into the form

$$\mathbf{A}\frac{\text{adj }\mathbf{A}}{\det \mathbf{A}} = \mathbf{A}$$

and compared with

$$\mathbf{A}\mathbf{A}^{-1} = \mathbf{I}$$

it becomes evident that the inverse of the matrix $\mathbf{A}$ is equal to its adjoint divided by its determinant

$$\mathbf{A}^{-1} = \frac{\text{adj }\mathbf{A}}{\det \mathbf{A}} \tag{A.16}$$

Observe that if $\det \mathbf{A} = 0$, the inverse of $\mathbf{A}$ cannot exist and is therefore *singular*. Thus the necessary condition for the matrix, $\mathbf{A}$, to be singular is for $\det \mathbf{A} = 0$.

Example A.9: Determine the inverse of the third-order symmetrical matrix matrix

$$\mathbf{a} = \begin{bmatrix} 3 & -2 & 0 \\ -2 & 4 & -1 \\ 0 & -1 & 6 \end{bmatrix}$$

SOLUTION In Ex. A.7, it was shown that the given matrix possesses a cofactor matrix

$$\mathbf{A}_c = \begin{bmatrix} 23 & 12 & 2 \\ 12 & 18 & 3 \\ 2 & 3 & 8 \end{bmatrix}$$

and the reader may verify that the given matrix has a determinant det $\mathbf{A} = 45$.

The given matrix is symmetrical as is the cofactor matrix. The adjoint (the transpose of the cofactor matrix) is also symmetrical and is equal to the cofactor matrix. Thus, by Eq. (A.16), the inverse is

$$\mathbf{A}^{-1} = \begin{bmatrix} 23/45 & 4/15 & 2/45 \\ 4/15 & 2/5 & 1/15 \\ 2/45 & 1/15 & 8/45 \end{bmatrix}$$

which is also observed to be symmetrical.

It is important to note that symmetrical matrices possess symmetrical transposes, symmetrical cofactor matrices, symmetrical adjoints, and symmetrical inverses.

The evaluation of the inverse can always be concluded with a check on its validity. In the example just concluded

$$\mathbf{A} = \begin{bmatrix} 3 & -2 & 0 \\ -2 & 4 & -1 \\ 0 & -1 & 6 \end{bmatrix} \begin{bmatrix} 23 & 12 & 2 \\ 12 & 18 & 3 \\ 2 & 3 & 8 \end{bmatrix} = \begin{bmatrix} 45 & 0 & 0 \\ 0 & 45 & 0 \\ 0 & 0 & 45 \end{bmatrix}$$

Example A.10: Determine the inverse of the nonsymmetrical second-order matrix

$$\mathbf{A} = \begin{bmatrix} 4 & -1 \\ 1 & 6 \end{bmatrix}$$

SOLUTION The matrix has a determinant

$$\det \mathbf{A} = 24 + 1 = 25$$

a cofactor matrix

$$\mathbf{A}_c = \begin{bmatrix} 6 & -1 \\ 1 & 4 \end{bmatrix}$$

and an adjoint

$$\text{adj } \mathbf{A} = \begin{bmatrix} 6 & 1 \\ -1 & 4 \end{bmatrix}$$

Its inverse is

$$\mathbf{A}^{-1} = \frac{\text{adj } \mathbf{A}}{\det \mathbf{A}} = \begin{bmatrix} 6/25 & 1/25 \\ -1/25 & 4/25 \end{bmatrix}$$

This can be verified by the reader, and it is observed that the inverse of a second-order determinant is obtained by swapping the elements that lie on the principal diagonal, changing the sign of the off-diagonal elements and then dividing all elements by the determinant.

Appendix

B

NOMENCLATURE

Roman Letter Symbols

$\mathbf{A}$	node-branch incidence matrix, dimensionless
A	cross-sectional area, m^2 geometric factor, dimensionless
a	channel aspect ratio, dimensionless coordinate at fin or spine tip, m elements of node-branch incidence matrix, dimensionless significant dimension of heat source, m
B	geometric factor, dimensionless
b	coordinate at fin or spine base, m fin or spine height, m
$\mathbf{C}$	matrix, dimensionless
C	combination of terms, dimensionless correlating coefficient, dimensionless skin friction coefficient, dimensionless with subscripts -1 and -2, arbitrary constant, dimensionless
c	specific heat, J/kg-K
d	diameter, m
det	determinant, dimensions vary
El	Elenbaas number, dimensionless
$\hat{E}$	effective channel emittance, dimensionless
e	edge of topological graph, dimensionless
$\mathcal{F}$	radiation surface factor, dimensionless
f	friction factor, dimensionless

G inverse hybrid parameter matrix, dimensions of elements vary

G body gravity function, dimensionless
irradiation, W/m^2
thermal conductance, W/m^2

Gr Grashof number, dimensionless

g acceleration of gravity, 9.81 m/s^2
elements of the inverse hybrid parameter matrix, dimensions vary

H hybrid parameter matrix

H height extension, m
microhardness, N/m^2

h elements of the hybrid parameter matrix, dimensions vary
heat-transfer coefficient, W/m^2-K

I modified Bessel function of first kind, dimensionless

i counter, dimensionless

J adjusted column vector, dimensionless

J radiosity, W/m^2

j counter, dimensionless
elements of adjusted column vector, dimensionless

K combination of matrices, dimensionless

K modified Bessel function of the second kind, dimensionless
radiation conductance, m^2

k counter, dimensionless
elements of the matrix, **K**, dimensionless
thermal conductivity, W/m-K

L	length, m
ℓ	number of fins, dimensionless
M	gas parameter, m heat sink utility function, m^{-1}
m	particular fin effective absolute surface slope, dimensionless fin performance parameter, m^{-1} mass, kg surface slope, dimensionless
n	particular fin exponent, dimensionless fin performance parameter, defined where used, m^{-1} number of entities, dimensionless
Nu	Nusselt number, dimensionless
Pr	Prandtl number, dimensionless
P	array parameter, dimensionless contact pressure, N/m^2 perimeter, m plate air parameter, m^{-4}
p	pressure, N/m^2
$\tilde{\mathbf{Q}}$	node heat source vector, W
$\hat{\mathbf{Q}}$	heat flow vector, W
Q	heat flow per unit length, W/m total heat flow, W optimization parameter, dimensionless
q	heat flow, W

Ra Rayleigh number, dimensionless

Re Reynolds number, dimensionless

R electrical resistance,
thermal resistance, K/W

r radial coordinate, m
radius, m
St Stanton number, dimensionless

$\mathbf{S}$ sign-correcting matrix, dimensionless

S surface area, m^2

s arc length, m
optimization parameter, dimensionless
pin fin array spacing, m

$\Delta\mathbf{T}$ matrix of branch temperatures, K

T temperature, °C or K

t time, s

U optimization parameter, dimensionless

u tranformed variable, dimensionless

V average velocity, m/s
dimensional constant, 0.4646 m^{-1}
volume, m^3

v optimization parameter, dimensionless
velocity, m/s

W width, m
Wronskian

w transformed variable, dimensionless

x height coordinate, m
thickness of individual layer, m

$\mathbf{Y}$ admittance parameter matrix, W/K

Y thermal admittance, W/K
distance between mean planes, m

y elements of the admittance parameter matrix, W/K
coordinate transverse to the flow direction, m
fin profile function, m

$\mathbf{Z}$ impedance parameter matrix, K/W

Z combination of terms, dimensionless
thermal impedance, K/W
combination of terms, dimensionless

z elements of the impedance parameter matrix, K/W
clear space between fins or plates, m

Greek Letter Symbols

α accommodation parameter, dimensionless
linear transformation relating dependent and independent variables
exponent (defined in Section 3.6), dimensionless
parameter defined in Eq. (3.54)
thermal diffusivity m^2/s

β fluid property parameter, dimensionless
coefficient of volumetric expansion, K^{-1}
exponent (defined in Section 3.6), dimensionless
optimization parameter, dimensionless
parameter defined in Eq. (3.54)

$\mathbf{\Gamma}$ Linear transformation matrix from fin base to fin tip

γ elements of the matrix $\mathbf{\Gamma}$, dimensions vary
exponent (defined in Section 6.8), dimensionless

Δ change or difference
determinant

δ thickness, m
thickness, in general, m
boundary layer thickness, m
with subscript, the side of a spine of rectangular cross section, m
with subscript, the major or minor diameter of a spine of elliptical cross section, m

ϵ eddy diffusivity, m^2/s
emissivity, dimensionless
ratio of heat source area to substrate area, dimensionless

η fin effeciency, dimensionless

$\hat{\Theta}$ temperature excess vector, K

Θ temperature excess vector, K

θ temperature difference, K or °C
temperature excess between fin surface and surroundings, K or ° C

κ optimization parameter, dimensionless

$\mathbf{\Lambda}$ mean free path of molecules, m
combination of modified bessel functions, dimensionless

λ solution of a differential equation, dimensions vary
fin taper ratio, dimensionless

μ dynamic viscosity, $N\text{-}s/m^2$
Thermal transmission ratio, W/K

ν optimization parameter, dimensionless
kinematic viscosity, m^2/s

ρ	fluid density, kg/m^3 radius ratio, dimensionless
σ	optimization parameter, dimensionless effective rms surface roughness, dimensionless surface roughness, dimensionless Stefan-Boltzmann constant, 5.669×10^{-8}, W/m^2-K^4
$\mathbf{T}$	Linear transformation matrix from fin tip to fin base
τ	elements of the matrix $\mathbf{T}$, dimensions vary shear stress, N/m^2
$\boldsymbol{\Upsilon}$	matrix (Section 7.2), dimensionless combination of modified bessel functions (Section 6.8), dimensionless
ϕ	fin taper angle, radians parameter defined by Eq. (3.53) dimensionless
ψ	combination of terms (Section 3.13) dimensionless a geometric parameter, dimensionless optimization parameter, dimensionless
$\boldsymbol{\Omega}$	an incidence matrix, dimensionless electrical resistance, ω
ω	elements of the incidence matrice, dimensionless

Roman Letter Subscripts

$\sqrt{A}$	value based on the square root of an area
a	apparent area fin or array tip entity augmented matrix
av	average value or condition

a, opt	optimum value for an array
b	quantity taken at a fin base bond layer fin base bond layer or bond resistance number of branches
c	chip cluster contact solid-to-solid conductance
co	contact quantity value at a component case value at a chip
cd	conduction quantity
co	contact quantity
cv	convection quantity
ctr	value at the center of a configuration
d	reference to a diameter
e	effective value equivalent value deignates an equivalent diameter
ex	external condition
f	fin friction
fl	fluid

g	generated quantity quantity at a gap an array-graph incidence matrix
h	horizontal or lateral pin-fin spacing thermal eddy diffusivity
id	ideal condition
in	inlet condition
int	internal condition
j	junction
ja	junction-to-ambient value
jc	junction-to-case value
ℓ	left end of configuration
m	momentum particular fin
max	maximum quantity
n	particular fin
o	average value characteristic value edge condition represents overall condition wall condition gap based quantity
out	outlet condition
p	constant pressure condition profile area reference to a plate

p, s	pin-grid array Nusselt number
r	radiation right end of a configuration condition based on a hydraulic radius
req	required value or condition
s	harmonic mean thermal conductivity environment or surroundings pin-grid array Nusselt number value taken at a surface source
sat	saturated condition
sf	surface-fluid combination
sk	sink (as in heat sink)
sp	spreading resistance
s, opt	optimum value for a single fin
T	total value thermal boundary layer
tr	transition
v	vertical pin fin array spacing
w	wall condition wetted perimeter
1	entity number 1
2	entity number 2

∞ free stream condition

Greek Letter Subscripts

ϕ phase change

Superscripts

T the transpose of a matrix

-1 the inverse of a matrix

$'$ first derivative

$''$ second derivative
heat flux in the transverse direction
modified Elenbaas number

∞ diffusive limit

$*$ pin fin channel Elenbaas number

AUTHOR INDEX

SUBJECT INDEX